PROGRESS IN HUMAN
BEHAVIOR GENETICS

PROGRESS IN HUMAN BEHAVIOR GENETICS

Recent Reports on Genetic Syndromes, Twin Studies, and Statistical Advances

Edited by

Steven G. Vandenberg

The Johns Hopkins Press

Baltimore, Maryland

Baltimore, Maryland 21218

Manufactured in the U.S.A.
Library of Congress Catalog Card Number 68-22675

PREFACE

After more than a quarter century of neglect, there has been a rapidly growing interest in the genetics of behavior, both in mice and men (as well as in other animals). In fact, most recently the increase is beginning to move beyond the traditional position in which heredity was seen as a factor competing with environmental influences as an explanation of individual differences. Instead, current studies are beginning to answer the question so aptly phrased in 1958 by Anne Anastasi: "Heredity, environment and the question 'How?' " in her paper in the *Psychological Review* (pages 197-208).

The papers in this volume are good examples of this advance. At the start there is a statement reminding us of the need to keep evolution, and therefore biological utility, in mind as the basic explanation for the continued presence of any trait in the behavioral repertoire of a species. The first part of the book then continues with explorations of the behavioral consequences of specified genetic anomalies. This is one way in which a specific pathway may be found leading from gene to behavior. In the second part of the book various new research methods with twins are demonstrated, and in the last part, several statistical methods are proposed for the unraveling of the mutual interrelations at the genetic (as well as on the environmental) level between a number of variables. These methods promise to furnish us in time with psychological tests specially tailored for use in studies of the modes of inheritance of psychological factors.

The chapters in this volume comprise part of the papers presented at the Second Invitational Conference on Human Behavior Genetics, held in Louisville, Kentucky, from April 30 to May 2, 1966. This conference was supported, in part, by a grant (1–R13–MH 12,638) from the National Institute of Mental Health, U.S. Public Health Service. Other aspects of the conference were supported by the U.S. Public Health Service through the grant for the Louisville Twin Study (HD–00843) and through a Career Development Award to Vandenberg (K3–MH–18,382). The University of Colorado has supported some of the final phases such as editing, proofreading, and related correspondence in the preparation of this book. Many of the papers were typed or retyped by Mrs. Mickey Gliesner, Mrs. Betty Matthews and Miss Linda McCarthy. I am grateful for all their help.

STEVEN G. VANDENBERG
University of Colorado

LIST OF CONTRIBUTORS

V. Elving Anderson
Dight Institute for Human Genetics
University of Minnesota

J. Bradford Block
University of Cincinnati
School of Medicine

R. Darrell Bock
University of Chicago

Anne Brown
University of Louisville
School of Medicine

Harvey F. Dingman
California State Department of Mental Hygiene
Pacific State Hospital, Pomona, Calif.
(Now at the University of Texas)

L. Erlenmeyer-Kimling
Department of Medical Genetics
New York State Psychiatric Institute

Daniel G. Freedman
Committee on Human Development
University of Chicago

Irving I. Gottesman
University of Minnesota

Axel Hoffer
Adult Psychiatry Branch
National Institute of Mental Health

Ronald C. Johnson
University of Colorado

Arnold R. Kaplan
Laboratory of Medical Genetics
Cleveland Psychiatric Institute

Einar Kringlen
Institute of Psychiatry
University of Bergen, Norway

John C. Loehlin
University of Texas

William Meredith
University of California, Berkeley

John Money
Department of Psychiatry and Behavioral Sciences & Department of Pediatrics
Johns Hopkins University
School of Medicine

Loren R. Mosher
Adult Psychiatry Branch
National Institute of Mental Health
(Now at Department of Psychiatry, School of Medicine, Yale University)

William Pollin
Adult Psychiatry Branch
National Institute of Mental Health

Glenn E. Roudabush
University of Pittsburgh
(Now at American Institute for Research, Palo Alto, California)

Sandra Scarr
Graduate School of Education
University of Pennsylvania

James Shields
MRC Psychiatric Genetics Research Unit
Maudsley Hospital, London, England

Felicia S. Siegel
Dight Institute for Human Genetics
University of Minnesota

Barbara J. Spillman
Adult Psychiatry Branch
National Institute of Mental Health

James R. Stabenau
Adult Psychiatry Branch
National Institute of Mental Health

Richard E. Stafford
Ohio State University

Steven G. Vandenberg
University of Colorado

Ronald S. Wilson
University of Louisville
School of Medicine

CONTENTS

PART I
Genetic Syndromes

Introduction

Several methods are available for the study of hereditary factors in human behavior. They may be summarized as family studies, studies of adopted children, the comparison of identical and fraternal twin concordance, the study of identical twins reared apart, the study of the effects of inbreeding, and the study of the effects of racial intermarriage. Another method consists of studying the psychological concomitants of specific gene substitutions or abnormal chromosome complements, or of other genetic abnormalities.

In this first part of the book, the latter approach is represented by a number of papers. Because they take as their starting point a known genetic anomaly, these studies are probably closer to the causal chain from abnormal genome through malfunctioning biochemistry to final atypical behavior, and should therefore give more pertinent information. The problem with this method is that it may at times be difficult to know what aspect of behavior to observe. One would hope that results from the other methods can help to answer this question, because they tell what aspects of behavior are, in part, controlled by heredity. Unfortunately there is no guarantee that the same genes that control normal variation in various behavioral traits will play an important role in genetic anomalies. It seems highly likely that by chance alone some of the loci involved in normal variation will also be involved in defects, but equally likely that a number of others will not be involved in such key roles.

On the other hand, it may be that normal variation is merely the result of the presence or absence of a large number of "abnormal" genes, most of which have small cumulative effects, so that all loci for normal variation also play a role in one or another abnormal conditions.

Regardless of one's view on the first question, it seems likely that studies of hereditary variation in "normal" populations may tell us which behavioral domains are especially responsive to hereditary variation, and which therefore form promising hunting grounds for students of genetic anomalies.

1

DANIEL G. FREEDMAN
Committee on Human Development
University of Chicago

AN EVOLUTIONARY FRAMEWORK FOR BEHAVIORAL RESEARCH

It has been my impression that behavior geneticists have a gimmick rather than a theory. We feel superior to other psychologists because we know that in diploid organisms individual differences are largely due to genetic variation and we continue to push this point wherever we can. This has served to upset the environmentalists' applecart, and strict environmentalism is now passé; but aside from this heuristic value to our work, we seem to be in the same boat as anyone else: up the creek without an over-all guiding theory.

Let me begin with my own case history. Starting as a clinical psychologist with a strong Gestalt-holistic bias, my Ph.D. thesis revealed to me, in a very dramatic way and not by design, the importance of genotype. I reared different breeds of dogs in two ways, hoping to prove a "purely" psychological hypothesis (Freedman, 1958); instead, I came away with striking breed-by-environment interactions, and I have been obsessed with such interactions ever since. Following the thesis, I began to look to geneticists for research leads. I visited Kopec at New York University with the notion of doing chromosome surgery on hamsters and relating this to behavior. It turned out to be perfectly possible. I then spent a year at the Institute for Medical Genetics in Uppsala where, among other things, the world of biochemical genetics was opened to me.

But whenever I became bored with reading and decided to do some work, I found myself involved in psychotherapy of twins, or studying babies (e.g., twins, blind babies, Mongoloid babies), or in some way dealing with humans very much as I had done as a clinician. I obviously still preferred to work with my subjects over a substantial period of time and through a developing relationship.

But something new had been added. On the basis of my new interest in genetics I had become an evolutionist, and the notion of adaptive function began filling my brain. In the arena of animal behavior, for example, it no longer concerned me, as it does so many animal psychologists, whether imprinting was traditional learning as opposed to a special kind of learning. It was clear that in the ground-nesting mallard, unless ducklings pursued the mother into the pond soon after hatching, the last mallard would have been eaten long ago. Imprinting is obviously something that has appeared under strong pressure of predators, and as Lorenz has said to those using barnyard chicks, "If you are going to study imprinting, study it in birds that imprint."

(Parenthetically, I prefer not to use the terms "innate" and "acquired," and instead I have found the simpler term "evolved behavior" much more congenial. Imprinting obviously involves both innate and acquired elements, and rather than get lost in a make-believe partitioning of these elements, I prefer to use the subsuming term, "evolved," or its synonym, "phylogenetically adaptive.")

We may now ask, what does this evolutionary thinking do for personality theory? Let me give examples. For several years I have been interested in the human smile and the fear of strangers, behaviors which I view as phylogenetically adaptive. Three years ago (Freedman, 1965) I presented data, since corroborated by further work, that these behaviors are significantly more concordant in identical twin infants than in same-sexed fraternals. Let me elaborate my thinking on this.

Since, from an evolutionary point of view, phylogenetically adaptive social signals must be matched by complementary receiving mechanisms, it has become clear to me that the baby's smile is meaningless without a sympathetic recipient or participant in that smile. Although my teacher, Kurt Goldstein, wrote this ten years ago (Goldstein, 1957), it has only recently taken on an evolutionary meaning for me. I now see that many evolved behavioral mechanisms in the infant have counterpart reactions in the caretaking adult.

For example, we will probably all agree that crying is a phylogenetically adaptive mechanism, and most newborn mammals, when out of the nest, start to cry. In dogs, one has only to watch the bitch's excited seeking to realize that her reaction constitutes an evolved mechanism complementary to the pup's cry. In the human, similarly, it can be demonstrated that within hours after birth, and before the first feeding, a crying infant will quiet when held and carried. Consider how this cessation of crying coordinates beautifully with the intense anxiety felt by the human parent until the infant is quieted. In this way the human baby does about as well as the macaque in getting next to the parent without having the ability to cling.

Let us consider, further, the two-month-old infant's persistent searching for the faces of adults and the wave of love an adult feels as eyes meet and the first smiles ensue. These feelings of love within the adult are data, too! (An infant smiles most readily at the full face view of an adult, and turning one's profile is like turning off a switch: the smile disappears and the baby searches with its eyes at about the level of your ear. In addition, there is considerable evidence, still largely unpublished, that babies prefer to look at models of the face rather than at various other competing configurations. See, for example, Fantz, 1961.)

A few weeks after smiling starts, the infant begins to coo at the beholding adult—try *not* to coo back at a vocalizing baby, as we have to do as experimenters, and see how unnatural it feels. The infant is now "talking" and we feel the irrestible urge to respond. I have little doubt but that these species-specific mutualities are the stuff social bonds are made of.

Consider further the clocking-in of laughter at about four months and the joy it gives us. Now the baby and caretaker can indulge in genuine mutual play; is there any reason to hold that the joy the adult feels is less of a mechanism than the laughter of the baby? As the first year progresses, a fear of strangers appears which draws the infant and caretaker even closer; by the time imitation and the first use of words start, late in the first year, social bonds are very strong and the child is an integral part of the lives of those about him.

I have thought most about infant-adult interactions, but evolved mechanisms are at work in all aspects of man's behavior. A particularly clear example to explore is man's constant engagement in dominance-submission testing, particularly among males (as in other primate species). One can see the competitive interplay among young boys in any school, and when the same behavior occurs at home vis à vis the father we refer to it as the working of the Oedipus complex. As in rhesus monkeys, the hierarchy often starts with rough and tumble play and becomes more serious with age. Try as we may not to engage in it, no matter what culture we are reared in this behavior always characterizes a human group. Reconstruction of the social order of our progenitors, Australopithecus, suggests that they lived as groups of hunters, and the establishment of dominance-submission hierarchies, since they lead to dynamically stabilized groups, still suits us well.

One has to look through evolutionary glasses to find the meaningful units of behavior. It is clear that paper and pencil tests given to twins will not in themselves reveal the processes of evolution and that one has to look through evolutionary glasses to find the meaningful units of behavior. For a personality theorist this leads to a view that people often act in mutual concert or discord, and are built to send and receive cues in the service of various evolved behaviors. It will require some ingenuity to decide on the proper units of behavior and to put these to a meaningful test.

Let me give two examples of ongoing studies which appear to follow from the above development. Since we hypothesized that the baby's cry and the adult's need to do something about it are complementary inherited mechanisms, we have set up the following experiment. A tape of various baby-cries and "control" noises was played to adult subjects, and concurrent psycho-physiological measures were taken. As we expected, women reacted more than men, and both men and women with children reacted more intensely than those without. Here we again see illustrated probable E $\times$ G interactions, and if we had used twins we could say more about that. This is, in fact, our planned next step.

In a second study, which seeks to examine experimentally the adaptive value of beardedness, we put beards on some figures in the Thematic Apperception Test and used the regular version as a control. We found that for our male respondents, but not for females, the bearded figures tended to come out in stories as more independent and higher in status. To eliminate the possibility of stereotyped responses, we plan as a second step to present the pictures subliminally, using a tachistoscope, so that conscious registration is avoided. In this technique the judgments and associations are made to a neutral "masking" figure instead. This method may be used with a wide variety of facial expressions purported to have phylogenetically adaptive function (Freedman, 1967), such as blushing with shame, reddening with anger, the direct vs. the indirect threat-stare, and so on. In this area of facial responsivity, one might well use a twin population to study genetic variation in the *elicitation* of these behaviors.

As a final word, I should like to offer an evolutionary definition of personality. It derives from our work with twins, some of whom we have now followed from birth through six years. We have never had trouble describing the nuances of personality in fraternal individuals, but we have found it nearly impossible to speak about identical individuals with the same rich detail. A little introspection revealed why. The fact that we knew two individuals whose variation was so very much the same confused and tongue-tied us. This led to the definition: *Personality amounts to an individual's unique variation on the basic hominid theme.*

Just as all of our phylogenetically adaptive structures are standard, yet variable, so too for behavior. This is what gives us our individuality, and it leads to the trouble we have in describing two identical children. It is as if we are either constructed, or else deeply habituated, to perceive and assess another's uniqueness. (One could make a case for this tendency being phylogenetically adaptive [Freedman, 1967].) We are solving this difficulty by the use of films. Each identical twin is viewed and rated, as if he were a singleton, by a separate investigator.

In closing, I hope the point is clear that if psychologists continue to focus on the individual, the self, personality, *or any other ontogenetically limited concept,* they will be committing a major mistake. It strikes me as a safe prediction that most correlations obtained with twin studies will dry and blow away with time, and only those that attain comprehensibility in the light of our evolved nature will remain.

References

Fantz, L. 1961. The origin of form perception. *Sci. Amer.* **204:** 66-72.

Freedman, D.G. 1958. Constitutional and environmental interactions in rearing four breeds of dogs. *Science* **127**: 585-86.

Freedman, D. G. 1965. An ethological approach to the genetical study of human behavior. In *Methods and Goals in Human Behavior Genetics,* ed. S. G. Vandenberg, pp. 141-61. New York: Academic Press.

Freedman, D. G. 1967. A biological view of man's social behavior. In *Social Behavior from Fish to Man,* ed. W. Etkin, pp. 152-88. Chicago: University of Chicago Press.

Goldstein, K. 1957. The smiling of the infant and the problem of understanding the "other." *J. Psychol.* **44:** 175-91.

2

V. ELVING ANDERSON and FELICIA SIEGEL
Dight Institute for Human Genetics
University of Minnesota

STUDIES OF BEHAVIOR IN GENETICALLY DEFINED SYNDROMES IN MAN

Our interest in this topic arose several years ago while preparing a review of genetics in mental retardation (Anderson, 1964). This review included a list of over 50 different conditions associated with mental retardation which can be shown to be genetic in origin. In some conditions a specific inborn error of metabolism is known. In others the pattern of inheritance is sufficient to establish a genetic basis, even though the physiological mechanisms are not understood. Newly defined syndromes are being added to the literature frequently (Waisman, 1966).

A variety of psychological tests have been employed for a few of these conditions, notably Down's syndrome and Turner's syndrome, which will be discussed extensively later in this symposium. In many of the other disorders a great deal of attention has been given to biochemical details, but relatively little to behavior.

Furthermore, the emphasis on "mental retardation" has tended to obscure the possible presence of other behavioral signs. The clinical reports, however, included comments on aspects of behavior not readily identified by IQ tests, such as attention span, hyperactivity, irritability, or emotional instability. Some affected children show psychotic-like behavior, and on occasion the presenting symptoms may resemble schizophrenia (Lippman, Perry, and Wright, 1958).

These observations based on the literature suggested the possible value of a more extensive study of behavior in these genetically conditioned syndromes. To the psychologist this approach would offer a set of subjects defined on objective laboratory criteria not related to behavior, and the potential opportunity to relate changes in behavior with concurrent biochemical findings. The values to the clinician might include a more

---

The work in progress reported here has been supported in part by a grant from the Graduate School of the University of Minnesota, and by grant HD 01396 from the National Institutes of Health.

adequate measure of the effectiveness of therapy and a better way to identify the constellation of problems in a specific child.

For the geneticist, behavior is an important aspect of physiological genetics, tracing the pathways from genes to traits. The approach suggested here is an illustration of the genotypic approach to behavior discussed by Scott and Fuller (1963). Merrell (1965) has more recently stressed the advantages offered by the analysis of the effects of single gene differences upon behavior. Finally, the interest expressed in this symposium is evidence that behavior genetics has reached the point where reasonably adequate techniques and models are available for the study of behavior in man.

Two general problems are involved, which might well be the subject of later discussion. (1) Should we expect qualitative or quantitative differences in behavior as the various syndromes are compared? Reitan (1959) has examined subjects with brain damage with a variety of tests, and his data indicated quantitative, but not qualitative, deviations from normal behavior. Perhaps similar results will characterize the genetic syndromes. (2) How can we establish the cause-effect relationships involved? Any deviant behavior observed in affected children could result from neurological damage or it could represent the child's awareness of the serious nature of his disease or his response to the necessary therapy. Scheinberg (1958), in his discussion of Wilson's disease, pointed out that both explanations must be kept open until decisive evidence can be obtained.

Which of the syndromes is likely to be most rewarding for the study of behavior? The following factors might enter into such a decision. (1) Survival should be high enough to permit following the subjects through childhood. (2) The mean IQ should be high enough to permit testing a reasonably wide range of behavior. (3) The frequency in the population should be high enough to permit the comparison of findings among subjects and institutions. (4) A more adequate physiological interpretation will be possible if the biochemical pathways are known. (5) If a means of treatment is available, the treated subjects will generally have higher IQ's, and the changes of behavior in response to therapy can be studied.

These points may be illustrated by some of the findings in galactosemia, Hartnup disease, histidinemia, and phenylketonuria. All four of these conditions are inherited as autosomal recessive traits, and the biochemical defects have been explored extensively. In untreated cases the degree of mental retardation is most severe in phenylketonuria, less marked in galactosemia, and relatively mild in histidinemia, with probably no retardation (but episodes of ataxia and psychological disturbances) in Hartnup disease. A method of treatment is not generally needed for histidinemia, but is available for the other three conditions. The survival rates and the mean IQ levels, at least for treated cases, are suitable for the purposes of psychological testing. Children with galactosemia and phenylketonuria are seen in most medical centers, while Hartnup disease and histidinemia are less common. The general features of the conditions are described by Hsia (1966).

Galactosemia results from an enzyme deficiency which leads to the accumulation of galactose-1-phosphate in the lens of the eye and other tissues. Many affected children, unless treated by a galactose-free diet, develop cataracts, enlargement of the liver and spleen, jaundice, and mental retardation. A few children with the enzyme deficiency show none of the classic signs or symptoms (Baker *et al.,* 1966).

Children who have had good dietary control perform significantly better on intelligence tests than those who have had poor dietary control or none at all (Hsia and Walker, 1961). Some with early diagnosis and treatment, however, may show a marked distractibility and lack of concentration accompanying lack of interest (Holzel, 1964).

Fishler *et al.* (1966) have followed 34 children with galactosemia over a period of eight years. About half of the children aged five or more years showed some degree of visual-perceptual difficulty. It was the impression of these authors that children with galactosemia seem to exhibit distinct characteristics of behavior not seen in disorders such as hypoglycemia or phenylketonuria. "The younger age level children are typically anxious and fearful in interpersonal contacts, shying away from people and their peers. They often manifest signs of emotional disturbance including thumb-sucking, nail-biting, or bed-wetting. . . . Older galactosemic children, especially at puberty, go through stormy periods of adjustment because of their inability to handle environmental impact."

The clinical manifestations of Hartnup disease are intermittent and resemble the signs found in pellagra. A red, scaly rash develops during the "attacks," and some patients show a severe but fully reversible cerebellar ataxia (Jepson, 1966). The attacks appear to be precipitated by an inadequate or irregular diet.

The underlying biochemical problem involves a defect in the transport of certain neutral alpha-amino acids (including tryptophan) across membranes in the intestines and kidney (Scriver, 1965). One of the results is an unusually high urinary level of a number of amino acids. Marked improvement in the dermatitis and the neurological signs usually follows treatment with oral nicotinamide.

Some of the earlier reported cases were mentally retarded, but this is not a consistent finding. Rodnight (1961) has emphasized the possible implications for wider problems of mental illness. Some cases of Hartnup disease presented as a confusional psychosis, a depression, or an anxiety state, and were initially treated as psychiatric illnesses. A more detailed study of such behavioral changes would appear desirable.

Histidinemia was first described in 1960. A deficiency of the enzyme histidase results in an elevated blood level of the amino acid histidine. Affected children show a "false-positive" reaction on the ferric chloride test of the urine for the diagnosis of phenylketonuria. The first reported cases appeared to have a characteristic speech disorder, but further study has suggested a general mild mental retardation rather than a specific malfunction of speech (Berlow *et al.,* 1965; Efron, 1965).

Two patients studied by a speech pathologist showed disorders of both articulation and language (Witkop and Henry, 1963). The children could not move the tongue independently from the mandible. As a result, "la, la, la" became "ya, ya, ya." In addition, two-phrase sentences "were scrambled nominally, syntactically, and serially with errors in pattern changing on each attempt." Other cases with histidinemia should be evaluated carefully by specialists in speech and other aspects of behavior.

Phenylketonuria (PKU) was first recognized in 1934 through the discovery of an unusual constituent in the urine from a mentally retarded brother and sister in Norway. Biochemical studies later showed an enzyme deficiency which led to a high plasma level of phenylalanine and other metabolites and a subsequent overflow of some of these through the kidney. A reduction of tyrosine and alterations in the metabolism of tryptophan and epinephrine are also involved (Lyman, 1963).

Phenylketonuria appears to produce two effects on the nervous system: (1) a permanent effect upon the developing brain, and (2) other effects upon brain function which are alterable through changing the level of phenylalanine intake. In untreated children the mean IQ is about 30. If a low-phenylalanine diet is initiated before four months of age, the mean IQ turns out to be in the 80-90 range (Berman *et al.*, 1966). If the diet is started much after one year of age there seems to be less beneficial effect on IQ measures. However, it has been reported that other changes might be observed, such as an increased attention span or a decrease in seizures (Bickel and Grüter, 1963).

Fuller (1967) has studied 112 phenylketonurics and a comparison group of outpatients with other conditions (most of whom were retarded). Separate scores were evaluated for the four Gesell test categories. Among the phenylketonurics the adaptive and language areas were most severely affected, while motor and personal-social development were least severely affected. The pattern of scores was more evenly distributed in the comparison group. Taking the phenylketonurics as a group, those placed earlier on dietary treatment showed less impairment. However, some individuals with PKU maintained on good dietary control since shortly after birth showed marked retardation, and some started on the diet after three years of age showed great improvement.

The variability in IQ and other aspects of behavior in these children has been puzzling, but some possible explanations can be advanced (Kleinman, 1964). It has been assumed that affected children are alike with respect to homozygosity for the PKU gene, but some heterozygotes may show an elevated serum level of phenylalanine (Anderson *et al.*, 1966). Other mutant genes may produce a "leaky" or partially functioning enzyme, a defective co-factor, or an enzyme inhibitor. Still other mutants may involve different enzymes affecting phenylalanine metabolism (Bessman, 1964).

In addition, the children may vary in genes at other modifying loci. The accumulated phenylalanine must be handled by other enzymes which may be present in adequate amounts in normal persons. In some children with PKU, however, even a minor deficiency in secondary enzyme systems may be taxed by the increased concentration of phenylalanine (or other metabolites) which must be handled. As a result of such genetic variation the relative concentration of secondary metabolites may vary widely from one phenylketonuric child to another.

Furthermore, there is growing evidence for genetic variation in the nervous system response of experimental animals to drugs (Meier, 1963). It is thus possible that phenylketonuric children with precisely the same blood levels of several secondary metabolites might nevertheless show differences in behavioral response.

A pilot study was initiated several years ago (Anderson, Shechtman, and Fisch, 1964) to study two questions: (1) How do the patterns of behavior in PKU differ from those in other syndromes? and (2) Which aspects of behavior are altered by dietary change? A detailed study was made of behavior changes in three children upon controlled alteration of diet. One boy was hospitalized for ten days, a second for fifteen days, and a girl was kept at home with visits to the hospital for observation. All three were about six years of age.

Each child was initially placed on a diet providing a total daily phenylalanine intake of 17 mg. per kilogram of body weight. The diet was similar to that used routinely for the children in their homes, and included a Lofenalac cornstarch pudding. On the fifth day a milk and egg cornstarch pudding with added L-phenylalanine was substituted, raising the total daily intake of phenylalanine to 100 mg. per kg. of body weight. Each child was tested three or more times using the Merrill-Palmer Scale or appropriate items selected from the test. Actometer ratios were obtained as measures of activity level (Schulman and Reisman, 1959). Fasting blood samples were used to determine the level of serum phenylalanine. Urine collections were made over 24-hour periods, and the samples were frozen for later biochemical study.

In two of the three children a decrease in IQ scores was associated with an increase in serum phenylalanine level. More significantly, variations in performance on the subtests appeared to reflect changes in speed of performance and in perception of error. Hyperactivity (i. e., undirected, random activity) and distractibility decreased on the low-phenylalanine diet. Perseveration was less apparent, tremors decreased, and critical ability seemed to improve.

These results provided evidence that certain aspects of behavior may change within a few days after alteration of diet. Two further problems became apparent, however: (1) What are the effects of learning with repeated testing over a short period of time? (2) Can the results from a short-term trial be used to predict the effects of a long-term use of a high-level diet? A child's body may possess compensatory mechanisms whereby some of the detrimental effects can be minimized and a high level diet more readily tolerated.

At this point we adopted two major changes in research strategy. (1) Tests were chosen which were appropriate for repeated trials and observations. For example, tasks using a modification of the Wisconsin General Test Apparatus have been given at approximately weekly intervals until a plateau is reached before a change in diet is instituted. (2) Instrumentation was employed to provide precise measurement of response latencies.

Testing is still underway, and the data are not ready for publication. It is possible, however, to review two different types of tests, each on a different population, and to discuss some preliminary impressions.

The first of these studies involves 12 phenylketonuric children who attend the PKU Clinic at the University of Minnesota Hospitals, under the medical supervision of Dr. Robert Fisch. They range in age from five to fourteen years and in IQ from 43 to 104. At the time of testing, seven were being maintained on a low-phenylalanine diet.

The Minneapolis Board of Education permitted selection of a control group from the elementary schools matched with the experimental group for age, IQ, sex, and race, with our further stipulation that a control subject should not come from a neglected home or have an obvious motor disability (such as cerebral palsy). This control series turned out to include two children with Down's syndrome, but the remaining ten are not known to have any identifiable clinical entity.

Data from a psychomotor test (developed by Dr. Auke Tellegen) on these two groups are now essentially complete. This involves automatic recording of responses to five tasks: tapping, two pegboards (large and small pegs), finger dexterity, and steadiness. Each task is performed twice with each hand. The score for tapping is the number of taps in a 12-second period. The next three tasks involve progressively finer motor control, and the individual is scored for the number of seconds required to complete the task. The score for steadiness is the total number of seconds that a stylus held by the subject is not in contact with the rim of a hole into which the stylus is inserted (for an 18-second period).

We have been able to secure additional data from a normative sample of school children from kindergarten through grade six, with six children from each sex from each grade. The children were selected by their teacher, with the only stipulation that they were to be of "average ability." These data show an essentially linear improvement with age for each task. The mean scores for both the PKU and the control groups were significantly different from the normative data on all tests.

The scores for each pair (PKU and control) were then compared for each test, and a judgment was made as to whether the score for the PKU child was better or worse than the score for the control child. The normative data were used to correct for any intra-pair differences in age at time of test. Sequential analysis based on pair differences indicates that the PKU sample is farther from normal than the control sample just at the five per cent level of significance. These results would appear to suggest that the observed motor dysfunction in PKU is not simply a general function of the level of retardation.

A preliminary analysis taking diet into account, however, suggests that almost all of the differences between the PKU and control groups are contributed by the pairs in which the PKU child is on an ad lib diet. Several possible explanations for this observation remain to be explored.

Two of the five PKU children with IQ's over 85 were within the range of the normative group for all five tests. One was nine years, ten months, at the time of testing, with an IQ of 92, while the other was five years, eight months, with an IQ of 95.

The second study involves five PKU children (four boys and one girl) who are patients at the State School and Hospital in Faribault, under the medical supervision of Dr. Heinz Bruhl. Their ages range from seven to fifteen years. Three are profoundly retarded (less than 25 IQ) and two are trainable (IQ 35-39).

These children were tested weekly over a period of six months, using a position discrimination and reversal task. The measures used were response latencies and errors. A candy reward was given for correct responses.

Three types of diets were used, each being maintained for at least six weeks: a regular ad lib diet, a low phenylalanine diet, and a low phenylalanine diet plus enough added phenylalanine to bring the blood level to the level maintained on the regular diet. The latter diet is essential to distinguish between the effect of phenylalanine per se and the many other nutritional differences between the first two diets.

This population has turned out to be suitable for several reasons (Bruhl *et al.,* 1964). Dietary control is feasible for any desired length of time. Blood and urine samples can be collected. Children of varying mental and chronological ages are available in the institutions. Furthermore, the test employed is feasible for use with children having a low tested IQ.

For every child the addition of phenylalanine to the diet has resulted in an increase in response latency. However, two of the boys clearly show an adaptation, in that within two weeks after a high phenylalanine diet has been initiated the response latency has dropped to the level observed while on a low-phenylalanine diet. A third boy has not shown this adaptation. His response latencies vary somewhat from test to test, but the mean on low phenylalanine was 0.96 seconds. This rose to 1.13 with added phenylalanine and to 1.25 on an ad lib diet. Two other children showed considerable variability and we have not yet tried to interpret their scores.

There are at least three possible explanations for the adaptation (or recovery) noticed in the two boys. (1) Biochemical mechanisms might become activated which have the effect of reducing the blood level of certain toxic metabolites. (2) The nervous system may become less responsive to the metabolites. (3) The child may learn to compensate for the behavioral limitation involved. The urine samples will be analyzed for evidence related to the first possibility.

Both of these points can be illustrated in the case history of S. B., a boy twelve years old with an IQ of about 70. The diagnosis of PKU was not made until age ten. At that time he cooperated in going on diet "if it kills me." Recently, it became of some importance to decide how essential the diet was. A test of size and color discrimination was administered for five trials over a two-month period before he was changed from a low-phenylalanine to an ad lib diet. A week later the mean response was considerably longer, but after an additional three weeks on the ad lib diet the response latency was back down to the level before the diet change. Thus, S. B. appears to show the adaptation phenomenon. On the motor coordination battery the only change upon alteration of diet was a two-fold increase in contact time of the steadiness probe. His handwriting showed larger letters, increased tremor, and less control. He frequently dropped the candy used as a reward in the discrimination task.

Several other tests are now in use or in preparation. A school behavior profile (including 65 questions) has been sent to the classroom teachers of those PKU and control children who are in school. A delayed response task using form-color stimuli is administered after delay intervals of zero, three, and twenty seconds. In order to investigate the more complex memory functions a short-term memory task has been developed by modifying the procedure of Atkinson, Hansen, and Bernbach (1964). The variable of attention or vigilance is being approached by a modification of the Continuous Performance Test (Rosvold *et al.,* 1956), using a carousel projector and selected visual stimuli from the Binet picture vocabulary.

SUMMARY

The careful study of behavior in genetically defined syndromes is of potential value for the fields of medicine, psychology, and genetics. Some of the findings were reviewed briefly for galactosemia, Hartnup disease, and histidinemia, and more extensively for phenylketonuria. Some tentative results were then presented from studies in progress directed toward two questions: (1) How do the patterns of behavior in phenylketonuria differ from those in other syndromes? and (2) Which aspects of behavior are altered by dietary change?

References

Anderson, J. A.; Fisch, R.; Miller, E.; and Doeden, D. 1966. Atypical phenylketonuric heterozygote. Deficiency in phenylalanine hydroxylase and transaminase activity. *J. Pediat.* **68:** 351-60.

Anderson, V. E. 1964. Genetics in mental retardation. In *Mental Retardation,* ed. H. A. Stevens and R. Heber, pp. 348-94. Chicago: University of Chicago Press.

Anderson, V. E.; Shechtman, A. M.; and Fisch, R. O. 1964. Behavior correlates of phenylketonuria upon alteration of diet. Paper presented at the American Society of Human Genetics. (Unpublished.)

Atkinson, R. C.; Hansen, D. N.; and Bernbach, H. A. 1964. Short-term memory with young children. *Psychon. Sci.* **1:** 255-56.

Baker, L.; Mellman, W. J.; Tedesco, T. A.; and Segal, S. 1966. Galactosemia: Symptomatic and asymptomatic homozygotes in one Negro sibship. *J. Pediat.* **68:** 551-58.

Berlow, S.; Arends, R.; and Harries, C. 1965. Studies in histidinemia. *Lancet* **85:** 241-46.

Berman, P. W.; Waisman, H. A.; and Graham, F. K. 1966. Intelligence in treated phenylketonuric children: a developmental study. *Child Developm.* **37:** 731-47.

Bessman, S. P. 1964. Some biochemical lessons to be learned from phenylketonuria. *J. Pediat.* **64:** 828-38.

Bickel, H., and Grüter, W. 1963. Management of phenylketonuria. In *Phenylketonuria,* ed. F. L. Lyman, pp. 136-72. Springfield, Ill.: C C Thomas.

Bruhl, H. H.; Arnesen, J. F.; and Bruhl, M. G. 1964. Effect of a low-phenylalanine diet on older phenylketonuria patients (long-range controlled study). *Amer. J. Ment. Defic.* **69:** 225-35.

Efron, M. L. 1965. Aminoaciduria. *New Engl. J. Med.* **272:** 1058-67, 1107-13.

Fishler, K; Koch, R; Donnell, G; and Graliker, B. V. 1966. Psychological correlates in galactosemia. *Amer. J. Ment. Defic.* **71:** 116-25.

Fuller, R. 1967. Psychological results in treated phenylketonuria. I. Gesell findings. In *Psychopathology of Mental Development,* ed. J. Zubin and G. A. Jervis, pp. 153-80. New York: Grune & Stratton.

Holzel, A. 1964. Ten year follow-up study cases of galactosemia. In *Neurometabolic Disorders in Childhood,* ed. K. S. Holt and J. Milner, pp. 83-87. Edinburgh: E. & S. Livingstone, Ltd.

Hsia, D. Y-Y. 1966. *Inborn Errors of Metabolism. Part I. Clinical Aspects,* 2d ed. Chicago: Medical Year Book Publishers.

Hsia, D. Y-Y. and Walker, F. A. 1961. Variability in the clinical manifestations of galactosemia. *J. Pediat.* **59:** 872-83.

Jepson, J. B. 1966. Hartnup disease. In *The Metabolic Basis of Inherited Disease,* 2d ed.; ed. J. B. Stanbury, J. B. Wingaarden and D. S. Fredrickson, pp. 1283-99. New York: McGraw-Hill.

Kleinman, D. S. 1964. Phenylketonuria. A review of some deficits in our information. *Pediatrics* **33:** 123-34.

Lippman, R. W.; Perry, T. L.; and Wright, S. W. 1958. The biochemical basis of mental dysfunction. II. Mental deficiency (amentia). *Metabolism* **7:** 274-80.

Lyman, F. L. 1963. *Phenylketonuria.* Springfield, Ill.: C C Thomas.

Meier, H. 1963. *Experimental Pharmacogenetics. Physiopathology of Heredity and Pharmacologic Responses.* New York: Academic Press.

Merrell, D. J. 1965. Methodology in behavior genetics. *J. Hered.* **56:** 263-66.

Reitan, R. M. 1959. Impairment of abstraction ability in brain damage: Quantitative versus qualitative changes. *J. Psychol.* **48:** 97-102.

Rodnight, R. 1961. Body fluid indoles in mental illness. *Int. Rev. Neurobiology* **3:** 251-92.

Rosvold, H. E.; Mirsky, A. F.; Sarason, I; Bransome, E. D., Jr.; and Beck, L. H. 1956. A continuous performance test of brain damage. *J. Consult. Psychol.* **20:** 343-50.

Scheinberg, I. H. 1958. Hereditary defects in protein synthesis as related to psychiatry. *Dis. Nerv. Syst.* **19:** 25-30.

Schulman, J. L. and Reisman, J. M. 1959. An objective measure of hyperactivity. *Amer. J. Ment. Defic.* **64:** 455-56.

Scott, J. P. and Fuller, J. L. 1963. Behavioral differences. In *Methodology in Mammalian Genetics,* ed. W. J. Burdette, pp. 283-96. San Francisco: Holden-Day.

Scriver, C. R. 1965. Hartnup disease: A genetic modification of intestinal and renal transport of certain neutral alpha-amino acids. *New Eng. J. Med.* **273:** 530-32.

Waisman, H. A. 1966. Some newer inborn errors of metabolism. *Pediat. Clin. No. Amer.* **13:** 469-501.

Witkop, C. J. and Henry, F. V. 1963. Sjögren-Larsson syndrome and histidinemia: Hereditary biochemical diseases with defects of speech and oral functions. *J. Speech Dis.* **28:** 109-23.

3

HARVEY F. DINGMAN
California State Department of Mental Hygiene
Pacific State Hospital
Pomona, California
(Now at the University of Texas)

PSYCHOLOGICAL TEST PATTERNS IN DOWN'S SYNDROME

INTRODUCTION

Mental retardation has become the source of considerable public concern, and substantial funds have been appropriated, on both national and local levels, to investigate this problem further. The continuing research on the genetics of mental retardation has received a new emphasis from the studies of Reed and Reed (1965) and Anderson (1964), as well as from the many studies of chromosomes that are being carried out with the mentally retarded (Robinson and Robinson, 1965).

Pollitt and Money (1964) have discussed the use of standardized test scores with mentally retarded individuals. The difficulty in utilizing standardized test scores with children has been re-emphasized in a recent monograph by Stott and Ball (1963), as well as in an earlier article by Meyers and Dingman (1960).

Following nearly a half century's efforts at differentiating the scores on mental tests into their appropriate components, an attempt is made by Meyers *et al.* (1964) to specify some independent aspects of mental development in children, both normal and retarded, at mental ages two, four, and six. In retrospect, this model has been judged to be consistent with the Structure of Intellect model proposed by Guilford (1956) in a series of publications.

The separate test scores used by Meyers *et al.* (1964) were compared with the diagnoses for the mentally retarded children and, while some consistent findings were obtained, it now seems more appropriate to compute the factor scores for each subject in the Meyers *et al.* (1964) sample

---

This investigation was supported by Public Health Service Research Grant No. MH-08667; Socio-Behavioral Study Center for Mental Retardation, Pacific State Hospital, Pomona, California; and Public Health Service General Research Support Grant No. 1-S01-FR-05632-01; Pacific State Hospital, Pomona, California.

and to make comparisons with such specific syndromes as Down's syndrome. Of course several different chromosomal abnormalities lead to Down's syndrome, and the label "Down's Syndrome" covers them all; but while these tabulations are presented for a gross group of chromosomal disorders, they still represent what is yet today a single psychiatric entity.

Rather than compare the patients with Down's syndrome to normal children, it was decided to make comparisons with patients who have specific disorders of a relatively verifiable nature and with patients who have as yet no satisfactory verifiable disorder other than mental deficiency. Finally, some anthropometric data available on part of the sample were correlated with the factor scores. Since some kinds of microcephaly are hereditary and since many patients with mongolism have very small heads, the test scores were compared with head size.

METHOD

Factor scores were computed for the mentally retarded subjects who were evaluated as part of the Meyers *et al.* (1964) project. These factor scores were computed using the normalizing equations and the Procrustes rotation which had been developed from the data of the normal sample; that is, these factor scores are deviations from the mean factor scores of the normal group using the rotations specified for the normal subjects at each age level. The diagnoses used for most of our statistical reports were taken from the computer files, and they are listed in Tables 3-1 and 3-2. These diagnoses are not the current American Association on Mental Deficiency diagnoses but rather are coding groups established by the Statistical Research Bureau of the California State Department of Mental Hygiene in a memorandum dated August 1, 1960. The body measurements included in the study were taken as part of a cross-sectional study of physical growth measures reported by Mosier, Grossman, and Dingman (1965) ; the techniques are fully reported there but are those that are usually considered standard.

RESULTS

Table 3-1 shows the nature of the tasks that define the four ability factors used in the present analysis. A detailed description of these tasks and instructions for their administration may be found in Meyers *et al.* (1964) . The four abilities are : (1) linguistic ability; (2) hand-eye psychomotor coordination; (3) perceptual speed; (4) figural reasoning.

Table 3-2 gives a summary of the results. There were 56 cases in the Mental Age 2 group, 50 in the Mental Age 4 group and 46 in the Mental

Age 6 group. Their distribution over the various categories is shown in the three columns on the right of the table.

The means for each of these groups on the four factor scores are reported in the three columns on the left of the table, while the three center columns present the standard deviations.

As can be seen from the tables, the patients with Down's syndrome uniformly have negative mean factor scores on linguistic ability, and in the two younger groups these deviations are of substantial size.

On the MA 6 scores for all subjects and all factors, few extremely large negative mean scores are found, indicating that the older ages of all subjects have given them an opportunity to catch up, particularly in psychomotor skills and linguistic reasoning. The positive figural reasoning scores at the younger age level indicate that the subjects are somewhat more advanced compared to normals, but at the older mental age level this advantage is lost, with patients with mongolism again sharing the common trend rather than being different from the others.

The correlations between the physical measurements and the test factor scores were generally negligible, with the following exceptions: when linguistic ability is correlated with total height (crown heel), a correlation of .86 is obtained whereas a correlation of only .37 is obtained when linguistic ability is correlated with head circumference, thus indicating that probably some general developmental phenomenon is more responsible for the growth of linguistic ability than just a specific deviation due to a general diagnostic category such as microcephaly. Thus, even though four factors similar to Guilford's (1956) Structure of the Intellect were obtained for research purposes and the results were tabulated by specific diagnostic categories, still no significant pattern appears obvious with respect to diagnostic category.

It has often been noticed that patients with mongolism have low linguistic ability scores, but, as can be seen from these data, their linguistic ability scores are not necessarily lower than those of other groups, and, when matched for mental age level, the deficit in linguistic ability tends to disappear. Thus, since the mongoloid patient tends to have limited IQ (Zeaman, 1962; Ross, 1962), one could expect that the mental age categories would tend to wash out those findings that are so often "clinically" noticed.

SUMMARY

Comparing factor scores from tests similar to Guilford's (1956) Structure of the Intellect and using factor equations calculated on a normal sample, it was discovered that there are no systematic differences between Down's syndrome patients and other mentally retarded patients, but that the differences seem to be due to developmental growth and not to genetic structure.

References

Anderson, V. E. 1964. Genetics in mental retardation. In *Mental Retardation,* ed. H. A. Stevens and R. Heber, pp. 348-94. Chicago: University of Chicago Press.

Guilford, J. P. 1956. The structure of intellect. *Psychol. Bull.* **53:** 267-93.

Meyers, C. E. and Dingman, H. F. 1960. The structure of abilities at the preschool ages: Hypothesized domains. *Psychol. Bull.* **57**: 514-32.

Meyers, C. E.; Dingman, H. F.; Orpet, R. E.; Sitkei, G.; and Watts, C. A. 1964. Four ability-factor hypotheses at three preliterate levels in normal and retarded children. *Monog. Soc. Res. Child Devel.,* Serial No. 96, **29:** 5.

Mosier, H. D.; Grossman, H. J.; and Dingman, H. F. 1965. Physical growth in mental defectives: A study in an institutionalized population. *Pediatrics* **36:** 3, suppl. 2.

Pollitt, E. and Money, J. 1964. Studies in the psychology of dwarfism. I. Intelligence quotient and school achievement. *J. Pediat.* **64:** 415-21.

Reed, Elizabeth W. and Reed, S. C. 1965. *Mental Retardation: A Family Study.* Philadelphia: W. B. Saunders.

Robinson, H. B. and Robinson, Nancy M. 1965. *The Mentally Retarded Child.* New York: McGraw-Hill.

Ross, R. T. 1962. The mental growth of mongoloid defectives. *Amer. J. Ment. Defic.* **66:** 736-38.

Stott, L. and Ball, Rachel S. 1963. *Evaluation of Infant and Pre-School Mental Tests.* Detroit: The Merrill-Palmer Institute.

Zeaman, D. and House, Betty J. 1962. Mongoloid MA is proportional to log CA. *Child Devel.* **33:** 481-88.

Table 3-1

Procrustes Rotated Factors for the Two-Year, Four-Year, and Six-Year Groups

Test Names	Factor Loadings
Two-Year Groups	
Linguistic Ability	
1. Pacific Expressive Vocabulary and Expressive Language Check List	63[a]
2. Pacific Receptive Vocabulary and Receptive Language Check List	65
3. Pacific Identification-By-Use	58
Hand-Eye Psychomotor	
1. Bead Stringing	55
2. Disk Stacking	82
3. Cube Stacking	37

Test Names	Factor Loadings
Perceptual Speed	
1. Form-Color-Size Matching	62
2. Form-Color Matching	77
3. Form Matching	60
Figural Reasoning	
1. Pacific Pattern Completion	32
2. Pacific Form and Picture Completion	72
3. Design Copying	43
Four-Year Groups	
Linguistic Ability	
1. Pacific Expressive Vocabulary	64
2. Pacific Receptive Vocabulary with Ammons FRPV	49
3. Response to Pictures and Monroe Ideational Fluency	56
Hand-Eye Psychomotor Coordination	
1. Bead Stringing	42
2. Disk Stacking	46
3. Cube Stacking	57
Perceptual Speed	
1. Pacific Color-Form Matching	69
2. Pacific Figure Matching	54
3. Pacific Design Discrimination	34
Figural Reasoning	
1. Pacific Object Classification	46
2. Pre-Raven Pattern Completion	36
3. Design Copying and Pacific Pattern Copying	21
Six-Year Groups	
Linguistic Ability	
1. Pacific Expressive Vocabulary	60
2. Pacific Receptive Vocabulary with Ammons FRPV	57
3. Monroe Ideational Fluency	45
Hand-Eye Psychomotor	
1. Bead Stringing	59
2. PMA Motor Test	48
3. Circle Dotting	60
Perceptual Speed	
1. PMA Picture Matching	58
2. PMA Figure Matching	41
3. Pacific Form Matching	20
Figural Reasoning	
1. IPAT Classification	45
2. Raven Matrices	49
3. Pacific Pattern Copying and Design Copying	30

[a] All decimals omitted.

TABLE 3-2
MEANS AND STANDARD DEVIATIONS OF FACTOR SCORES BY FACTOR AND DIAGNOSIS FOR MENTALLY RETARDED PATIENTS WITH MA'S OF 2, 4, AND 6

	Mean Factor Scores			Standard Deviation			Number of Cases		
Diagnosis Category	MA 2	MA 4	MA 6	MA 2	MA 4	MA 6	MA 2	MA 4	MA 6
Linguistic Ability									
52 (Mongolism)	−2.17	−3.85	− .16	1.41	1.86	.83	16	15	8
53 (DCA)	−2.58	—	.88	2.30	—	1.44	3	—	3
55 (Infection)	−2.43	−3.02	1.13	2.94	1.08	1.42	2	5	3
56, 57 (Trauma)	− .86	−4.48	.75	1.09	3.08	1.32	5	3	6
59, 62 (Glandular, Organic)	−2.88	—	—	1.52	—	—	2	—	—
63 (Hereditary)	−2.53	—	.14	1.94	—	1.38	15	—	11
51 (Familial)	−2.45	−1.43	.74	2.21	.57	.84	2	2	4
64 (Undifferentiated)	−1.98	−3.58	.56	1.60	1.71	1.04	11	8	9
53, 58, 59 (DCA, Epilepsy, Glandular)	—	−3.07	—	—	1.46	—	—	3	—
62, 63 (Organic, Hereditary)	—	−2.90	—	—	.85	—	—	4	—
58 (Epilepsy)	—	—	− .98	—	—	1.67	—	—	2
Hand-Eye Psychomotor									
52 (Mongolism)	1.98	− .39	− .45	1.16	2.96	1.26	16	15	8
53 (DCA)	1.63	—	− .42	1.60	—	1.67	3	—	3
55 (Infection)	1.10	.78	−1.02	1.68	2.82	.12	2	5	3
56, 67 (Trauma)	1.09	−2.21	.26	.80	2.43	.74	5	3	6
59, 62 (Glandular, Organic)	2.24	—	—	1.14	—	—	2	—	—
63 (Hereditary)	2.02	—	.16	2.09	—	.73	15	—	11
51 (Familial)	1.88	−1.13	− .56	1.75	2.76	.56	2	2	4
64 (Undifferentiated)	1.81	− .70	.23	1.00	1.30	1.83	11	8	9
53, 58, 59 (DCA, Epilepsy, Glandular)	—	−1.24	—	—	2.70	—	—	3	—
62, 63 (Organic, Hereditary)	—	5.77	—	—	4.46	—	—	4	—
58 (Epilepsy)	—	—	.32	—	—	.95	—	—	2
Perceptual Speed									
52 (Mongolism)	− .13	2.86	− .58	.85	1.36	1.33	16	15	8
53 (DCA)	1.10	—	−1.26	1.65	—	1.30	3	—	3

55 (Infection)	.82	1.76	−1.33	2.68	1.91	2.39	2	5	3
56, 67 (Trauma)	− .24	4.16	−1.27	1.64	1.56	1.55	5	3	6
59, 62 (Glandular, Organic)	− .19	—	—	1.65	—	—	2	—	—
63 (Hereditary)	.25	—	.008	.66	—	1.36	15	—	11
51 (Familial)	− .35	1.14	− .25	2.15	1.07	1.87	2	2	4
64 (Undifferentiated)	− .17	2.49	−1.21	.72	.64	.83	11	8	9
53, 58, 59 (DCA, Epilepsy, Glandular)	—	2.08	—	—	1.56	—	—	3	—
62, 63 (Organic, Hereditary)	—	−1.88	—	—	2.61	—	—	4	—
58 (Epilepsy)	—	—	−1.61	—	—	2.02	—	—	2
Figural Reasoning									
52 (Mongolism)	1.41	−1.34	− .03	.92	1.08	.10	16	15	8
53 (DCA)	1.03	—	− .85	1.05	—	.32	3	—	3
55 (Infection)	1.26	− .99	− .98	1.50	1.26	1.77	2	5	3
56, 67 (Trauma)	.32	−2.12	− .89	.91	1.07	.77	5	3	6
59, 62 (Glandular, Organic)	1.27	—	—	.42	—	—	2	—	—
63 (Hereditary)	1.21	—	− .54	.99	—	1.06	15	—	11
51 (Familial)	1.42	−1.08	−1.24	.25	1.19	.84	2	2	4
64 (Undifferentiated)	1.07	−1.69	− .82	1.60	.83	1.49	11	8	9
53, 58, 59 (DCA, Epilepsy, Glandular)	—	−1.21	—	—	1.15	—	—	3	—
62, 63 (Organic, Hereditary)	—	−1.41	—	—	1.04	—	—	4	—
58 (Epilepsy)	—	—	− .98	—	—	1.71	—	—	2

KEY TO CODE OF DIAGNOSIS

52 = Mongolism
53 = With developmental cranial anomaly (presumably due to defective fetal development)
55 = Due to central nervous system infection (post natal) involving cerebrum
56, 57 = Due to trauma during birth; and due to trauma after birth
59, 62 = With glandular disorder; and with other organic nervous disease (hereditary)
63 = Other forms (hereditary causes)
51 = Familial mental retardation and familial-cultural retardation
64 = Undifferentiated
53, 58, 59 = With developmental cranial anomaly (presumably due to defective fetal development); due to epilepsy; and with glandular disorder
62, 63 = With other organic nervous disease (hereditary); and other forms (hereditary causes)
58 = Due to epilepsy

4

JOHN MONEY
Department of Psychiatry and Behavioral Sciences and Department of Pediatrics
The Johns Hopkins University School of Medicine

COGNITIVE DEFICITS IN TURNER'S SYNDROME

Among human beings, experimental inbreeding for behavioral traits must for ethical reasons be replaced by the procedures of fieldwork genetics. One such procedure, not yet widely employed in behavioral genetics, is the behavioral study of a clinical population which is known to be genetically distinct, and on some criterion homogeneous. One such population on which I wish to report is that of Turner's syndrome.

In the chromatin negative type of Turner's syndrome the karyotype is 44 + XO and in the chromatin positive type it is a mosaic (44 + XO/44 + XX) or there may be a translocation or deletion of chromosomal material which has similar effects (Bartalos and Baramki, 1967).

Individuals with Turner's syndrome are phenotypic females who are staturally dwarfed (rarely more than 5 feet tall in adulthood) and, until treated, sexually infantile at puberty, secondary to gonadal agenesis. There are many other somatic stigmata with which they may be affected, severally or singly, including congenital cardiac defect, webbed neck, epicanthal folds, micrognathia, peripheral lymphedema and pigmented moles.

It is erroneous to associate the behavioral attributes of over-all mental deficiency with Turner's syndrome (Money, 1963; 1964) as has frequently been done since the syndrome was first described in 1938. There is, however, a proneness to a specific cognitive deficit, first recognized in a verbal-nonverbal IQ disparity by Shaffer (1962) working in my psychohormonal unit. On the Wechsler Intelligence Scales, this deficit shows up in three factorially derived scores (Cohen, 1957; 1959). The specific factor score for Perceptual Organization (Block Design + Object Assembly) is low, and often extremely low when compared with the Verbal

Supported by Research Grant No. HD–00325 and Research Career Development Award No. HD–K3, 18635, The National Institute of Child Health and Human Development, The United States Public Health Service.

Comprehension score (Information + Similarities + Comprehension + Vocabulary). A third specific factor score which Cohen labeled Freedom from Distractibility has to do with numerals and calculation. It may also be low, relative to the verbal factor, though not so low as the perceptual or space-form factor.

This mild dyscalculia may be related to defective performance on an alliteration test (Word-fluency of the SRA Primary Mental Abilities Test) which also requires focalized attention and stereotyped adherence to a rule within a time limit (Alexander and Money, 1965). In this test the subjects were required to write in five minutes all words they could think of beginning with *s* and many of them did poorly.

The space-form blindness or visual-constructional dysgnosia associated with Turner's syndrome has been further substantiated by failures on other tests, namely, the Bender Visual-Motor Gestalt Test; the Benton Visual Retention Test; the Draw-a-Person Test (Goodenough-Harris); the Draw-a-Floor-Plan exercise; and the Space subtest of the SRA Primary Mental Abilities Test (Alexander and Money, 1965).

In view of the difficulty with shapes, it is not surprising that girls with Turner's syndrome also show difficulty with right-left directional discrimination (Alexander, Walker, and Money, 1964). In the teenage years and later, one expects this discriminatory skill to be well developed, but in the Turner's syndrome sample it was not. The patients, as a group, did very poorly on a Roadmap Test of Direction Sense, newly standardized for the purpose (Money, Alexander, and Walker, 1965). This roadmap test requires orientation to right and left simultaneously with orientation toward and away from the subject on a flat surface resembling part of a city map. The patient's deficiency on the roadmap showed up also in Benton and Kemble's Right-Left Discrimination Battery, which tests recognition of Right-Left on the person facing oneself; they were also weak on the Orientation Test of the Detroit Tests of Learning Aptitude, which requires imaginary locomotion and rotation of the body in space, without losing track of the direction one would finally be facing (Alexander and Money, 1966).

These three types of deficit—space-form dysgnosia, directional-sense dysgnosia, and mild dyscalculia—are the only ones that have emerged from the tests so far given. There has been no evidence of finger agnosia, color blindness, or color agnosia, dysgraphia, dyslexia, or aphasia. In fact, reading is the academic forte of these patients and they may be superior in it.

Directional sense difficulty with mild dyscalculia is suggestive of Gerstmann's syndrome, but finger agnosia and dysgraphia—which classically are included in the tetrad of symptoms in that disease—are missing. Because of the prominence of space-form disability, one thinks alternatively of a parietal lobe syndrome, and in particular of a developmental parietal lobe functional deficit, perhaps involving the nondominant more than the dominant hemisphere.

It is consistent with a hypothesis of right parietal lobe involvement that the personality of patients with Turner's syndrome (Hampson, Hampson, and Money, 1955; Shaffer, 1963) is phlegmatic and lacking in initiatory verve. Langworthy (1964) mentions a woman with right parietal lobe damage who seemed less alert, and more passive, placid, and indifferent to things that had formerly aroused her attention and emotion. Girls with Turner's syndrome display a placid stolidity and resignation to the special demands which life imposes on them by reason of the dwarfism, pubertal failure before treatment, sterility, and other physical disabilities. Their tolerance of these indignities is in marked contrast to the emotional disturbance that rather frequently characterizes the mothers.

There is no available direct evidence to implicate a specific relationship between the chromosomal deficit and the behavioral features of Turner's syndrome. Nonetheless, there is good presumptive evidence that one may attribute to the genetic defect the final responsibility for the behavioral as for the other stigmata of the syndrome. In this indirect way it is thus possible to attribute a specific and restricted deficit of intellect, as contrasted with general mental deficiency, to a genetic defect—and specifically to one involving an X chromosome.

In summary, many patients with Turner's syndrome (irrespective of sex-chromatin type) exhibit a degree of space-form dysgnosia, directional sense dysgnosia and mild dyscalculia that suggests a developmental right parietal lobe anomaly. It is presumably related to the chromosomal defect basic to the syndrome.

References

Alexander, D. and Money, J. 1965. Reading ability, object constancy and Turner's syndrome. *Percept. Mot. Skills* **20:** 981-84.

Alexander, D. and Money, J. 1966. Turner's syndrome and Gerstmann's syndrome: Neuropsychologic comparisons. *Neuropsychologia* **4:** 265-73.

Alexander, D.; Walker, H. T., Jr.; and Money, J. 1964. Studies in direction sense: I. Turner's syndrome. *Arch. Gen. Psychiat.* **10:** 337-39.

Bartalos, M. and Baramki, T. A. 1967. *Medical cytogentics.* Baltimore: Williams and Wilkins.

Cohen, Jacob. 1957. A factor-analytically based rationale for the Wechsler Adult Intelligence Scale. *J. Consult. Psychol.* **21:** 451-57.

Cohen, Jacob. 1959. The factorial structure of the WISC at ages 7–6, 10–6, and 13-6. *J. Consult. Psychol.* **23:** 285-99.

Hampson, J. L.; Hampson, J. G.; and Money, J. 1955. The syndrome of gonadal agenesis (ovarian agenesis) and male chromosomal pattern in girls and women. *Bull. of The Johns Hopkins Hosp.* **97:** 207-26.

Langworthy, O. 1964. Only half aware: A review. *Amer. J. Psychiat.* **121:** 116-22.

Money, J. 1963. Cytogenetic and psychosexual incongruities with a note on space-form blindness. *Amer. J. Psychiat.* **119:** 820-27.

Money, J. 1964. Two cytogenetic syndromes: Psychologic comparisons. I. Intelligence and specific-factor quotients. *J. Psychiat. Res.* **2:** 223-31.

Money, J. and Alexander, D. 1966. Turner's syndrome: Further demonstration of the presence of specific cognitional deficiencies. *J. Med. Genet.* **3:** 47-48.

Money, J.; Alexander, D.; and Walker, H. T., Jr. 1965. *A Standardized Road-Map Test of Direction Sense.* Baltimore: Johns Hopkins Press.

Shaffer, J. 1962. A specific cognitive deficit observed in gonadal aplasia (Turner's syndrome). *J. Clin. Psychol.* **18:** 403-6.

Shaffer, J. W. 1963. Masculinity-femininity and other personality traits in gonadal aplasia (Turner's syndrome). In *Advances in Sex Research,* ed. H. B. Beigal, pp. 219-32. New York: Hoeber.

5

ARNOLD R. KAPLAN
Laboratory of Medical Genetics
Cleveland Psychiatric Institute

PHYSIOLOGICAL AND PATHOLOGICAL CORRELATES OF DIFFERENCES IN TASTE ACUITY

INTRODUCTION

There have been numerous pedigree and population studies of taste thresholds for the bitter phenylthiourea ("PTC") type anti-thyroid compounds containing the characteristic H–N–C=S grouping. A population's taste threshold distribution for this class of compounds, which includes 6-n-propylthiouracil, 1-methyl-2-mercaptoimidazole, etc., tends to approach a bimodal curve. Taste thresholds for the majority of other compounds tend to follow monomodal and approximately "Gaussian" distributions.

Early classical studies (Fox, 1931 and 1932; Blakeslee and Salmon, 1931; Blakeslee, 1932; Snyder, 1931 and 1932) indicated that taste sensitivity for phenylthiourea has a bimodal population distribution. The literature has largely supported the hypothesis that taste *in*sensitivity (i.e., occurrence of an individual's taste acuity level on the insensitive mode of the population's bimodal distribution) is the effect of a homozygous pair of recessive genes. The hypothesis was independently derived by Snyder (1931) and Blakeslee and Salmon (1931). Reports of discrepancies in twin and pedigree data have, however, prevented definite confirmation of the hypothesis (Ardashnikov *et al.,* 1936; Rife, 1938; Harris and Kalmus, 1951; Das, 1956 and 1957; Kalmus, 1957; Merton, 1958; Dencker, Hauge, and Kaij, 1959; Verkade, Wepster, and Stegerhoek, 1959; Sutton, de Lamadrid, and Esterer, 1962).

These studies have been supported by the State of Ohio, Department of Mental Hygiene and Correction, Division of Mental Hygiene, and by research grants from the Council for Tobacco Research—U.S.A. and from the National Institutes of Health (HD 00591-02 and GRS 05563). The author is particularly indebted to the following collaborators in cited investigations: Roland Fischer, Frances Griffin, Edward V. Glanville, Bertram Fleshler, and Wilma Powell. The present report was also facilitated by the helpful cooperation of Edward N. Hinko, Sue Tullius, and Anita Kirk Kaplan.

The various studies have involved use of the compound in one of three forms—in solution, as a crystal, or as phenylthiourea-impregnated filter paper. Hartmann (1939) demonstrated that both the crystal and paper tests were significantly less reliable than the solution test. Harris and Kalmus (1949) devised a more reliable methodology to determine taste threshold, which involves using "a few c.c." of each sample in a tumbler, a double blind placebo, and a final sorting-out procedure. Merton (1958), Kalmus (1958), and Leguèbe (1960) described extensive genetic studies with data based on the improved methodology. Sinnot and Rauth (1937) used distilled water instead of tap water and a between-sample mouth rinse. Fischer, Griffin, England, and Pasamanick (1961) used 6-n-propylthiouracil instead of phenylthiourea, and they observed that a volume of 5.0 ml. was necessary for a reliable taste test. They also reintroduced the use of distilled instead of tap water and the use of a between-sample mouth rinse. The major advantage of using 6-n-propylthiouracil is the odorless character of solutions which are half-saturated or less concentrated. Phenylthiourea solutions, on the other hand, have a distinct odor (Skude, 1963), which affects detection of the difference and determination of apparent taste threshold.

METHODS

In our laboratory, taste thresholds were determined according to modifications (Fischer *et al.*, 1961) of the procedure described by Harris and Kalmus (1949). Their basic procedure of double blind placebo and final sorting out has proved to be the most reliable of the published methods for determining taste thresholds (Fischer, Griffin, and Kaplan, 1963; Kaplan *et al.*, 1963; Fischer and Griffin, 1964). Serial dilutions of the compounds were prepared by dissolving each substance in distilled water, in concentrations ranging from 7.32×10^{-7} M to 6.00×10^{-3} M. The most concentrated solution of PROP (6-n-propylthiouracil), number 14, consists of 1.0212 grams of the compound dissolved in 1.0 liter of distilled water; that of quinine (l-quinine sulfate), number 13, consists of 1.11744 grams of the compound in 1.0 liter of distilled water; and the highest concentration of hydrochloric acid is a 0.012 M solution, number 15. Each solution number represents twice the concentration of its preceding solution number, as shown in Table 5-1.

A subject's threshold, expressed as a solution number, is defined as the lowest concentration at which the samples of solution are correctly differentiated from the placebos. Each threshold is first estimated by providing the subject with solutions in progressively doubled concentrations until a

TABLE 5-1
CONCENTRATION OF TASTE SOLUTIONS IN MOLARITY

Solution No.	Molarity	Solution No.	Molarity
15	1.20×10^{-2}	7	4.69×10^{-5}
14	6.00×10^{-3}	6	2.34×10^{-5}
13	3.00×10^{-3}	5	1.17×10^{-5}
12	1.50×10^{-3}	4	5.86×10^{-6}
11	7.50×10^{-4}	3	2.93×10^{-6}
10	3.75×10^{-4}	2	1.46×10^{-6}
9	1.88×10^{-4}	1	7.32×10^{-7}
8	9.38×10^{-5}		

definite taste is reported. The threshold is then determined by final sorting out of eight cups, four containing solution and four containing distilled water placebo (5.0 ml. per cup), presented in a double-blind manner. A subject is required to utilize a mouth rinse with distilled water after tasting the contents of each cup. Due to the strong affinity of phenylthiourea type compounds for taste receptor cells, the thresholds for subsequently tasted compounds may be higher than when determined without prior tasting of a phenylthiourea type compound. Therefore, the compounds were tested in the following sequence: hydrochloric acid, quinine, and PROP. The testing of a subject for all three compounds required about 45 minutes. Data regarding age, smoking habits, drug and hormone therapy, menstruation, and pregnancy were obtained from questionnaires completed by the subjects.

SUBJECTS

Most of our subjects were volunteers contacted through the Cleveland Area Twin Registry and Mothers-of-Twins Clubs in the greater Cleveland area. Additional volunteers were obtained from the staffs and students at our own and neighboring institutions, and from the general public in response to publicity which solicited such volunteers. Ulcer patients were contacted through cooperation with staff members of the Western Reserve University School of Medicine, Cleveland Metropolitan General Hospital, and the Crile Veterans Administration Hospital.

PHENOTYPES AND GENOTYPES

Taste thresholds for most substances have a continuous and normal or nearly "Gaussian" distribution in the population. Such patterns have been described for population distributions of taste thresholds for hydrochloric acid, quinine sulfate, and over two dozen other studied compounds (Fischer and Griffin, 1964). Population distributions of thresholds for phenylthiourea type compounds such as PROP, however, manifest a bimodal tendency.

Figure 5-1 is a schematic presentation of the distribution of a multifactorial trait in a population where the factors are cumulative and where the trait's distribution is continuous and normal. The factors determining taste thresholds for quinine and those for hydrochloric acid are evidently multiple and cumulative.

Figure 5-2 compares schematically the distribution in a population of a trait based on two different monofactorial genetic mechanisms. One indicates a continuous and normal distribution of the trait, the result of cumulative gene action; the other indicates the kind of distribution that occurs when the gene action is not cumulative. In this case, the dominant homozygote and the heterozygote each manifest the trait, while the recessive homozygote is different, and there is no clearly observed intermediate group. In such cases, as our measuring techniques are improved, controlled, and made more specific for the genetic factor, the curve of population distribution more clearly approaches discontinuity between the modes. As other variables affect the parameter being measured, however, the antimode becomes less distinct. Taste thresholds for a phenylthiourea type compound such as PROP exhibit such a population distribution. This is characteristic of the class comprising dozens of antithyroid com-

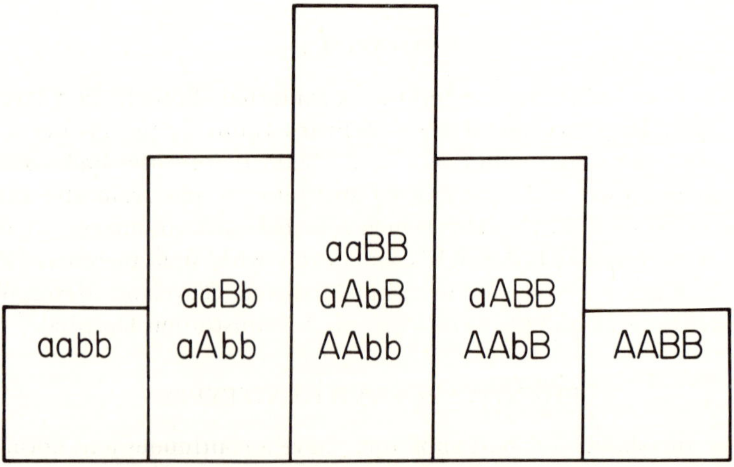

FIG. 5-1. SCHEMATIC TO SHOW CONTINUOUS DISTRIBUTION IN A POPULATION OF A TRAIT BASED ON MULTIPLE FACTORS THAT ARE CUMULATIVE IN EFFECT.

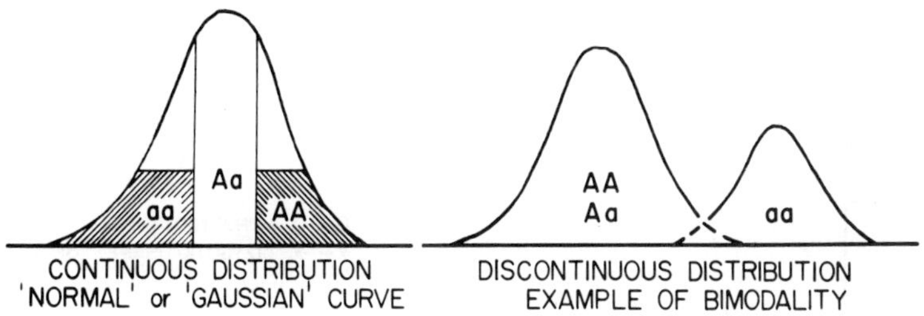

FIG. 5-2. SCHEMATIC TO SHOW A CONTINUOUS DISTRIBUTION IN A POPULATION OF A TRAIT BASED ON GENETIC FACTORS AT A SINGLE LOCUS, WHICH ARE CUMULATIVE IN EFFECT (LEFT); AND A DISCONTINUOUS, BIMODAL DISTRIBUTION IN A POPULATION OF A TRAIT BASED ON GENETIC FACTORS AT A SINGLE LOCUS WHICH ARE DOMINANT AND RECESSIVE RATHER THAN CUMULATIVE IN EFFECT (RIGHT).

pounds, natural and synthetic, and all bitter (Fischer and Griffin, 1964). Evidently, taste sensitivity for the compounds involves a single pair of genetic factors (Kalmus, 1958; Leguèbe, 1960; Merton, 1958), but numerous other variables also affect the trait. Control of these other variables is essential for any genetic study to be clear. Several of these variables have been studied and are discussed in the present paper. The incompleteness and imperfection of our controls may be reflected by the incompleteness of the antimodes in the distribution curves for the variables. The bimodal classification according to the population distribution based on the measured phenotype or trait does not completely correspond to the genotypic distribution, and there is some overlap of phenotypes.

TWIN AND SIBLING STUDIES

Intrapair differences in taste thresholds have been compared for hydrochloric acid, quinine, and PROP. Comparative distributions of intrapair threshold differences in monozygotic and same-sex dizygotic twin pairs have been graphed (Kaplan and Fischer, 1965). Figure 5-3 indicates the distribution of intrapair differences in thresholds for hydrochloric acid, based on 25 monozygotic (16 female and 9 male) and 26 dizygotic (15 female and 11 male) pairs. The distributions were very similar for two categories of same-sex twins.

Figures 5-4 and 5-5 indicate the distributions of intrapair differences in thresholds for quinine and PROP, respectively. The graphs are based on 69 monozygotic (48 female and 21 male) and 45 dizygotic (29 female and 16 male) pairs. The distribution of intrapair differences for quinine (Figure 5-4) indicates a greater difference between the two twin categories than that observed for hydrochloric acid. The difference of PROP intra-

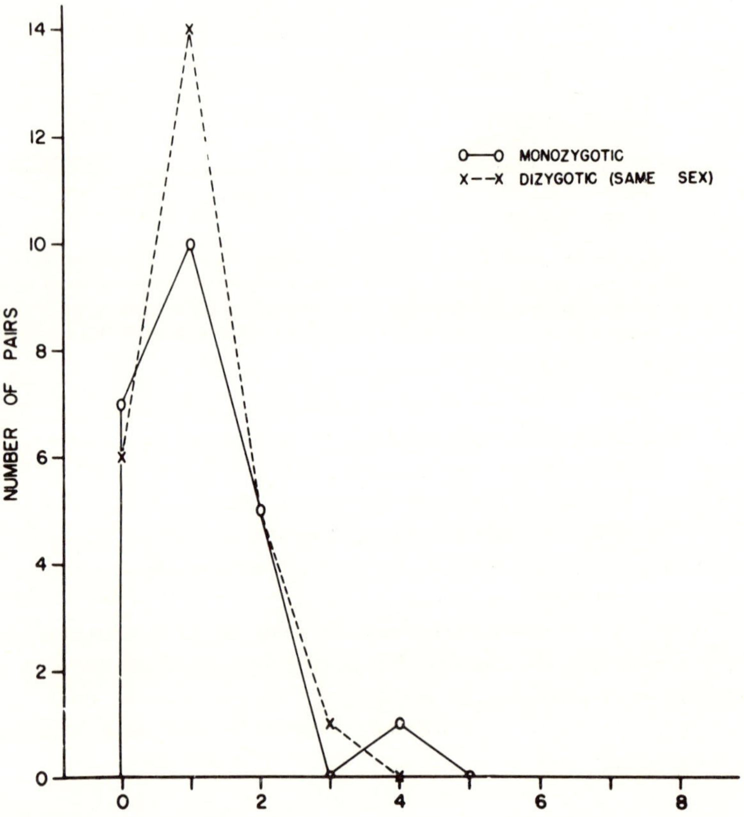

FIG. 5-3. COMPARATIVE DISTRIBUTION OF INTRAPAIR TASTE-THRESHOLD DIFFERENCES FOR HYDROCHLORIC ACID IN 25 MONOZYGOTIC (16 FEMALE, 9 MALE) AND 26 DIZYGOTIC (15 FEMALE, 11 MALE) TWIN PAIRS.

pair threshold differences between the two twin categories (Figure 5-5) was the greatest of the three compounds. The relative importance of genetic factors in taste threshold differences was also suggested by a comparison of the average intrapair differences observed in the same monozygotic and dizygotic twins, respectively, for each of the three compounds: for hydrochloric acid, 1.00 and 1.04; for quinine, 1.22 and 1.64; for PROP, 0.75 and 2.87. There was very little average intrapair difference between the pairs of genetically identical and those of the same-sex fraternal twins for hydrochloric acid. A much greater difference of average intrapair differences was observed between the two twin categories for quinine, and the greatest difference was observed for PROP. Intrapair variance of taste

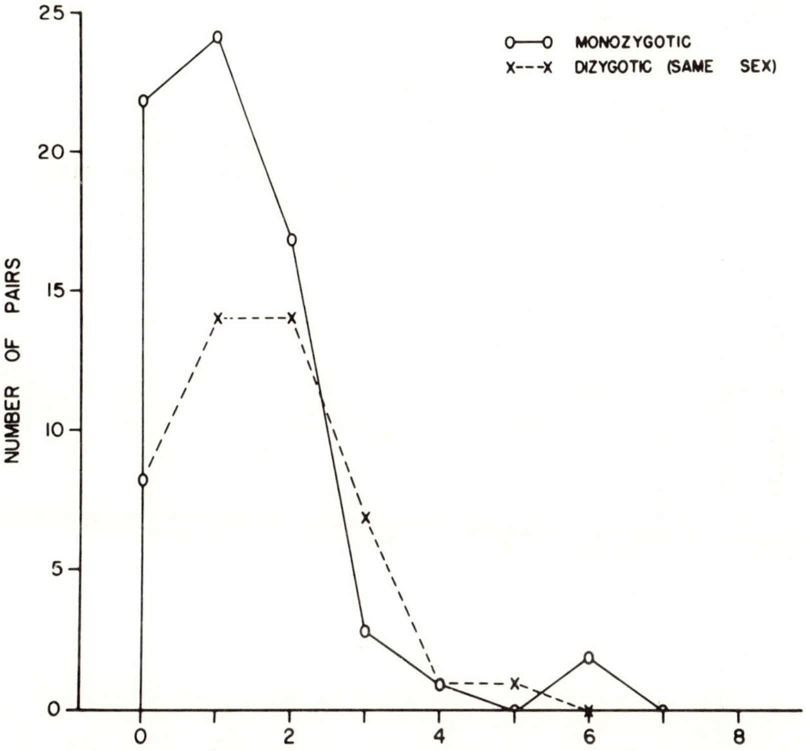

FIG. 5-4. COMPARATIVE DISTRIBUTION OF INTRAPAIR TASTE-THRESHOLD DIFFERENCES FOR QUININE IN 69 MONOZYGOTIC (48 FEMALE, 21 MALE) AND 45 DIZYGOTIC (29 FEMALE, 16 MALE) TWIN PAIRS.

thresholds for hydrochloric acid, quinine, and PROP, has been determined for larger numbers of monozygotic twin, dizygotic twin, and non-twin sibling pairs (Kaplan, *et al.,* 1967).

The numbers of respective types of pairs studied were 26, 45, and 142 for hydrochloric acid; 75, 70, and 191 for each of the other two substances.

A constant of 1.0 was added to each intrapair difference in raw scores, since the variances of the raw difference scores were not independent of their means. Each sum was then transformed to its reciprocal, making the variances of the groups homogeneous and independent of their means. For each compound, a standard (Z) score was calculated from the reciprocal of each intrapair difference score. This was done in order to permit comparisons among the three compounds, which could not be carried out directly due to their different scales. Each standard score

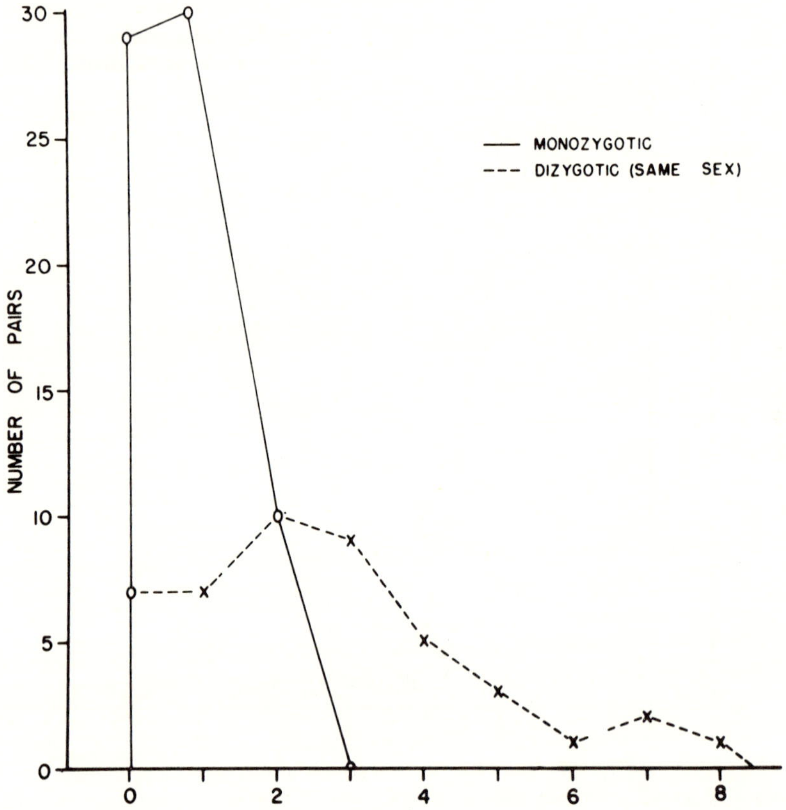

FIG. 5-5. COMPARATIVE DISTRIBUTION OF INTRAPAIR TASTE-THRESHOLD DIFFERENCES FOR PROP IN 69 MONOZYGOTIC (48 FEMALE, 21 MALE) AND 45 DIZYGOTIC (29 FEMALE, 16 MALE) TWIN PAIRS.

represents the difference between the overall mean and the individual score, divided by the standard deviation. A constant of 2.0 was added to each standard score to eliminate negative values. The standard scores, based on intrapair threshold difference scores, were significantly smaller for the monozygotic twin pairs than for the dizygotic twin pairs or for the sibling pairs (P less than .01) for each of the three solutions. There were no significant differences between the dizygotic twin pairs and the nontwin sibling pairs (P more than .10). The data are tabulated in Tables 5-2, 5-3, and 5-4. The variance analysis of standard scores based on intrapair score differences in monozygotic twin, dizygotic twin, and sibling pairs indicates genetic involvement regarding each of three compounds tested.

TABLE 5-2
DISTRIBUTION OF TWIN AND SIB PAIRS ACCORDING TO SEX, WITH THEIR TASTE THRESHOLD DATA FOR HYDROCHLORIC ACID (HCl), L-QUININE SULFATE (QUININE), AND 6-N-PROPYLTHIOURACIL (PROP)

			Number of Pairs	Mean Threshold	Standard Deviation	Mean Intrapair Difference
HCl	MZ		26	10.12	1.17	1.00
	DZ	same sex	26	10.06	1.26	1.04
		opposite sex	19	10.64	1.53	1.61
		Total	45	10.30	1.40	1.27
	SIB	same sex	86	10.27	1.14	1.27
		opposite sex	56	10.05	1.34	1.25
		Total	142	10.19	1.22	1.26
Quinine	MZ		75	5.37	1.76	1.29
	DZ	same sex	51	5.58	1.84	1.69
		opposite sex	19	5.71	1.41	1.84
		Total	70	5.56	1.90	1.73
	SIB	same sex	113	5.10	1.68	1.43
		opposite sex	78	5.34	2.21	1.99
		Total	191	5.20	1.89	1.66
PROP	MZ		75	9.51	2.18	0.84
	DZ	same sex	51	9.70	2.76	2.71
		opposite sex	19	10.10	2.25	1.63
		Total	70	9.81	2.65	2.41
	SIB	same sex	113	9.25	2.41	2.37
		opposite sex	78	9.79	2.26	2.30
		Total	191	9.45	2.33	2.34

TABLE 5-3
MEAN (X), VARIANCE (σ^2), AND NUMBER OF PAIRS (N), BASED ON RECIPROCALS OF SUMS OF EACH INTRAPAIR DIFFERENCE PLUS CONSTANT (1.0)

		DZ ♀ ♀	DZ ♂ ♂	DZ ♀ ♂	SIB ♀ ♀	SIB ♂ ♂	SIB ♀ ♂	MZ ♀ ♀	MZ ♂ ♂
HCl									
	X	61.1	52.2	47.6	60.2	51.3	61.3	52.1	77.8
	σ^2	624.4	644.0	869.0	719.0	793.0	946.6	611.0	694.4
	N	15	11	19	45	41	56	17	9
Quinine									
	X	47.3	48.6	44.4	50.0	55.4	46.9	55.6	62.3
	σ^2	723.4	701.4	703.0	641.0	775.0	757.2	771.0	1056.0
	N	34	17	19	65	48	78	52	23
PROP									
	X	42.6	40.1	50.0	43.9	43.4	44.0	66.6	68.8
	σ^2	882.0	970.0	823.0	849.0	889.0	829.0	831.0	820.0
	N	34	17	19	65	48	78	52	23

TABLE 5-4
Z SCORES BASED ON RECIPROCALS OF INTRAPAIR DIFFERENCE SCORES. THE MEAN (X), VARIANCE (σ^2), AND NUMBER OF PAIRS (N) CALCULATED FROM STANDARD SCORES OF EACH SUBJECT FOR EACH COMPOUND. ALL NEGATIVE VALUES WERE ELIMINATED BY ADDITION OF A CONSTANT (2.0) TO EACH SCORE

		DZ ♀ ♀	DZ ♂ ♂	DZ ♀ ♂	SIB ♀ ♀	SIB ♀ ♂	SIB ♂ ♂	MZ ♀ ♀	MZ ♂ ♂
HCl									
	X	2.12	1.81	1.74	2.09	1.78	2.13	1.80	2.71
	σ^2	.78	.81	1.06	.90	.99	1.19	.76	.87
	N	15	11	18	45	41	56	17	9
Quinine									
		1.93	1.97	1.81	1.85	2.21	1.86	2.31	2.65
	σ^2	1.10	.99	.97	.77	1.08	.90	1.09	1.98
	N	15	11	18	45	41	56	17	9
PROP									
	X	1.83	2.00	2.21	2.03	1.90	1.87	2.84	2.45
	σ^2	1.19	1.47	.87	1.02	.91	.79	.91	1.08
	N	15	11	18	45	41	56	17	9

CORRELATIONS BETWEEN THRESHOLDS FOR DIFFERENT COMPOUNDS

The PROP thresholds of monzygotic twins with differing thresholds were examined to determine whether the co-twins with the higher quinine thresholds also tended to have higher PROP thresholds. The quinine thresholds of 53 pairs of monozygotic twins in our population differed from each other by one or more thresholds. In 14 of the pairs, the same individuals manifested the higher threshold for each of the two substances; in 22 pairs, both co-twins had the same PROP threshold despite their differences in quinine threshold; and in 17 pairs the co-twin with the lower quinine threshold had the higher PROP threshold. In fact, there was no significant correlation of intrapair differences for any two of the compounds. Apparently, independent factors influence the intrapair taste threshold differences observed in monozygotic twins for the two bitter-tasting drugs, quinine and PROP.

When individual scores, not twin differences, are used a different picture arises. Taste thresholds for the three compounds investigated were significantly and positively correlated with each other. Analyses based on the threshold data for 308 individuals indicated the following correlation coefficients: between PROP and quinine, $r = 0.444 \pm .046$ (P less than .01); between quinine and hydrochloric acid, $r = 0.35 \pm .05$ (P less than .01); between PROP and hydrocloric acid, $r = 0.166 \pm .055$ (P less than .05).

Dietary Preferences Related to Taste Thresholds

The flavor of a particular food is the product of the interacting effects of various tastes and odors, and therefore might not be expected to be highly correlated with taste threshold for one particular compound. Taste threshold for PROP, however, has been found to be correlated with thresholds for a wide variety of other compounds, including those tasting bitter, salty, sour, and sweet (Fischer and Griffin, 1964). Food preferences and rejections are related to various factors, including social custom, experience, and physiological state. Individual differences in taste acuity for particular compounds have also been correlated with food likes and dislikes. Fischer *et al.* (1961) found that, in a group of 48 college students, proportions of foods disliked from a given list of 118 specific items were correlated with taste thresholds for quinine and probably also for that of PROP; but they found no correlations with taste thresholds for sucrose, sodium chloride, or hydrochloric acid. More recently, we studied 187 adults (51 males, 136 females) ranging in age from 22 to 66 years and with an average age of 38 (Glanville and Kaplan, 1965a). The sample included 39 husband-wife pairs, 16 pairs of MZ twins, and 10 pairs of DZ twins. Our questionnaire was designed to minimize the influence of social custom through use of a list of only a few carefully selected foods which are in wide use locally and are commonly prepared in several different forms. Each subject was rated according to preference for "mild," "moderate," or "strongly tasting" preparations.

The questionnaire was divided into two parts. Part A involved a list of widely used foods which are commonly prepared in a variety of ways, providing a graded series ranging from mild through intermediate to strong tasting. Subjects were advised that the listed foods and drinks could each be prepared in a number of different ways, and that the preferred way should be checked in each case from a group of choices. The listed alternatives were as follows, except that the choices were randomized: (1) Coffee: with more than one spoon of cream/with one spoon of cream/black, no cream. (2) Coffee: with more than one spoon of sugar/with one spoon of sugar/no sugar. (3) Cheese: American/Longhorn or Swiss/blue cheese. (4) Cheddar cheese: mild/medium/sharp. (5) Salad dressing: mild/oil and vinegar/Roquefort or blue cheese. Scores were determined as follows: questions (1) and (2), both of which involve coffee, were treated as single question. One point was given for a selection of: more than one spoon of either cream or sugar, or one each of cream and sugar. One spoon of cream or one of sugar only was given two points, and black coffee without sugar was scored as three points. For questions (3), (4), and (5), one point was given for the first choice, two points for the second, and three for the third. The minimum score was therefore four points (mildest choice) and the maximum was 12 (strongest tasting choice).

Part B of the questionnaire involved a list of five foods. Subjects were asked to answer "yes" or "no" to questions inquiring whether they liked: grapefruit juice, lemon juice, sauerkraut, vinegar, and horse-radish. Verbal explanations of the questionnaire were given if required and subjects were asked to indicate those foods not customarily consumed. The entire food preference questionnaire was given to 181 subjects. Of this number, 15 could not complete Part B because they were unfamiliar with one or more of the foods listed. Incomplete forms were not counted, leaving a total of 166 completed scores on Part B. Six subjects could not complete Part A for the same reason, but an additional 12 subjects were tested for Part A only (since Part B was added after the investigation had begun). The total number who completed Part A was therefore 187. Table 5-5 shows the mean taste thresholds with their standard errors for quinine and PROP, for subjects who preferred the mild, medium, and strongly tasting choices in questions (1) through (5) of Part A. Taste sensitivity for quinine and PROP apparently is more important in determining the answers to questions (3), (4), and (5), than in determining preferences regarding the ways coffee may be prepared (the first two questions).

TABLE 5-5
MEAN TASTE THRESHOLDS WITH THEIR STANDARD ERRORS FOR SUBJECTS SELECTING MILD, MEDIUM AND STRONG TASTING CHOICES IN QUESTIONS 1-5 ON THE FOOD QUESTIONNAIRE. THE NUMBER IN EACH GROUP IS SHOWN WITHIN PARENTHESIS

	Question No.	Choice of foods tasting Mild		Medium		Strong	
Quinine	1 and 2	(100)	5.87 ± 0.19	(42)	5.65 ± 0.32	(45)	5.82 ± 0.27
	3	(78)	5.21 ± 0.19	(82)	6.17 ± 0.19	(27)	6.33 ± 0.48
	4	(84)	5.23 ± 0.20	(46)	5.96 ± 0.25	(57)	6.49 ± 0.24
	5	(52)	5.58 ± 0.29	(84)	5.66 ± 0.20	(51)	6.24 ± 0.24
PROP	1 and 2	(100)	9.29 ± 0.25	(42)	9.28 ± 0.41	(45)	10.24 ± 0.37
	3	(78)	8.68 ± 0.26	(82)	10.20 ± 0.27	(27)	10.11 ± 0.54
	4	(84)	8.74 ± 0.25	(46)	9.52 ± 0.32	(57)	10.77 ± 0.26
	5	(52)	8.84 ± 0.37	(84)	9.68 ± 0.25	(51)	10.06 ± 0.38

The total scores on Part A of the food questionnaire for the total sample have been tabulated with the taste thresholds for quinine in Table 5-6 and for PROP in Table 5-7. The correlation coefficients between the thresholds for quinine and PROP and the food preference scores are shown in Table 5-8. There are highly significant positive correlations between high scores on the food questionnaire (that is, preference for strongly tasting foods) and high taste thresholds (that is, low taste sensitivity) for both quinine and PROP. Table 5-8 also shows that the thresholds for quinine and PROP are themselves correlated with each other.

TABLE 5-6
TASTE THRESHOLDS FOR QUININE AND SCORES ON PART A OF FOOD-PREFERENCE QUESTIONNAIRE

Quinine threshold	Questionnaire score (Part A) 4	5	6	7	8	9	10	11	12	Total
13					1					1
12						1				1
11										
10		1			1	1				3
9					5	1		1		7
8		1	4	2	2		3	1	2	15
7	1	5	2	9	6	6	1	3	3	36
6	3	9	7	10	9	6	3		1	48
5	4	4	4	5	12	3	2	1		35
4		5	1	7	3			2		18
3	5	3	1	2	2	2	1			16
2		1	2	1	1				1	6
1	1									1
Total	14	29	21	36	42	20	10	8	7	187

TABLE 5-7
TASTE THRESHOLDS FOR PROP AND SCORES ON PART A OF FOOD-PREFERENCE QUESTIONNAIRE

PROP threshold	Questionnaire score (Part A) 4	5	6	7	8	9	10	11	12	Total
>14					5			1	3	9
14		2			3	1		1		7
13				2	2	3			1	8
12	2		1	5	4	4	1	1		18
11		3	2	4	4	1	2			16
10		1	4	7	5	4	2		2	25
9	1	6	3	8	5	1	3	2		29
8	2	12	6	7	7	4	1	1		40
7	5	1	2	3	4	1	1	3		20
6	2	3	3		2	1				11
5	1				1					2
4		1								1
3										
2										
1	1									1
Total	14	29	21	36	42	20	10	9	6	187

TABLE 5-8
TOTAL AND PARTIAL CORRELATION COEFFICIENTS WITH THEIR STANDARD ERRORS BETWEEN THRESHOLD FOR QUININE AND PROP AND THE SCORE ON PART A OF THE FOOD-PREFERENCE QUESTIONNAIRE.
(DATA FROM TABLES 6 AND 7)

Correlation coefficients	
Quinine/food score	+0.262 ± 0.068
PROP/ food score	+0.366 ± 0.064
Quinine/PROP	+0.540 ± 0.052

Results from Part B of the food preference questionnaire were analyzed by calculation of the linear regression coefficients of taste thresholds on the number of foods disliked. The coefficient with quinine is 0.319 ± 0.114, and that with PROP is 0.480 ± 0.146. Both of these values are significantly different from zero, and support the conclusion that there is a positive association between increasing number of foods disliked and increasing taste sensitivity (i.e., decreasing taste threshold). Both males and females showed a positive correlation between preference for mild tasting foods and relatively sensitive perception of the taste of quinine and PROP. The males showed a somewhat higher correlation than the females, but the difference was not statistically significant.

The 16 pairs of monozygotic twins in the sample were all married and the co-twins had lived apart for varying lengths of time. Their scores on Part A of the food-preference form were found to be highly correlated (0.70, significant for N = 16 at the 0.01 level). The scores on Part A of the questionnaire for 10 pairs of dizygotic twins, eight pairs of whom lived apart, showed an insignificant correlation of 0.18.

The sample included 39 husband-wife pairs. The members of these pairs are not closely related genetically and there was no significant intrapair correlation between their thresholds for quinine and for PROP despite their having shared a common home environment for varying lengths of time. A significant positive correlation was observed between the scores on the food questionnaire of husbands and wives. On Part A of the questionnaire, the correlation coefficient was found to be 0.482 (significant for N = 39 at the 0.01 level). The correlation between the indicated food preferences of spouses observed in our data, despite absence of a correlation between their taste thresholds, demonstrates that social factors, as well as individual taste thresholds, are important for the determination of individual food preferences.

TABLE 5-9
MEAN TASTE THRESHOLDS FOR QUININE TABULATED ACCORDING TO AGE

Age Group (Years)	Males threshold (Solution number)			Females threshold (Solution number)		
	N	Mean	σ_d	N	Mean	σ_d
1-5	2	5.00		3	5.67	
6-10	37	5.38	1.72	46	5.48	1.87
11-15	34	4.85	2.28	50	5.04	1.56
16-20	78	5.05	1.73	98	4.89	1.91
21-25	72	5.18	1.69	45	5.29	1.41
26-30	22	5.23	1.82	21	5.76	2.26
31-35	23	5.83	1.72	21	5.86	1.80
36-40	13	7.38	2.36	21	5.91	1.69
41-45	20	6.45	2.44	30	5.80	1.61
46-50	3	6.67		18	6.67	1.71
51-55	4	7.00		15	6.13	2.84
Total	308			368		

TABLE 5-10
MEAN TASTE THRESHOLDS FOR PROP TABULATED ACCORDING TO AGE

Age Group (Years)	Males threshold (Solution number)			Females threshold (Solution number)		
	N	Mean	σ_d	N	Mean	σ_d
1-5	2	8.50		3	10.67	
6-10	37	9.62	2.27	46	9.52	2.47
11-15	34	9.59	2.23	50	9.66	2.24
16-20	78	9.28	2.25	98	9.37	2.62
21-25	72	9.43	2.51	45	9.42	2.37
26-30	22	9.45	2.40	21	9.05	2.58
31-35	23	9.74	2.53	21	10.38	2.04
36-40	13	10.00	2.86	21	9.52	2.56
41-45	20	10.45	2.33	30	9.63	2.77
46-50	3	13.00		18	10.28	1.93
51-55	4	11.00		15	9.73	2.96
Total	308			368		

AGE, SEX, SMOKING, AND TASTE THRESHOLDS

In earlier studies, it was observed that the proportion of smokers was lower in groups of sensitive quinine tasters, and higher among the insensitive tasters (Krut, Perrin, and Bronte-Stewart, 1961; Fischer, Griffin, and Kaplan, 1963). These observations appeared to be analogous to those indicating large numbers of reported food dislikes associated with low taste threshold for quinine and small numbers of food dislikes associated with high quinine taste threshold (Fischer, Griffin, and Kaplan, 1963), as well as the differences in preferences for mildly or strongly tasting foods (discussed above).

Harris and Kalmus (1949) observed that taste sensitivity for phenylthiourea solutions decreased at a rate of one doubling in concentration (i.e., increase in one threshold) for each 20-year age span. Kalmus and Trotter (1962) observed a decline in phenylthiourea taste sensitivity corresponding to a mean annual increase of 0.03 threshold. They also noted a more rapid deterioration in men than in women. Leguèbe (1960) observed that women were approximately 0.8 and 0.5 threshold more sensitive tasters than men for phenylthiourea and quinine, respectively. He observed that age was related to threshold differences only for the insensitive tasters ("nontasters") of phenylthiourea, and that age was not related to differences in quinine threshold. Previously, Byrd and Gertman (1959) had reported no significant age-related difference in taste threshold for quinine.

Extensive investigations of taste threshold distributions have demonstrated an age-related exponential decline in sensitivity (i.e., increase in threshold) for quinine and PROP, tabulated by age for 308 male and 368 female subjects. Table 5-11 indicates the comparative rates of decrease in taste sensitivity for the male and female subjects. Figures 5-6 and 5-7 illustrate for quinine and PROP respectively, the increase in taste threshold (i.e., decrease in sensitivity) associated with increase in age, and the more rapid changes occurring in males.

TABLE 5-11
COMPARATIVE RATES OF DECREASE IN TASTE SENSITIVITY IN MALES AND FEMALES AGED 16 TO 55

Substance	Sex	A[a]	B[b]
PROP	M	0.052 ± 0.011	19
	F	0.026 ± 0.013	38
Quinine	M	0.066 ± 0.013	15
	F	0.043 ± 0.009	23
Hydrochloric	M	0.071 ± 0.028	14
Acid	F	0.018 ± 0.014	55

Notes: [a] Decrease in sensitivity in thresholds per year, together with the standard error.
[b] Number of years required to increase mean score by one threshold, according to the rate indicated in A.

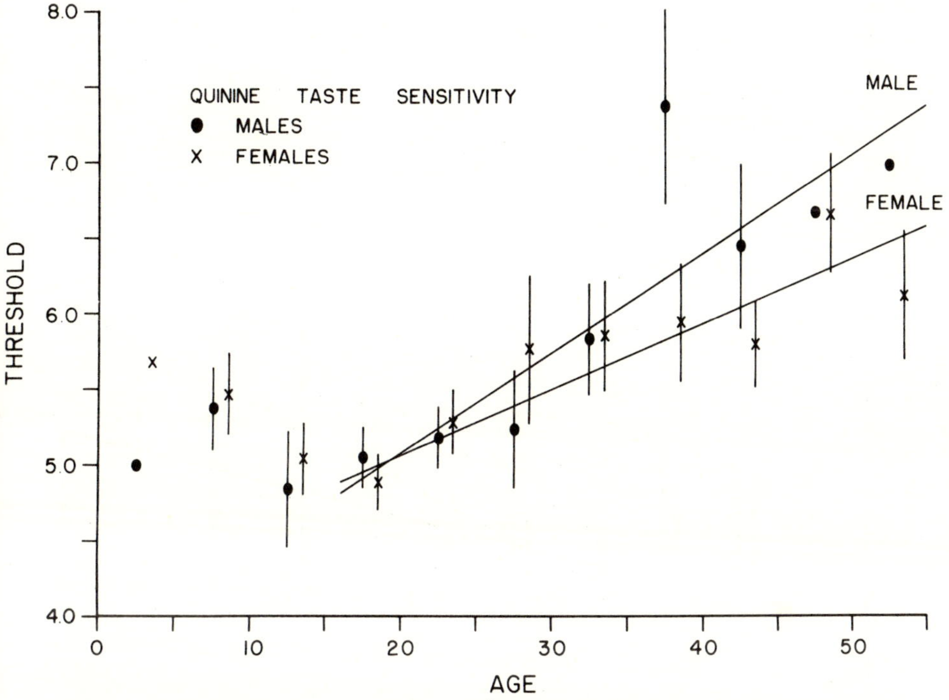

FIG. 5-6. INFLUENCE OF AGE ON TASTE THRESHOLD FOR 1-QUININE SULFATE. ● MALES; X FEMALES. THE MEAN SCORE AND STANDARD ERROR FOR EACH AGE GROUP ARE SHOWN, TOGETHER WITH THE FITTED REGRESSION LINE FOR THE SCORES OF Ss AGED 16 TO 55 YEARS. THRESHOLDS ARE BASED ON SOLUTION NUMBERS. VERTICAL LINES REPRESENT $\pm$ 1 σ.

The above studies were continued until sufficient numbers of subjects had been investigated to allow categorization according to smoking habits as well as age and sex (Kaplan, Glanville, and Fischer, 1965). The latter sample included 84 males and 184 females who had never smoked regularly, as well as 78 males and 49 females who had regularly smoked 20 or more cigarettes per day at the time they were tested. Tables 5-12 and 5-13 indicate the distribution of taste thresholds for quinine and PROP, respectively, according to age and sex and smoking habits (i.e., nonsmokers vs. subjects smoking 20 or more cigarettes per day). The mean thresholds for quinine and PROP, distributed according to age in the "heavy smokers" and, separately, in the nonsmokers, are shown in Figures 5-8 and 5-9 for the females, and in Figures 5-10 and 5-11 for the males. These four illustrations show the mean thresholds with their standard errors, together with the fitted regression lines. The regression coefficients calculated for taste threshold on age are shown in Table 5-14.

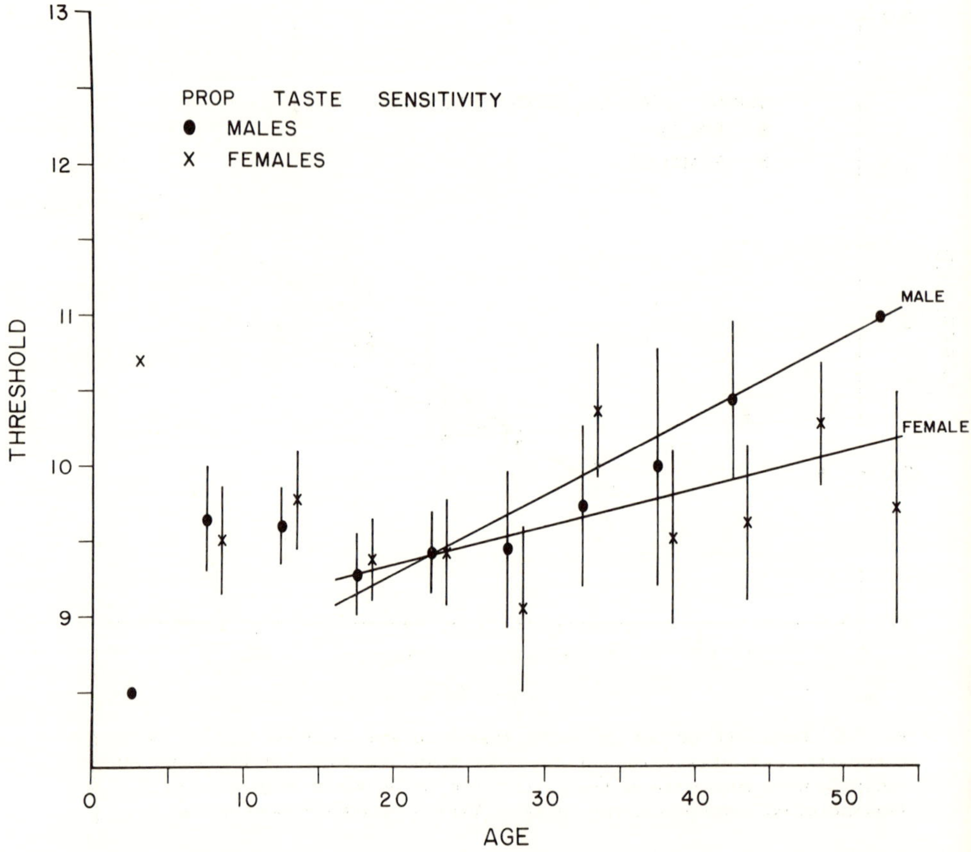

FIG. 5-7. INFLUENCE OF AGE ON TASTE THRESHOLD FOR PROP (6-N-PROPYLTHIOURACIL). ● MALES; X FEMALES. THE MEAN SCORE AND STANDARD ERROR FOR EACH GROUP ARE SHOWN, TOGETHER WITH THE FITTED REGRESSION LINE FOR THE SCORES OF Ss AGED 16 TO 55 YEARS. THRESHOLDS ARE BASED ON SOLUTION NUMBERS. VERTICAL LINES REPRESENT $\pm$ 1 σ.

Separate analyses of quinine and PROP threshold distributions for each sex indicated again that the heavy cigarette smokers were significantly less sensitive than the nonsmokers. In each case, the statistical significance was beyond the 5 percent level of chance probability.

Among nonsmokers, the regression coefficients did *not* differ significantly from zero, and there was no significant difference between the males and females either in mean threshold or in the apparent effect of age upon taste. There was, however, a highly significant decline in taste sensitivity among the heavy smokers with increasing age. The rates of decline and the mean scores were not significantly different in the two sexes after the samples were controlled for smoking habits.

TABLE 5-12
MEAN TASTE THRESHOLDS FOR QUININE AND PROP, TABULATED ACCORDING TO AGE FOR MALE NON-SMOKERS AND HEAVY SMOKERS OF CIGARETTES (I.E., SUBJECTS WHO SMOKED 20 OR MORE PER DAY WHEN TASTE-TESTED)

	Males					
	Non-smokers' threshold (Solution number)			Heavy smokers' threshold (Solution number)		
Age group (Years)	N	Mean	σ_d	N	Mean	σ_d
Quinine						
16–20	30	5.27	1.64	6	4.50	1.05
21–25	25	5.80	2.45	16	5.50	1.46
26–30	6	4.50	1.87	9	7.00	1.58
31–35	9	4.33	1.50	11	6.82	1.47
36–40	6	6.00	1.95	11	7.45	1.97
41–45	3	4.67		13	7.69	2.38
46–50	2	5.00		9	8.11	1.90
51–55	3	7.00		3	7.33	
Total	84			78		
PROP						
16–20	30	8.83	2.53	6	8.83	1.60
21–25	26	10.28	2.13	16	9.25	2.41
26–30	6	9.33	2.58	9	10.33	2.45
31–35	9	8.11	2.67	11	11.36	1.91
36–40	6	10.83	2.64	11	9.91	2.95
41–45	3	11.00		13	11.46	2.26
46–50	2	11.50		9	13.11	1.83
51–55	3	9.67		3	12.33	
Total	84			78		

Previous studies of the influence of age and sex on taste sensitivity have been based on populations which were heterogeneous regarding smoking habits. The separate examination of the influence of age and sex on taste sensitivity in nonsmokers and in "heavy smokers" of cigarettes has indicated no significant difference in taste threshold for quinine and PROP related to age or sex in the nonsmokers. The regression lines for the non-smokers and "heavy smokers" diverge near or within the 16–20 year age range, and threshold differences become increasingly pronounced with advancing age. Since "heavy smokers" in the different age groups had been smoking for different average lengths of time, the divergence of the regression lines could be interpreted as being due to a cumulative effect of heavy smoking. The previously reported sex differences, based on population data which were analyzed without regard to smoking habits, might be ascribed to the occurrence of different proportions of smokers in women and men. Men, as a group, smoke more than

TABLE 5-13
MEAN TASTE THRESHOLDS FOR QUININE AND PROP, TABULATED ACCORDING TO AGE FOR FEMALE NON-SMOKERS AND HEAVY SMOKERS OF CIGARETTES (I.E. SUBJECTS WHO SMOKED 20 OR MORE PER DAY WHEN TASTE-TESTED)

	Females					
	Non-smokers' threshold (Solution number)			Heavy smokers' threshold (Solution number)		
Age group (Years)	N	Mean	σ_d	N	Mean	σ_d
Quinine						
16–20	64	5.06	1.92	11	5.73	2.05
21–25	23	5.13	1.32	5	5.00	2.00
26–30	14	5.36	1.78	9	6.33	2.83
31–35	12	6.00	1.81	8	6.50	1.31
36–40	21	5.33	1.66	8	6.75	1.49
41–45	25	5.08	1.38	5	7.00	1.73
46–50	18	5.61	1.82	2	10.00	
51–55	7	5.57	1.72	1	7.00	
Total	184			49		
PROP						
16–20	64	9.14	2.80	11	9.36	2.54
21–25	23	9.35	2.25	5	10.20	2.17
26–30	14	8.86	2.35	9	10.56	2.24
31–35	12	9.42	1.93	8	11.00	2.20
36–40	21	7.76	2.05	8	11.00	1.85
41–45	25	8.24	2.03	5	11.80	2.78
46–50	18	9.94	2.46	2	13.50	
51–55	7	8.57	2.44	1	8.00	
Total	184			49		

women (Matarazzo and Saslow, 1960; Horn, 1963). In our own sample, collected without regard for smoking habits, women comprised 68.7 per cent of our nonsmokers, but only 38.6 per cent of our "heavy smokers." Aging is not associated with significant increase in taste threshold unless combined with smoking. The sex differences sometimes observed in population surveys apparently are the result of differences in smoking habits between the two sexes.

MENSTRUATION AND TASTE THRESHOLDS

Variations in numerous physiological and psychological characteristics have been correlated with phases of the menstrual cycle, but reports on associated changes in sensitivity of sensory perception are rare. The relative concentrations of certain hormones are known to be capable of affecting taste thresholds (Henkin, Gill, and Bartter, 1963). Fluctuations

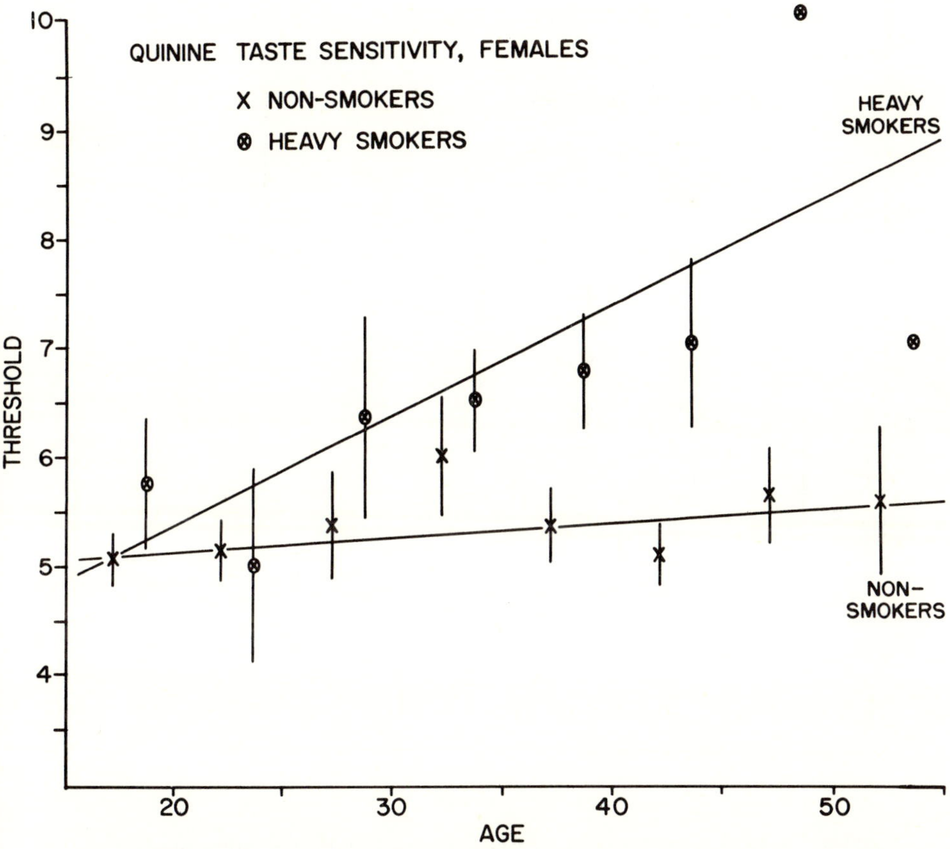

FIG. 5-8. INFLUENCE OF AGE ON TASTE THRESHOLD FOR QUININE IN FEMALE HEAVY SMOKERS AND NON-SMOKERS. THE MEAN SCORE AND STANDARD ERROR FOR EACH GROUP ARE SHOWN, TOGETHER WITH THE FITTED REGRESSION LINE FOR THE SCORES OF Ss AGED 16 TO 55 YEARS. THRESHOLDS ARE BASED ON SOLUTION NUMBERS. VERTICAL LINES REPRESENT ± 1 STANDARD ERROR.

in taste sensitivity at the time of menstruation may be correlated with fluctuations in endocrine balance at this time. Hoyme (1955) suggested that hormonal changes, such as those associated with the menstrual cycle, could modify taste sensitivity for phenylthiourea. Our observations, based on repeated tests of the same female subjects and on differences observed in female monozygotic twins, confirmed Hoyme's impression (Kaplan *et al.,* 1964a). Beiguelman (1964), however, observed no difference in taste threshold for phenylthiourea associated with the menstrual cycle. His study involved 100 subjects who were each tested twice, once during

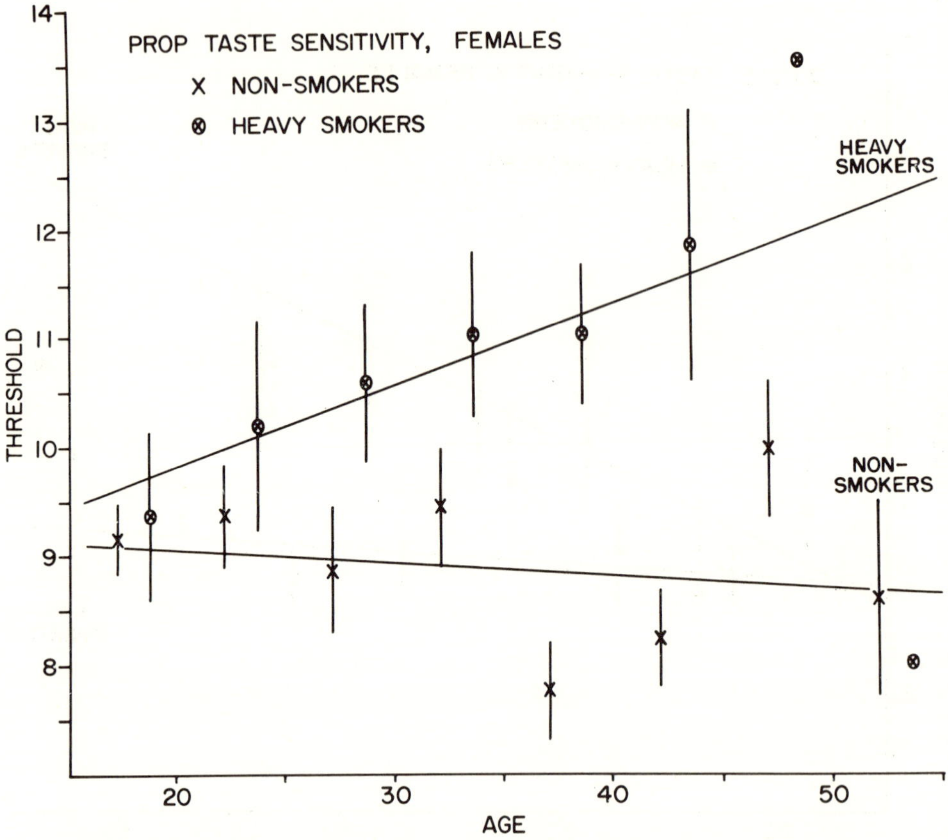

FIG. 5-9. INFLUENCE OF AGE TASTE THRESHOLD FOR PROP (6-N-PROPYLTHIOURACIL) IN FEMALE HEAVY SMOKERS AND NON-SMOKERS. THE MEAN SCORE AND STANDARD ERROR FOR EACH GROUP ARE SHOWN, TOGETHER WITH THE FITTED REGRESSION LINE FOR THE SCORES OF Ss AGED 16 TO 55 YEARS. THRESHOLDS ARE BASED ON SOLUTION NUMBERS. VERTICAL LINES REPRESENT ± 1 STANDARD ERROR.

the menstruation phase and again between periods. The wide variation which we have observed in individual response indicates the necessity for utilizing extreme caution in generalizing from limited amounts of taste threshold data. Repeated testing under controlled conditions provided a reliable method of examining the relationship between the menstrual cycle and constancy of taste thresholds.

Taste thresholds for quinine and PROP were determined throughout one or more menstrual cycles in a group of nineteen subjects (Glanville and Kaplan, 1965b and 1965c). Taste tests were carried out at the same time each day, three days per week, for periods ranging from four to nine

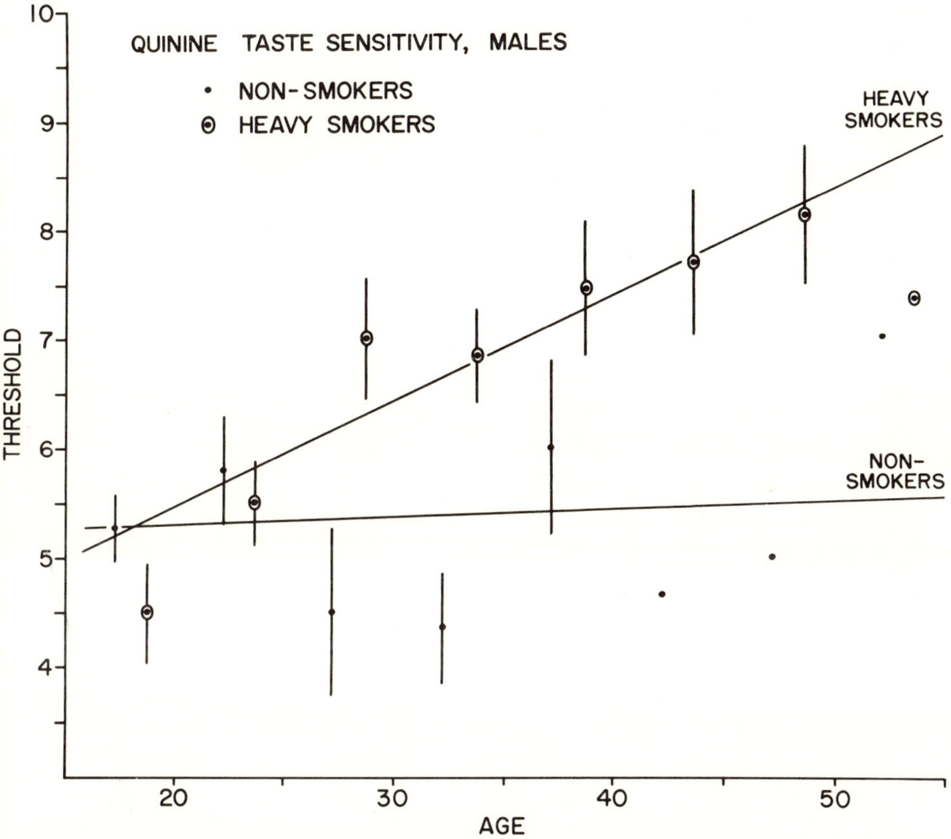

FIG. 5-10. INFLUENCE OF AGE ON TASTE THRESHOLD FOR QUININE IN MALE HEAVY SMOKERS AND NON-SMOKERS. THE MEAN SCORE AND STANDARD ERROR FOR EACH GROUP ARE SHOWN, TOGETHER WITH THE FITTED REGRESSION LINE FOR THE SCORES OF Ss AGED 16 TO 55 YEARS. THRESHOLDS ARE BASED ON SOLUTION NUMBERS. VERTICAL LINES REPRESENT ± 1 STANDARD ERROR.

weeks. The subjects were apparently healthy and normal females and all but two were student nurses. Their ages ranged from 19 to 27 (average 20.7). Subjects were requested not to take drugs during the experiment. Four other subjects were excluded from the analysis because of illness or medication. The days in the menstrual cycle were numbered forward and backward from the first day of menstruation. In those instances in which two cycles were recorded, the days were counted from the first day of menstruation in the second cycle. Sensitivities within three phases of the cycle were compared: the pre-menstrual (days —9 to —5), menstrual (days —1 to +4) and post-menstrual (days +6 to +10). The average

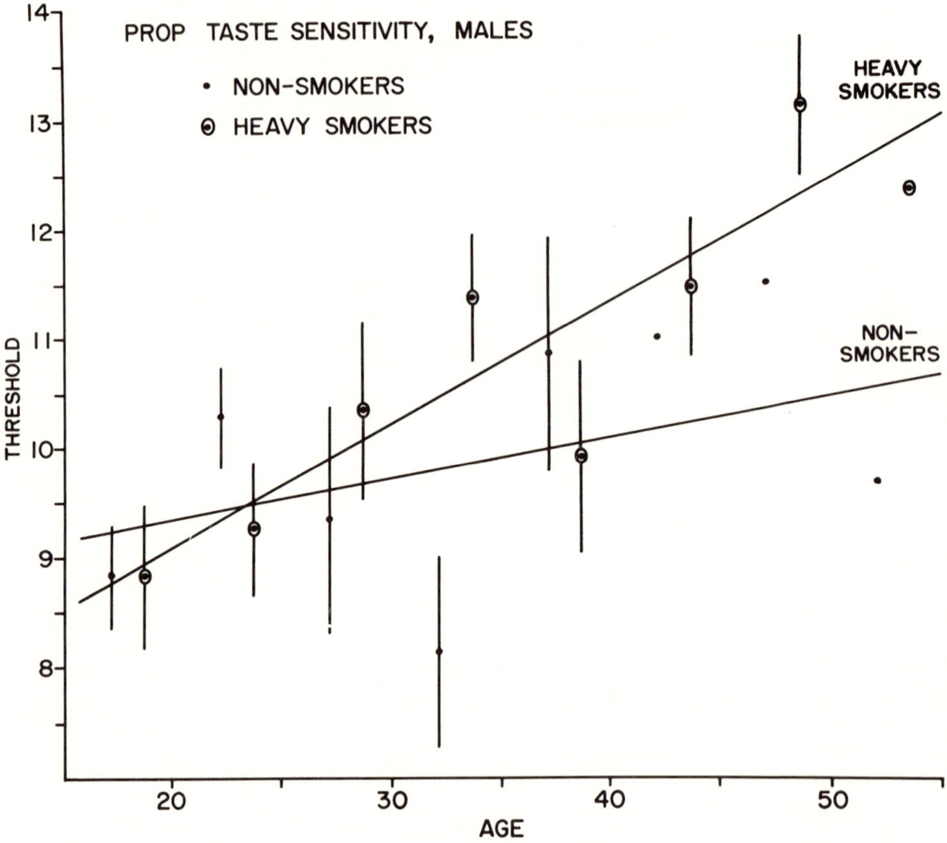

FIG. 5-11. INFLUENCE OF AGE ON TASTE THRESHOLD FOR PROP (6-N-PROPYLTHIOURACIL) IN MALE HEAVY SMOKERS AND NON-SMOKERS. THE MEAN SCORE AND STANDARD ERROR FOR EACH GROUP ARE SHOWN, TOGETHER WITH THE FITTED REGRESSION LINE FOR THE SCORES OF Ss AGED 16 TO 55 YEARS. THRESHOLDS ARE BASED ON SOLUTION NUMBERS. VERTICAL LINES REPRESENT ± 1 STANDARD ERROR.

scores of each subject tested during these phases are shown in Table 5-15. The scores of the initial two tests of each subject were excluded from the analysis, as several showed an improvement with practice. The variances of the scores are also given in Table 5-15. When these values were calculated, the scores on days —3 to +5 were excluded, so that the variances refer to the intermenstrual phases.

Several subjects continued to improve after the first two tests and therefore the most conservative comparison was between the menstrual and post-menstrual phases of the cycle. The final column of Table 5-15 shows the differences in scores between these two phases. In the majority

TABLE 5-14
REGRESSION COEFFICIENTS OF TASTE THRESHOLDS FOR INCREASING AGE.
NUMBER OF SUBJECTS:
MALE NON-SMOKERS = 84 MALE HEAVY SMOKERS = 78
FEMALES NON-SMOKERS = 184 FEMALE HEAVY SMOKERS = 49
(HEAVY SMOKERS EACH SMOKED 20 OR MORE CIGARETTES DAILY AT TIME OF TEST)

		Regression coefficient with standard errors Non-smokers	Heavy smokers
Quinine	Males	+0.007 ± 0.023	+0.097 ± 0.020
	Females	+0.012 ± 0.011	+0.100 ± 0.029
PROP	Males	+0.038 ± 0.029	+0.114 ± 0.023
	Females	+0.014 ± 0.015	+0.075 ± 0.034

of individuals, thresholds tended to be lower (that is, more sensitive) during the menstrual period. Analysis by the Wilcoxon test for matched pairs and signed ranks showed that these differences were statistically significant. In the two-tailed test, for PROP, $P < 0.05$; for quinine, $P = 0.05$; and, for both combined, $P = 0.003$. The average increase in taste sensitivity during menstruation was 0.68 standard thresholds for PROP and 0.45 for quinine. When the scores for PROP and quinine were considered together, 18 per cent showed a decreased sensitivity during the menstrual phase compared with the post-menstrual; 16 per cent showed no change; and 66 per cent increased in sensitivity. An increase of 0.5 standard thresholds or more was shown by 47 per cent of the total sample; 18 per cent had an increase of one threshold or more; and eight per cent increased by two thresholds or more.

The responses of the subjects were markedly heterogeneous, as shown in Figure 5-12. The most extreme change was shown by the subject designated as No. 13. Six days before the onset of menstruation, her scores for both compounds changed markedly and reached maximum sensitivity on day —3, returning to their usual values by day +6. Her fluctuation in taste sensitivity, as indicated by the differences in menstrual and intermenstrual scores, represent a 1,024-fold change in PROP sensitivity and a 362-fold change in quinine sensitivity. This subject was tested over two cycles and, in both, showed the same change immediately before the onset of menstruation. The change, if any, shown by other subjects was less dramatic. The majority of the subjects reached maximum sensitivity after the onset of menstruation on days +1 to +5.

In large populations, as discussed above, taste thresholds for quinine tend to approach a normal distribution, but the curve of distribution for PROP thresholds tends toward bimodality. Individuals have been classi-

TABLE 5-15
MEAN TASTE THRESHOLD SCORES FOR PROP (P) AND QUININE (Q) DURING THREE 5-DAY PHASES OF THE MENSTRUAL CYCLE
(DAYS −3 TO +5 IN CYCLE WERE EXCLUDED FROM CALCULATIONS OF VARIANCE)

Subject	Variance (thresholds)	Pre-menstrual days −9 to −5	Menstrual days −1 to +4	Post-menstrual days +6 to +10	Difference (post-menstrual minus menstrual)
2	P 0.88	10.0	7.5	8.25	+0.75
	Q 0.27	5.5	4.25	5.5	+1.25
3	P 0.09	4.5	4.25	4.25	0.0
	Q 0.18	4.75	4.25	4.5	+0.25
4	P 0.15	8.0	7.75	7.75	0.0
	Q 0.10	5.5	5.5	6.0	+0.5
5	P 0.25	10.0	9.0	9.83	+0.83
	Q 0.09	5.67	5.0	5.5	+0.5
6	P 0.47	8.0	7.5	7.25	−0.25
	Q 0.24	3.25	2.67	2.75	+0.08
7	P 0.73	10.0	8.5	8.75	+0.25
	Q 0.17	2.25	1.75	2.25	+0.5
8	P 0.28	10.75	8.75	10.25	+1.5
	Q 0.12	3.25	2.75	3.5	+0.75
9	P 0.56	8.75	8.25	7.75	−0.5
	Q 0.14	5.75	5.25	5.75	+0.5
11	P 0.34	7.75	7.0	7.5	+0.5
	Q 0.26	4.75	5.5	5.0	−0.5
12	P 0.37	6.25	5.75	5.0	−0.75
	Q 0.46	5.25	4.5	3.5	−1.0
13	P 2.96	3.25	0.75	7.17	+6.42
	Q 2.92	0.75	1.75	3.67	+1.92
14	P 0.03	6.5	6.83	7.0	+0.17
	Q 0.19	6.25	5.5	5.75	+0.25
15	P 0.07	12.25	12.5	12.5	0.0
	Q 0.07	5.75	6.25	6.25	0.0
16	P 0.12	8.0	7.0	7.75	+0.75
	Q 0.27	4.5	4.5	3.75	−0.75
17	P 0.15	5.25	3.5	5.5	+2.0
	Q 0.19	5.0	2.5	5.5	+3.0
19	P 0.64	10.75	10.5	10.5	0.0
	Q 0.38	6.0	6.0	6.25	+0.25
20	P 0.79	4.25	4.25	5.5	+1.25
	Q 0.65	1.25	0.5	1.0	+0.5
21	P 0.27	11.5	11.0	11.0	0.0
	Q 0.17	5.5	4.75	4.5	−0.25
22	P 0.16	13.25	12.25	12.5	+0.25
	Q 0.22	6.75	5.75	6.5	+0.75
Mean (n = 19)	P	8.36	7.53	8.21	+0.68
	Q	4.61	4.15	4.60	+0.45

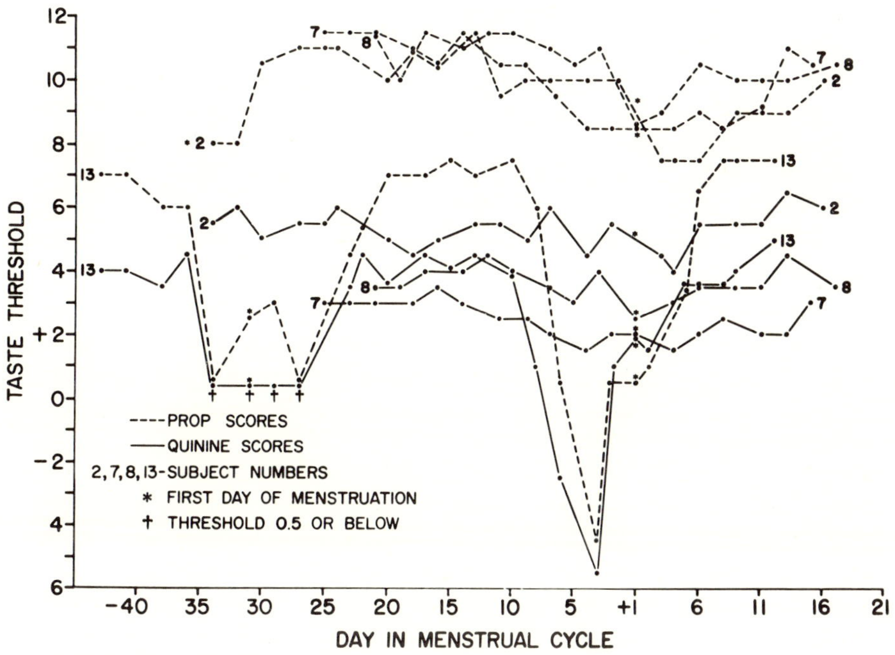

FIG. 5-12. TASTE THRESHOLD SCORES OF 4 SUBJECTS FOR 6-N-PROPYLTHIOURACIL (PROP) AND QUININE DURING ONE OR MORE MENSTRUAL CYCLES. THE DAYS IN THE CYCLE ARE NUMBERED FORWARD AND BACKWARD FROM THE FIRST DAY OF MENSTRUATION.

fied as "tasters" or "non-tasters" of PROP according to whether their thresholds occurred below or above the antimode between thresholds 9 and 10 (Fischer and Griffin, 1964). Both "tasters" and "nontasters" showed increased sensitivity at the time of menstruation. It is interesting that subject No. 2 (Figure 5-12 and Table 5-15) was initially classified as a "nontaster," but during menstruation she would clearly qualify as a "taster." She also changed from inability to taste phenylthiourea paper (Carolina Biologicals, Inc.) to being able to taste it as bitter during menstruation. The degree of change associated with menstruation does not appear to be associated with the position of the individual's taste threshold on the population distribution curve.

In our studies, information was obtained from each subject with a simple questionnaire relating to menstrual cycle regularity, symptoms experienced, and the time at which the symptoms were most severe. The subjects were asked to rate symptoms on a four-point scale: none, slight, moderate, and severe. The percentages of subjects who reported symptoms as moderate or severe were as follows: pain, 47 per cent; headache,

0 per cent; irritability, 58 per cent; depression, anxiety, nervousness, or tension, 37 per cent. In addition, 47 per cent reported some swelling of various parts of the body. Pain was most severe on the first day in all but one of the subjects who experienced it, but other symptoms were most commonly experienced immediately prior to the onset of menstruation. These figures are comparable to those obtained by Coppen and Kessel (1963a and 1963b) for a much larger and somewhat older population. No significant correlation was detected between increased taste sensitivity at menstruation and the occurrence of any of these symptoms.

Inquiries were also made into the length and regularity of the menstrual cycle. Subjects were asked to rate themselves as regular or irregular and to state the maximum and minimum interval between menstrual periods in recent months. Cycles varying in length by more than six days were classified as irregular. For the group in the present study there was complete agreement between the subjects' self-rating and the latter method of evaluation. No correlation was found between change in sensitivity and length of the cycle at the time of the tests. A significant positive correlation was detected, however, between increased taste sensitivity during menstruation and irregularity in length of the cycle as defined above. Table 5-16 shows a comparison of the change in sensitivities at the time of menstruation, compared with the postmenstrual phase, in subjects with regular and irregular cycles. Eleven subjects were classified as regular and eight as irregular. The magnitude and direction of the changes were compared by the Mann-Whitney U test. In the two-tailed test, comparing the scores of the two groups for PROP, $P = .090$; and for quinine, $P = .046$. When the data for PROP and quinine were combined, $P = .010$. This would indicate an association between irregularity in length of the cycle and increased taste sensitivity at or immediately prior to the time of menstruation.

TASTE DIFFERENCES BETWEEN DUODENAL AND GASTRIC ULCER PATIENTS

Taste thresholds for hydrochloric acid, quinine, and PROP were determined in 121 subjects with duodenal and/or gastric ulcers (Kaplan *et al.*, 1964b). The distributions of the 68 duodenal and 34 gastric ulcer subjects are similar with respect to age and sex (Table 5-17). In addition, the members of a group of 19 subjects who had both types of ulcers were taste-tested. This group of patients was too small to allow any reliable statistical interpretation. A group of 83 nonhospitalized volunteers (65 males, 18 females) was also taste-tested during the same time period, for

TABLE 5-16
COMPARISON OF CHANGES IN TASTE SENSITIVITY AT THE TIME OF MENSTRUATION IN SUBJECTS WITH REGULAR AND IRREGULAR MENSTRUAL CYCLES
(PROBABILITY VALUES, P, HAVE BEEN CALCULATED BY THE MANN-WHITNEY U TEST AND REFER TO THE TWO-TAILED TEST. FOR THE COMBINED DATA, P = .010)

Change in threshold	PROP		Quinine	
	Regular	Irregular	Regular	Irregular
1.75 or over		2		2
1.5 to 1.74	1			
1.25 to 1.49		1		1
1.0 to 1.24				
0.75 to 0.99	1	2	2	
0.5 to 0.74	1		2	3
+0.25 to 0.49	2		2	1
0.0 to 0.24	3	3	2	
−0.01 to 0.25	1			1
0.26 to 0.5	1		1	
0.55 to 0.75	1		1	
0.75 to 1.0			1	
Totals	11	8	11	8
	P = .090		P = .046	

comparative purposes. The age and sex distribution of this group is comparable to that of the ulcer groups. Although not confirmed by x-ray studies, the volunteers reported no history of ulcer.

Comparisons of taste thresholds were made between the subjects in the two-ulcer categories, as well as between those in each single ulcer category and the sample of nonhospitalized volunteers. The threshold scores for quinine and hydrochloric acid yielded continuous, monomodal distributions in these populations, and their statistical analysis is based on a standard t-test. Threshold scores for PROP, however, are distributed bimodally according to the previous studies on large populations. Therefore, the PROP thresholds were analyzed by nonparametric statistical tests, the Mann-Whitney U Test and the Kolmogorov-Smirnov two-sample test, each of which requires no assumption regarding the shape of the population distribution curve.

TABLE 5-17
AGE AND SEX OF SUBJECTS IN THE PEPTIC ULCER STUDY

	Duodenal	Gastric	Volunteer
Males	53	25	65
Females	15	9	18
Age (average)	48	53	41

No significant differences were indicated in the hydrochloric acid taste thresholds, and the distribution curves were similar for all groups. No significant differences were indicated in the quinine taste thresholds, and the distribution curves were similar for all groups.

The most striking result was a significant difference between the gastric and duodenal ulcer subjects in taste threshold for PROP, the duodenal group being the more sensitive tasters. Analyses of the differences in threshold distribution indicated that the greater PROP taste sensitivity of the duodenal ulcer subjects compared to the gastric ulcer subjects is statistically significant at a level where the P value is less than 0.01, according to both the Mann-Whitney and Kolmogorov-Smirnov nonparametric tests. Chi square evaluation of the differences between the duodenal and gastric ulcer samples, based on the assumption of a dichotomy occurring between taste threshold numbers 9 and 10, also yields a significance level in which P is less than 0.01 (Tables 5-18, 5-19, and 5-20). No significant differences in PROP taste sensitivity were found to occur between hospitalized patients and outpatients within any of the ulcer groups or the total ulcer sample.

These data support the concept that gastric and duodenal ulcers have differing constitutional backgrounds and that these differences may involve the specific genetic factors associated with differences in taste acuity for PROP and related drugs.

TABLE 5-18

DISTRIBUTION OF TASTE THRESHOLD SCORES FOR PROP IN THE VARIOUS ULCER CATEGORIES AND SAMPLE OF NONHOSPITALIZED VOLUNTEERS

Taste threshold in solution no.	Gastric ulcer	Duodenal ulcer	Both ulcer	Total ulcer	Non-hospitalized volunteers
Over 14	3	1	1	5	5
14	1	3	3	7	6
13	7	6	1	14	4
12	2	7	2	11	13
11	5	5	1	11	11
10	6	4	2	12	13
9	5	13	2	20	8
8	2	14	3	19	10
7	1	10	2	13	7
6	2	4	1	7	4
5					2
4		1	1	2	
3					
2					
1					
Total	34	68	19	121	83

TABLE 5-19

TABULATION OF "SENSITIVE" AND "INSENSITIVE" PROP TASTERS ACCORDING TO DICHOTOMY BETWEEN THRESHOLD NUMBER 9 AND 10, AS OBSERVED IN LARGE POPULATION SAMPLES (FISCHER AND GRIFFIN, 1961) AND X^2 EVALUATIONS OF COMPARATIVE DISTRIBUTIONS

	Sensitive tasters	Insensitive tasters	Total	X^2	P
Gastric ulcer	10	24	34	8.243	<0.01
Duodenal ulcer	42	26	68		
Gastric ulcer	10	24	34	0.3667	>0.1
Nonhospitalized volunteers	31	52	83		
Duodenal ulcer	42	26	68	7.977	<0.01
Nonhospitalized volunteers	31	52	83		
Gastric ulcer	10	24	34	1.038	>0.1
Both ulcer	9	10	19		
Duodenal ulcer	41	27	68	1.027	>0.1
Both ulcer	9	10	19		
Nonhospitalized volunteers	31	52	83	0.298	>0.1
Both ulcer	9	10	19		

TABLE 5-20

TASTE THRESHOLDS FOR QUININE AND HYDROCHLORIC ACID IN THE ULCER GROUPS AND IN THE NONHOSPITALIZED VOLUNTEER GROUP

(MEAN SCORE, STANDARD DEVIATION (SD) AND NUMBER OF SUBJECTS IN EACH GROUP ARE SHOWN. ALL COMPARISONS WERE STATISTICALLY INSIGNIFICANT)

Subjects	Number	Quinine		Hydrochloric acid	
		Mean	SD	Mean	SD
Gastric	34	6.059	1.883	11.265	1.601
Duodenal	68	6.029	1.869	10.721	1.629
Both	19	6.579	2.293	10.895	1.912
Total ulcer	121	6.124	1.939	10.901	1.670
Nonhospitalized volunteers	83	6.385	2.023		
Nonhospitalized volunteers	49			11.267	0.986

TASTE THRESHOLDS AND BIOLOGICAL ACUITY

Fischer and Griffin (1964) found that subjects who were very sensitive tasters and very insensitive tasters of both quinine and PROP tended to be sensitive and insensitive tasters, respectively, of numerous other compounds. The solutions which they tested and found to conform to this pattern included the following compounds: phenylthiourea, thioacetamide, thiourea, L-ergothioneine, Chlorpromazine (Smith, Kline, & French), Mellaril (Sandoz), Triflupromazine (Squibb), methylene blue, Tofranil (Geigy), desmethylimipramine (Geigy), acetamide, urea, L-phenylalanine, DL-phenylalanine, D-amphetamine sulfate, L-amphetamine sulfate, DL-dopa, sucrose, sodium chloride, potassium chloride, hydrochloric acid, Niamid (Pfizer), Antistine (Ciba). Their subjects who were very sensitive and very insensitive tasters of quinine and PROP tended to be, respectively, sensitive and insensitive tasters of drugs in general. The very sensitive and very insensitive tasters of the bitter-tasting drugs, quinine and PROP, also tended to be, respectively, sensitive and insensitive tasters of compounds whose primary taste qualities are sweet, salty, and sour, as well as bitter. Thus, taste acuity for both quinine and PROP were associated with high and low taste acuity in general.

Previous studies have indicated that there may be a relationship between a population's distributions of taste thresholds for different forms of a compound, and the comparative systemic activities of those different forms. Subjects generally display lower thresholds for l-quinine, and higher ones for quinidine or d-quinine (Fischer and Griffin, 1963). The former analog displays the higher toxicity of the two. The oral LD_{50} (i.e., lethal dose for 50 percent) of the "l" form has been found to be 214.8 ± 25.1 mg/kg for the mouse (Pfeiffer, 1956), whereas the corresponding value for the "d" form was 535 mg/kg (Schallek, 1952). Most subjects tested displayed lower thresholds for D-amphetamine, higher ones for L-amphetamine (Fischer and Griffin, 1964). The LD_{50} rating for L-amphetamine after very slow intravenous injection (1.0 cc. in two minutes) was 79.2 ± 8.5 mg/kg in the mouse, compared to only 5.0 ± 1.3 mg/kg for D-amphetamine (Fischer and Griffin, 1963 and 1964). The latter form, for which most subjects display lower taste thresholds, was 16 times as active as the former. In the above two cases, quinine vs. quinidine and L-amphetamine vs. D-amphetamine, the drug which elicited the greater taste sensitivity in man also manifested the lower or more powerful lethal dose in mice (Fischer and Griffin, 1963). The oral dosage of psilocybin which induced measurable neurological effects, as shown by finger-tapping tests, in a very sensitive taster of quinine and PROP produced no such changes in a very insensitive taster of quinine and PROP (Fischer, Griffin, and Pasamanick, 1965). Lower taste threshold (i.e., higher taste sensitivity) appears to be associated with a higher general systemic reactivity.

TASTE THRESHOLD AND PERSONALITY

The complete Wechsler Adult Intelligence Scale (WAIS) was administered to 27 college students whose taste thresholds had previously been determined for quinine and PROP. The group included four very insensitive tasters, with quinine thresholds of 7 or more and PROP thresholds of 11 or more; and five very sensitive tasters, with quinine thresholds of 4 or less and PROP thresholds of 9 or less. These categories are based on the observations of Fischer and Griffin (1963) that members of those two classes tend to generally be sensitive or insensitive tasters, respectively, for many other compounds as well as quinine and PROP. An intermediate group, with quinine thresholds of 5 or 6 and PROP thresholds equal to or more than 5 but not more than 13, showed less consistent responses for the different compounds. The investigator (D. Saunders of the University of Colorado) who administered the WAIS tests and then determined personality profiles, based on the subtest patterns, was not aware of his subjects' taste thresholds. The results indicated that the insensitive tasters of quinine and PROP ($n=4$) showed a "compensated" pattern of scores on the WAIS, while the sensitive tasters ($N=5$) showed a WAIS pattern described as "internalized" (Fischer, Griffin, and Pasamanick, 1965). In other words, the insensitive tasters (who may be relatively low in general systemic reactivity) yielded WAIS score profiles characteristic of subjects who easily maintain contact with other individuals, whereas the very sensitive tasters (who may be relatively high in general systemic reactivity) yielded patterns characteristic of relatively introverted subjects.

SUMMARY

Taste thresholds were determined with a modified form of the double blind placebo and sorting method of Harris and Kalmus. Intrapair threshold differences for hydrochloric acid, 1-quinine sulfate, and 6-n-propylthiouracil were investigated in pairs of monozygotic twins, dizygotic twins, and non-twin siblings. The intrapair differences were significantly less in the monozygotic twin pairs than in either the dizygotic twin or sibling pairs, for each of the three substances and regardless of sex. Significant positive correlations were observed between the individuals' threshold scores for the different compounds. There was no significant positive correlation, however, between the intrapair differences for any two compounds. These observations are consistent with the hypothesis that independent factors are involved in each of the three substances.

Sensitive tasters of quinine and 6-n-propylthiouracil reported relatively large numbers of food dislikes and a preference for mild-tasting foods, compared to insensitive tasters. The sensitive tasters included a relatively high proportion of nonsmokers, compared to the insensitive tasters, who included a relatively high proportion of heavy smokers. Taste threshold was not found to be related to age or sex when smoking habits were experimentally controlled, but heavy smokers were observed to manifest decreased taste sensitivity with increased age. Phases of the menstrual cycle were sometimes associated with significant changes in taste thresholds, and most of the subjects tested repeatedly showed an increased sensitivity during the period of menstruation. Apparently, high taste sensitivity for both quinine and 6-n-propylthiouracil may be associated with high general systemic reactivity, and vice versa. This observation appears to be relevant both with regard to the aspects of drug activity and subject responsiveness. There also appears to be a possible association with personality differences, very sensitive tasters having displayed relatively "internalized" WAIS score patterns, compared to relatively more "compensated" patterns in very insensitive tasters. Individual differences in taste acuity patterns, related to both genetic and nongenetic variables, have also been associated with significant differences in peptic ulcer pathology. Duodenal ulcer patients, as a group, were significantly more sensitive to the taste of 6-n-propylthiouracil, but not of quinine or hydrochloric acid, than either gastric ulcer patients or nonpatient controls.

BIBLIOGRAPHY

Ardashnikov, S. N.; Lichtenstein, E. A.; Martynova, R. P.; Soboleva, G. V.; and Postnikova, E. N. 1936. The diagnosis of zygosity in twins (Three instances of difference in taste acuity in identical twins.) *J. Hered.* **27**: 465-68.

Beiguelman, B. 1964. Taste sensitivity to phenylthiourea and menstruation. *Acta genet. med. gemell.* **13**: 197-99.

Blakeslee, A. F. 1932. Genetics of sensory thresholds: taste for phenylthiocarbamide. *Proc. Nat. Acad. Sci.* **18**: 120-30.

Blakeslee, A. F. and Salmon, M. R. 1931. Odor and taste blindness. *Eug. News.* **16**: 105-09.

Byrd, E. and Gertman, S. 1959. Taste perception in the aged. *Geriatrics* **14**: 381-86.

Coppen, A. and Kessel, N. 1963 a. Menstrual disorders and personality. *Brit. J. Psychiat.* **109**: 711-21.

——. 1963 (b). Menstrual disorders and personality. *Acta Psychotherapeutica* **11**: 174-80.

Das, S. R. 1956. A contribution to the heredity of the P.T.C. taste character based on a study of 845 sib-pairs. *Ann. Hum. Genet.* **20**: 334-43.

——. 1957. Inheritance of the P.T.C. taste character in man: an analysis of 126 Rarhi Brahmin families of West Bengal. *Ann. Hum. Genet.* **22**: 200-12.

Dencker, S. J.; Hauge, M.; and Kaij, L. 1959. An investigation of the PTC taste character in monochorionic twin pairs. *Acta Genet.* **9**: 236-44.

Fischer, R. and Griffin, F. 1963. Quinine dimorphism: a cardinal determinant of taste sensitivity. *Nature* **200**: 343-47.

——. 1964. Pharmacogenetic aspects of gustation. *Arzneim.-Forsch. (Drug Research)* **14**: 673-86.

Fischer, R.; Griffin, F.; England, S.; and Garn, S. M. 1961. Taste thresholds and food dislikes. *Nature* 191: 1328.

Fischer, R.; Griffin, F.; England, S.; and Pasamanick, B. 1961. Biochemical-genetic factors in taste polymorphism and their relation to salivary thyroid metabolism in health and mental retardation. *Med. Exp.* **4**: 356-66.

Fischer, R.; Griffin, F.; and Kaplan, A. R. 1963. Taste thresholds, cigarette smoking, and food dislikes. *Med. Exp.* **9**: 151-67.

Fischer, R.; Griffin, F.; and Pasamanick, B. 1965. The perception of taste: some psychophysiological, pathophysiological, and clinical aspects. In *Psychopathology of Perception,* ed. *P. H. Hoch and J. Zubin* pp. 129-63. New York: Grune & Stratton.

Fox, A. L. 1931. Tasteblindness. *Science* **73**: Supp. 14, (April 17).

——. 1932. The relation between chemical constitution and taste. *Proc. Nat. Acad. Sci.* **18**: 115-20.

Glanville, E. V. and Kaplan, A. R. 1965a. Food preference and sensitivity of taste for bitter compounds. *Nature* **205**: 851-53.

——. 1965b. Taste perception and the menstrual cycle. *Nature* **205**: 930-31.

——. 1965c. The menstrual cycle and sensitivity of taste perception. *Amer. J. Obst. Gyn.* **92**: 189-94.

Harris, H. and Kalmus, H. 1949. The measurement of taste sensitivity to phenylthiourea (P.T.C.). *Ann. Eugen.* **15**: 24-45.

——. 1951. The distribution of taste thresholds for phenylthiourea of 384 sib pairs. *Ann. Eugen.* **16**: 226-30.

Hartmann, G. 1939. Application of individual taste difference towards phenylthio-carbamide in genetic investigations. *Ann. Eugen.* **9**: 123-35.

Henkin, R. I.; Gill, J. R., Jr.; and Bartter, F. C. 1963. Studies on taste thresholds in normal man and in patients with adrenal cortical insufficiency: the role of adrenal cortical steroids and of sodium concentration. *J. Clin. Invest.* **42**: 727-35.

Horn, D. 1963. Behavioral aspects of cigarette smoking. *J. Chron. Dis.* **16**: 383-95.

Kalmus, H. 1957. Defective color vision, P.T.C. tasting and drepanocytosis in samples from fifteen Brazilian populations. *Ann. Hum. Genet.* **21**: 313-17.

——. 1958. Improvements in the classification of the taster genotype. *Ann. Hum. Genet.* **22**: 222-30.

Kalmus, H. and Trotter, W. R. 1962. Direct assessment of the effect of age on P.T.C. sensitivity. *Ann. Hum. Genet.* **26**: 145-49.

Kaplan, A. R. and Fischer, R. 1965. Taste sensitivity for bitterness: some biological and clinical implications. In *Recent Advances in Biological Psychiatry,* ed. J. Wortis, Vol. 7, pp. 183-96. New York: Plenum Press.

Kaplan, A. R.; Powell, W.; Fischer, R.; and Marsters, R. 1964a. Reexamination of genetic aspects of taste thresholds for phenylthiourea type compounds. In *Genetics Today,* S. J. Geerts, Vol. I, pp. 292-93. London: Pergamon Press.

Kaplan, A. R.; Fischer, R.; Glanville, E. V.; Powell, W.; Kamionkowski, M.; and Fleshler, B. 1964b. Differential taste sensitivities in duodenal and gastric ulcer patients. *Gastroenterology,* **47**: 604-09.

Kaplan, A. R.; Fischer, R.; Karras, A.; Griffin, F.; Powell, W.; Marsters, R.; and Glanville, E. V. 1967. Taste thresholds in twins and siblings. *Acta genet. med. gemell.* **16**: 229-43.

Kaplan, A. R.; Glanville, E. V.; and Fischer, R. 1965. Cumulative effect of age and smoking on taste sensitivity in males and females. *J. Geront.,* **20**: 334-37.

Krut, L. H.; Perrin, M. J. and Bronte-Stewart, B. 1961. Taste perception in smokers and non-smokers. *Brit. Med. J.* **1**: 384-87.

Leguèbe, A. 1960. Génètique et anthropologie de la sensibilité à phénylthiou-carbamide. *Bull. Inst. Roy. Sci. Nat. Belg.* **26**: 1-27.

Matarazzo, J. D. and Saslow, G. 1960. Psychological and related characteristics of smokers and non-smokers. *Psychol. Bull.* **57**: 493-513.

Merton, B. B. 1958. Taste sensitivity to PTC in 60 Norwegian families with 176 children. Confirmation to the hypothesis of single gene inheritance. *Acta Genet.* **8**: 114-28.

Pfeiffer, C. C. 1956. Optical isomerism and pharmacological action, a generalization. *Science* **124**: 29-30.

Rife, D. C. 1938. Contribution of the 1937 national twins convention to research. *J. Hered.* **29**: 83-90.

Schallek, W. 1952. Quinidine-like activity of thephorin. *J. Pharma. Exp. Ther.* **105**: 291-98.

Sinnot, J. J. and Rauth, J. E. 1937. The effect of smoking on taste thresholds. *J. Gen. Psychol.* **19**: 151-53.

Skude, G. 1963. Some factors influencing taste perception for phenylthiourea (P.T.C.). *Hereditas* **50**: 203-10.

Snyder, L. H. 1931. Inherited taste deficiency. *Science* **74**: 151-52.

———. 1932. Studies in human inheritance. IX. The inheritance of taste deficiency in man. *Ohio J. Sci.* **32**: 436-40.

Sutton, H. E.; de Lamadrid, E. G.; and Esterer, M. B. 1962. The hereditary abilities study: genetic variation in human biochemical traits. *Amer. J. Hum. Genet.* **14**: 64-82.

Verkade, P. E.; Wepster, B. M.; and Stegerhoek, L. J. 1959. Investigations on taste blindness with thiocarbamides. Intra-pair discrepancy of taste in pairs of identical twins. *Acta genet. med. gemell.* **8**: 361-68.

6

IRVING I. GOTTESMAN
University of Minnesota
JAMES SHIELDS
Medical Research Council Psychiatric Genetics Research Unit,
Maudsley Hospital, London

IN PURSUIT OF THE SCHIZOPHRENIC GENOTYPE

Fifty-six years ago Eugen Bleuler (1911), discussing the cause of the schizophrenias, commenced with the observation that they were commonly familial. "However," he continued, "if an adherent of an 'infectious theory' of this disease should choose to say that there is no hereditary factor in schizophrenia but merely an infection from some common source, or if someone else cares to assume that the modifications of the psychic or physical factors produced by communal living produce such accumulations of disease in a given family group, we would be unable to produce any proof to the contrary. Such skeptics could observe that in many cases, even after the most thorough study, no evidence of any hereditary *Anlage* and no individual predisposition (such as a seclusive, withdrawn character structure) has ever been proven."

"And yet heredity does play its role in the etiology of schizophrenia, but the extent and kind of its influence cannot as yet be stated. In order to be able to accomplish something more than what has already been done on this question of heredity, we first of all would need a *workable concept of heredity*" (1950, p. 337, italics added). Bleuler made these observations five years before Rüdin (1916) pioneered in the effort to apply the rediscovered principles of Mendelian genetics to the incidence of schizophrenia in relatives of probands. The first twin study of schizophrenia that was sound in regard to both genetics and problems of sampling was Luxenburger's, which was not conducted until 1928. Some dozen systematic twin studies of schizophrenia (Gottesman & Shields, 1966b) have been conducted to date. Erlenmeyer-Kimling, Rainer, and Kallmann (1966) have summarized the numerous sibling and family studies con-

---

The research reported here was supported in part by USPHS Grant MH-13117-02, and was in part carried out at the MRC Psychiatric Genetics Research Unit, Maudsley Hospital, London (Dr. Eliot Slater, Director).

ducted since 1916 (cf. Chap. 7). Where do we stand on the hereditary aspects of schizophrenia in the light of all these data? Do we yet have a workable concept of heredity that accounts for our observations?

So long as we adhere to simple Mendelian models of dominance and recessiveness and construe schizophrenia as a homogeneous entity, we have no workable concept of heredity. It has been known all along that simple Mendelian ratios are not found among the relatives of schizophrenic probands (cf. Gregory, 1960). Arguing by analogy with advances in research on mental deficiency, we find some forms of schizophrenia, individually of rare occurrence, that might fit classical Mendelian models. In doing so we would be permitting ourselves the non-parsimonious possibility of genetic and environmental heterogeneity. As with mental deficiency, such a use of multiple simple models would leave most schizophrenias unexplained, and so far no Mendelian sub-groups have been identified. Also, if there were some need to preserve the ideas about the genetic homogeneity of schizophrenia, it could be done by introducing the concept of "incomplete manifestation." Again, this is at the risk of being labeled Procrustean and non-parsimonious. It is perhaps worth noting that as scientists we are committed to parsimony because of convention and not because it guarantees our carving Nature at the joints. One monogenic theory of this kind which appears to account for many of the empirical observations on the familial incidences of schizophrenia is that of Slater (1958). According to his model, all homozygotes for a specific gene and a portion of heterozygotes would be affected, and virtually all schizophrenics would be heterozygotes.

Based on our own work (Gottesman and Shields, 1966a, 1966b; Shields, 1967) with a sample of schizophrenic twins to be described shortly, and encouraged by the ideas and findings of others, we should like to consider the merits of treating schizophrenia as a threshold character (Grüneberg, 1952) whose appearance is predictable from a diathesis-stress model (Rosenthal, 1963). Let us suppose that the diathesis is polygenically determined, and that what is inherited is a constitutional predisposition to schizophrenia termed variously the schizoid constitution (Kretschmer, 1948), with which an introverted, shut-in personality and a leptosomic body build were associated; a characterological defect of a specific kind (Essen-Möller, 1941); a neural integrative defect (Meehl, 1962); or, an independent tendency to manifest symptoms of a schizophrenic nature (Tsuang, 1967). (The workers cited did not necessarily embrace a polygenic diathesis.) Descriptions of the diathesis on more than one level, e.g., biochemical or cell membrane (cf. Grenell, 1962), may be essential in formulating research strategies. Furthermore, while some of the genetic influences may exert themselves early in development, others may be augmented or released only after specific psychosomatic states have been reached.

Polygenic theory, with or without a specific major gene, can go a long way toward giving us the workable concept of heredity hoped for by Bleuler. The likelihood of manifestation as overt schizophrenia, as well as the form and severity of the disorder, would depend on how many of the genes in the system were present, together with the remainder of the genetic background, and the environmental stress factors. The theory can contribute to explaining the distribution of Kraepelinian subtypes and the clinical variability across diagnostic boundaries seen in pairs of affected relatives. So-called process and reactive schizophrenias would be viewed as extremes on a continuum of severity, with severity determined in the first instance by the number of polygenes. Index cases who were head-injured, or severely traumatized psychologically, or who rapidly recovered would be viewed as having fewer of the polygenes and consequently fewer of their siblings or other first-degree relatives would be expected to be decompensated schizophrenics. Kay and Roth (1961) espoused this model for explaining the similarity between late paraphrenia and schizophrenias with earlier onsets. They reasoned that late paraphrenics, having fewer genes in the system, required such gerontological stress as social isolation and deafness in combination with previous personality eccentricities to produce decompensation. Polygenic inheritance could account for the observed irrational, schizoid, and "borderline" personalities in the parents and siblings of index cases. Finally, with this kind of theory we would expect the MZ co-twins of schizophrenics to be abnormal more often than their DZ counterparts. In particular, we would expect the nonpsychotic MZ co-twins to be more deviated along dimensions genetically related to schizophrenia. Abnormalities in DZ co-twins, when they occurred, would be more variable within pairs.

Rosenthal favored a diathesis-stress theory of schizophrenia over either monogenic-biochemical theories or life-experience theories, not only in the case of the Genain quadruplets but probably for most cases of schizophrenia. However, he also criticized the favored model, with justification, for its exasperating looseness. He called for research strategy to focus on clarifying the nature of the predispositions and stressors as well as their interaction, since they are usually only vaguely conceived or formulated (Rosenthal 1963, p. 509).

Whatever else may enter into the inheritance of the liability to schizophrenia, normal variation in introversion is generally believed to play an appreciable part. Summing up about the premorbid personalities of schizophrenics, Rosenthal said, "The best evidence available suggests that hereditary factors influence the introversive-extroversive expression of personality, that a disproportionate number of schizophrenics are drawn from the pool of introverts, but that only a small fraction of introversive persons ever become hospitalized schizophrenics" (Rosenthal, 1963, p. 534). Insofar as the inherited diathesis is reflected primarily in the previous personality one might expect acute onsets of florid schizophrenia with no evidence of predisposition to be unlikely or rare. In 1911 Bleuler observed that the onset was usually insidious and that ". . . whenever we have a thorough case history, it is an exception if we are not able to detect the previous, earlier signs of disease . . . . there are early character anomalies which can be demonstrated by careful case histories in more than half the individuals who later became schizophrenic: the tendency to seclusion, withdrawal, together with moderate or severe degrees of irritability. They already stood out as children because they were unable to play with others and followed their own ways instead" (Bleuler, 1950). He thought that the autistic traits were actually the first symptoms of the disease and not merely expressions of the predisposition to it. Bleuler also recognized the unreliability of informants in pinpointing the actual onset of the disorder.

Before we attempt to explore in more detail than has been customary, the application of a polygenic model to schizophrenia, let us briefly outline some of the findings in our ongoing twin study and, even more briefly show why we believe the vast majority of the twin studies conducted to date are replications of the same experiment (Gottesman and Shields, 1966b).

THE MAUDSLEY-BETHLEM SCHIZOPHRENIC TWIN STUDY

Through the foresight of Dr. Eliot Slater, Director of the Medical Research Council's Psychiatric Genetics Unit, a register of all twins seen at the Maudsley and Bethlem Royal Joint Hospital has been kept from 1948 onwards. On admission to the in-patient, out-patient, or children's services, every patient is routinely asked whether he was born a twin. In March, 1964, when our present series was closed, there were 392 adult patients on the register with twins of the same sex surviving to the age of 15. We believe ascertainment to be virtually complete.

Of the index twins (probands), 47 (12 per cent) had received an official hospital diagnosis of schizophrenia at the time of discharge. This is close to the 11 per cent of all adult Maudsley patients thus diagnosed. Since most of the remaining twins on the register had been followed up by the Unit, some were known to have received subsequent diagnoses of schizophrenia when re-hospitalized elsewhere. From 1963 to 1965 we followed up all other twins in whose case initial symptoms suggested the possibility of a future schizophrenic decompensation. In all, we added a further 21 probands who had subsequently received a hospital diagnosis of schizophrenia, making a total of 68. Six twins were omitted who had either been reared in non-European environments or about whom information was insufficient. Our 62 probands came from 57 pairs, since in 5 pairs both twins had been registered at the Maudsley. Our 57 pairs of twins were obtained from a starting material estimated at 45,000 psychiatric patients.

Zygosity was diagnosed by a combination of blood grouping, fingerprint analysis and resemblance in appearance. The distribution of the sample by sex and zygosity is given in Table 6-1 and supports the view that it is a representative one.

TABLE 6-1
GOTTESMAN AND SHIELDS SAMPLE BY ZYGOSITY AND SEX

	MZ	SS DZ	Total
Female	11	16	27
Male	13	17	30
Total	24	33	57

This is the first schizophrenic twin series which has no excess of females. In terms of probands we have 31 males and 31 females. The median age on last information for our twins was 37 with a range of 19 to 65. Of the 48 MZ twins, 42 (88 per cent) have been seen personally by one or both of us; 4 have been seen by other Unit members or by psychologists on our behalf, and the 2 remaining were especially seen by their family physician before disappearing from sight. Of the 66 DZ twins, we have seen 49 (74 per cent); 3 have been seen by others connected with the twin study. Of the remaining 14, 4 were dead, 4 were abroad, 1 untraced and 5 uncooperative at the time of follow-up. All discordant pairs of twins have been followed for over 3 years from the onset of illness in the proband, one of them for as long as 16 years.

Our sample cannot be put forward, anymore than any other, as completely representative of the domain of schizophrenia. Based on 16 years' consecutive admissions to out-patient facilities and to a short-stay, in-patient service, and including probands who at first appeared to be neurotic or have personality disorders, it may be said to make better provision for cases with a good prognosis than previous schizophrenic twin samples which were loaded with classical types of dementia praecox.

Table 6-2 presents our results in terms of hospital diagnoses uninfluenced by our hindsight. We have established four reasonably objective and reliable grades of similarity. The grades vary along a continuum of severity of psychopathology rather than along a specifically schizotypic dimension. Grade 1 consists of pairs in which both twins have been hospitalized and diagnosed as schizophrenic. Grade 2 co-twins have had a psychiatric hospitalization but have been diagnosed as other than schizophrenic. As it happens, none of the illnesses of Grade 2 co-twins had notable schizophrenic features. Grade 3 co-twins are otherwise psychiatrically abnormal, as determined by such criteria as out-patient psychiatric care, being in the care of a GP for a clear psychiatric problem, a neurotic or psychotic looking MMPI profile, or, in three cases, being manifestly abnormal on interview. Grade 4 co-twins were within normal limits at last information.

Concordance with respect to a hospital diagnosis of schizophrenia was 42 per cent in our 24 MZ pairs and 9 per cent in 33 same sexed DZ pairs. This is simple direct pairwise concordance and the figures have not been corrected for age. Counting double index cases twice, our MZ rate goes up to 50 per cent (14/28) and our DZ rate to 12 per cent (4/34). There were no important sex differences in concordance. The figures, of course, will be subject to change depending on how many probands and co-twins are discarded, using different diagnostic criteria. One of several methods we are attempting that will permit uncontaminated diagnoses of all twins involves the submission to Dr. Slater and other judges of case histories for each twin which do not mention zygosity or any information about the other twin in a pair.

Notice in Table 6-2 that only 21 per cent of MZ co-twins could be classed as normal compared to 55 per cent of the DZ co-twins and that

TABLE 6-2
CONCORDANCE FOR SCHIZOPHRENIA AND RATES OF PSYCHIATRIC HOSPITALIZATION AND MARKED ABNORMALITY IN THE TWINS OF SCHIZOPHRENICS [a]

Grade	MZ		DZ	
	N	%	N	%
1 [b]	10	42	3	9
1+2	13	54	6	18
1+2+3	19	79	15	45
Normal	5	21	18	55
Total	24	100	33	100

[a] All figures uncorrected for age.
[b] Chi square = 6.63, $p < .01$, one-tailed.

out of 19 MZ pairs in which both were abnormal the second twin was schizophrenic in ten pairs (53 per cent) compared with 3/15 (20 per cent) of DZ pairs. Following the lead of Kallmann and Rosenthal, we next analyzed the relationship between severity of schizophrenia in the proband and the degree of concordance for schizophrenia among the co-twins. Our calculations are in terms of probands rather than pairs at this point. Using criteria for severity such as length of hospitalization and outcome, we found that concordance was significantly higher when illness in the index case was severe. Table 6-3 gives our findings. We shall mention here only the effect of outcome on concordance. A detailed analysis of severity and concordance has been made by Gottesman (1968).

Outcome was assessed according to whether on follow-up the Maudsley index case had been out of the hospital for at least six months and was working or running a home. There were 12 cases with good outcomes and 16 with poor among the 28 MZ index cases. Only two out of 12 co-twins in the good outcome group had been diagnosed as schizophrenic, compared with 12 out of 16 in the poor outcome group: 17 per cent and 75 per cent respectively. Notice that the concordance rate for severe schizophrenia is similar to those found by earlier twin researchers for their samples which were mainly from resident or consecutive admissions to state type hospitals. MZ concordance in Kallmann's (1946) series was 100 per cent when the schizophrenia in the proband took a deteriorating course; it was only 26 per cent for MZ pairs with little or no deterioration in the first twin. Inouye (1963) in his study of Japanese schizophrenic twins obtained concordance rates of 74 per cent in MZ pairs where the index case was classified as chronic progressive schizophrenia, 86 per cent when called relapsing schizophrenia, but only 39 per cent in those MZ index pairs where the proband had been diagnosed as a "mild chronic" or transient schizophrenia.

TABLE 6-3
EFFECTS OF SEVERITY (GAINFULLY EMPLOYED AND OUT OF HOSPITAL MORE THAN 6 MONTHS) ON CONCORDANCE

Co-twin Status	MZ Proband [a]			DZ Proband		
	"Mild"	"Severe"	Total	"Mild"	"Severe"	Total
Schizophrenic	2	12	14	0	4	4
Non-Schizophrenic	10	4	14	16	14	30
% Concordance	17	75	50	0	22	12

[a] Chi square = 7.15, $p < .01$, one-tailed.

Such data as these on the relationship between severity and concordance can be explained by a polygenic theory. We could infer that a proband with a good outcome had had few of the genes in the system, and we would expect his co-twin to have a much lower probability of decompensating than the co-twin of a severe schizophrenic. Most instances of schizophrenia, from the most mild to the most severe, could then be regarded as biologically related.

COMMENTS ON RELATED STUDIES IN THE LITERATURE

Our sample included 14 unselected pairs of identical twins where only one had received a diagnosis of schizophrenia. This fact by itself is enough to rule out genetic factors as a *sufficient* cause of schizophrenia as well as to raise the question of etiological heterogeneity. The MZ concordance rate which we report lies between those of the earlier studies and the recent ones from Finland and Norway. We do not claim that our study gives the *true* concordance rate for schizophrenia. If we have learned anything from previous work it is that it is unreasonable to expect this from any study. What is valuable, is to discover in what ways concordance may change as a function of population, sampling, and the preferred statistical methods of the investigator.

Some of the criticisms of the earlier twin studies are statistical in nature. Reported concordance rates vary according to whether *pairs* have been counted (no pair more than once), or whether co-twins of defined *index cases* have been counted, in which case some pairs may have been counted twice. Concordance rates also vary according to whether, and how, correction has been made for age. Some studies may have had too many double index cases or have been over-corrected for age. But even if results are reported in terms of pairs and without any age correction, MZ rates in the earlier studies lie between 58 per cent and 69 per cent, those for DZ pairs between zero and 18 per cent (see Table 6-5 below).

A second source of error, and one of which twin investigators have been well aware since Luxenburger's work of 1928, is that of the preferential inclusion of concordant pairs. Luxenburger and Essen-Möller in their studies went to the length of checking the birth registers for many thousands of patients in order to be sure of not missing any who were twins. This method is not possible in most countries, and other investigators have proceeded as best they could. Some cases of twins were no doubt missed, discordant pairs more easily than concordant. However, in none of the studies under review was there an over-representation of MZ pairs such as would lead one to expect a gross bias in favor of similarity.

A third reason, or set of reasons, for suspecting concordance rates to be too high is the possibility of unconscious bias in diagnosis when the same investigator determines zygosity and also makes the psychiatric diagnosis of both members of a pair of twins. Errors can be reduced by using blood groups, fingerprints, and other objective data in deciding zygosity, by reporting findings in terms of hospital diagnoses as well as in terms of any changed diagnoses which the investigator has found it necessary to make, and above all by presenting case histories. Of the older studies only those of Essen-Möller and Slater provided case histories. Kallmann left himself more open to the kind of criticism under discussion. However, information which he provided to Kety in 1958 (Shields, Gottesman, & Slater, 1967) exonerates this indefatigable collector of twin and family data on schizophrenia from any gross bias arising from contaminated diagnoses.

A fourth reason why concordance rates may be too high is that females may be more often concordant than males, and that the samples studied contain an excess of females. In those studies that investigated both sexes and reported them separately, MZ concordance rates are 68 per cent in female pairs and 50 per cent in male pairs. It can be shown, however, that the difference is mainly accounted for by two studies (Rosanoff and Slater) in which there is an over-representation of females and so may be largely an artifact. One sex difference, however, that does appear consistently in all studies that included opposite-sexed pairs (except Kringlen, 1966) is that concordance is lower in opposite-sexed pairs than in DZ pairs of the same sex. This points to the operation of environmental factors.

One of Rosenthal's most interesting points was to suggest that a sampling bias may have occurred through selecting cases from the resident population of mental hospitals. If concordance were related to chronicity of illness, the rates found would be exaggerated. Essen-Möller, who sampled from consecutive admissions, found a lower concordance than Kallmann and Slater, who had started with a resident population and then added twins from subsequent admissions. The resident-versus-consecutive hypothesis can be put to the test by re-analyzing the Slater twin series according to how the cases entered it. Table 6-4 shows that when this was done (Gottesman & Shields, 1966b) no tendency was observed for concordance to be higher in the resident sub-sample, even though it contained a greater excess of females, than in the sub-sample of subsequently admitted twins. It must be pointed out, however, that there was probably little essential difference in severity of illness between the resident and the subsequently admitted index twins.

It would be cavalier to suggest that the variation among the twin studies of schizophrenia conducted to date can be accounted for in terms of

TABLE 6-4
CONCORDANCE BY SOURCE OF SLATER'S INDEX PAIRS, SEXES COMBINED

Source of Index Pairs	MZ	Combined DZ	All Twins
Resident	17/26 (65%)	4/71 (6%)	21/97 (22%)
Consecutive	7/11 (64%)	6/41 (15%)	13/52 (25%)
Total	24/37 (65%)	10/112 (9%)	34/149 (23%)

the inaccuracy and carelessness in the earlier work. In Table 6-5 we present what we consider to be reasonable estimates of the concordance rates in the various twin studies when age corrections have been removed and when all data are presented in terms of simple direct pairwise concordance.

Is it nevertheless possible that the MZ:DZ difference in concordance is largely accountable for in environmental terms? The most plausible environmental explanation is that factors such as identification, weak ego formation and confusion of identity predispose MZ twins in particular to developing schizophrenia. The simplest answer is that neither twins as such, nor MZ twins in particular, are more often schizophrenic than singletons. Studies of normal twins brought up apart (Shields, 1962; Juel-Nielsen, 1965) show that personality resemblances in MZ twins brought up together (e.g., Gottesman, 1963b, 1965) are not due simply to their being brought up by the same mother or to mutual identification.

PSYCHOMETRIC EVIDENCE OF PSYCHOPATHOLOGY IN THE MZ AND DZ CO-TWINS OF SCHIZOPHRENICS

Another line of evidence to support the polygenic basis for the etiology of schizophrenia comes from the objective picture we are able to obtain from the Minnesota Multiphasic Personality Inventory (MMPI) administered to 73 per cent (35/48) of the MZ twins and 67 per cent (44/66) of the DZ twins in our Maudsley-Bethlem twin study.[1] In a number of respects our MMPI data are unsatisfactory but their uniqueness and heuristic value far outweigh their shortcomings. The preliminary findings we shall report show up despite a number of obstacles to data homogeneity. We cannot be sure that untested or dead or illiterate twins are like those tested. The MMPI's were obtained from 28 pairs of twins and 23 unpaired twins; the unpaired twins may be either probands or co-twins. Since the MMPI profile varies with clinical condition, the MMPI's

[1] We are indebted to Beatrice Rouse for technical assistance in the data analysis.

TABLE 6-5

SUMMARY TABLE OF TWIN STUDIES OF SCHIZOPHRENIA [a]

		Concordance						Sampling						
Investigator	Country	MZ Pairs	%	DZ SS Pairs	%	DZ OS Pairs	%	Resident vs. Consec. Admit.	Long Stay vs. Short Stay	Is Severity Related to Concordance	Sex with Higher Concordance	Sample Sex Surplus	Hospital vs. Author Diagnosis	Blood and/or Fingerprints In Zyg. Dx.
Kallmann														
Preadolescent (1956)	USA	15/17	88	8/35*	23	*		R+C	L	?	?	M	A	Yes
Adult (1946)	USA	120/174	69	34/296	11	13/221	6	R+C	L	Yes	neither	F	A	No
Slater (1953)	UK													
Resident Sample		17/26	65	4/35	11	0/36	0	R	L	Yes	F	F	A	Yes
Consecutive Sample		7/11	64	4/23	17	2/18	11	C	L	Yes	neither	neither	A	Yes
Essen-Möller (1941)	Sweden	7/11**	64	4/27**	15	—		C	L	No	neither	neither	A	Yes
Rosanoff (1934)	USA	25/41	61	7/53	13	3/48	6	R	L+S?	?	F	F	H	No
Inouye (1961)	Japan	33/55	60	2/11	18	0/6	0	R+C	L+S	Yes	neither	F	A	Yes
Luxenburger (1928)	Germany	11/19	58	0/13	0	0/20	0	R+C	L+S	Yes?	neither	neither	A	No
Gottesman & Shields (1966)	UK	10/24	42	3/33	9	—		C	S	Yes	neither	neither	H	Yes
Harvald & Hauge (1965)	Denmark	2/7	29	2/31	6	1/28	4	neither	n.a.	?	?	?	H	Yes
Kringlen (1966)	Norway	14/50	28	6/94	6	6/81	7	R+C	L	No	neither	neither	H	Yes
Tienari (1963)	Finland	0/16	0	—		—		neither	n.a.	No	***	***	A	Yes

[a] By permission of Academic Press. From Progress in Experimental Personality Research, 1966. Vol. 3. © Academic Press.
* DZ pairs not broken down by type and include OS (opposite sex) pairs.
** Includes psychoses with schizophrenic-like features and Kaij (1960) follow-up. On other criteria MZ concordance ranges from 0%—86%.
*** Tienari did not include female probands.

should ideally have been administered at a uniform stage of the illnesses of all schizophrenic subjects, such as first admission or subsequent relapse, and at a similar stage in all pairs, such as a time when one twin is decompensated, the other not. Furthermore, concordant pairs should be tested more than once, so that twins may also be compared when both are in phases of remission, acute relapse, or chronicity. However, no control was possible over the clinical condition of the twins. One of our DZ co-twins was tested 26 years after his first admission and had spent virtually all that time in mental hospitals. Other twins were tested at various times in the course of their illness, sometimes in hospital, most often not. Probands in varying degrees of remission, either spontaneous or with an unknown degree of support from phenothiazine medication, will decrease the amount of pathology detectable by the MMPI and make contrasts with co-twins less impressive. Those probands who may be judged as having been incorrectly diagnosed as schizophrenic at the time of hospitalization are still in the sample. Their eventual removal along with their co-twins should increase the accuracy of the MMPI data.

Out of 24 MZ pairs we managed to test 19 probands (A-twins) and 16 co-twins (B-twins),[2] and out of 33 DZ pairs we tested 19 probands and 25 co-twins. Figure 6-1 shows the mean MMPI profiles for the MZ probands and co-twins; 7 of the latter had a diagnosis of schizophrenia. The more highly elevated profile of the probands is clearly psychotic in shape and not unexpected for a schizophrenic population. The MZ co-twin mean profile has only one score clearly outside of normal limits and that is on the Schizophrenia Scale with T = 70. The similarity of shape is conveyed by the fact that the four leading scales of the co-twin profile are found among the five leading scales of the MZ proband profile, and the configuration of the validity scales, L, F, and K, is similar and typical for a psychiatric sample. From the similarity in shape we inferred similarity in personality structure between probands and their identical twins.

Figure 6-2 shows the mean MMPI profiles for 19 DZ probands and 25 DZ co-twins; one of the co-twins had a diagnosis of schizophrenia. Again the highly elevated profile of the probands is clearly psychotic and very much like that of the MZ probands, suggesting that the two groups of twins are matched on many aspects of their schizophrenia tapped by the MMPI. The DZ co-twin mean profile is quite unremarkable and within normal limits. The slight elevation on the Depression Scale, T = 61, can be accounted for by the age of the twins. The validity scales do not have

[2] Twin A is the proband or, in pairs where both twins are probands, the first twin to enter the series. In the present section on the MMPI the terms proband and co-twin are used in the sense of A-twin and B-twin respectively.

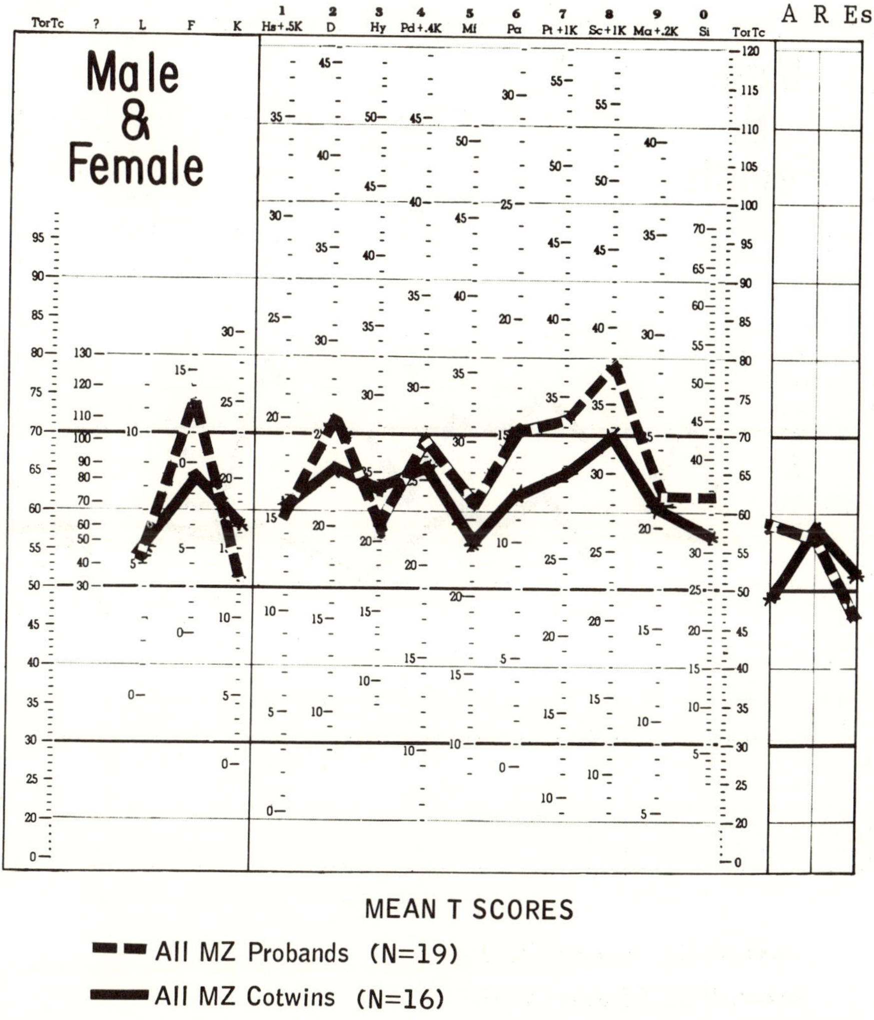

FIG. 6-1. MEAN MMPI PROFILES OF 19 SCHIZOPHRENIC MZ PROBANDS AND 16 MZ CO-TWINS.

the characteristic inverted V of psychiatric patients. From the dissimilarity in shape we inferred dissimilarity in personality structure between probands and their fraternal twins. A great deal of crucial information is lost from MMPI profiles by averaging, as we shall show below, with selected but not atypical individual twin pair profiles.

The same profile information but broken down by sex and zygosity is given in Figures 6-3, 6-4, 6-5, and 6-6. The female MZ and DZ probands

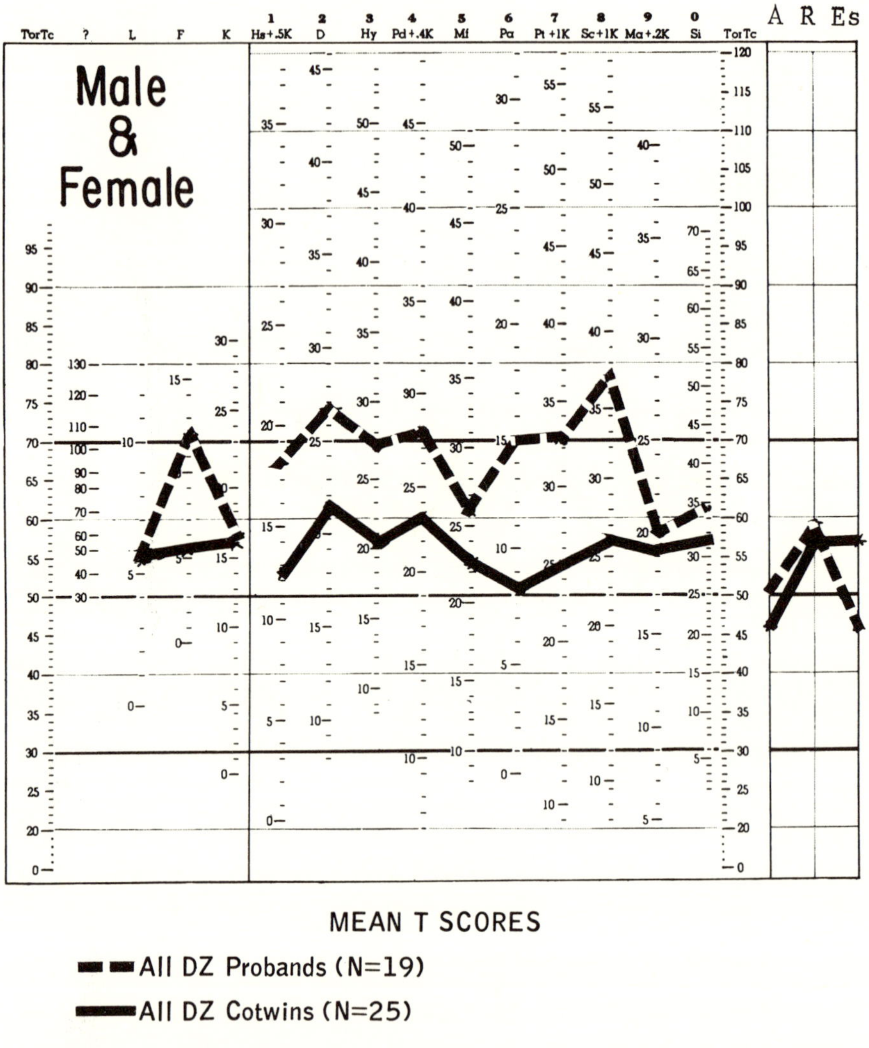

FIG. 6-2. MEAN MMPI PROFILES OF 19 SCHIZOPHRENIC DZ PROBANDS AND 25 DZ CO-TWINS.

appear to be well matched as do the two male samples. However, the MMPI picks up a sex difference that makes the females as a group look much more like paranoid schizophrenics and the males look like chronic undifferentiated schizophrenics (Marks and Seeman, 1963; Gilberstadt and Duker, 1965). The MZ male co-twins are the most disturbed of the four groups of co-twins, followed by MZ female co-twins and then the DZ co-twins. Again, much information is lost by the averaging process.

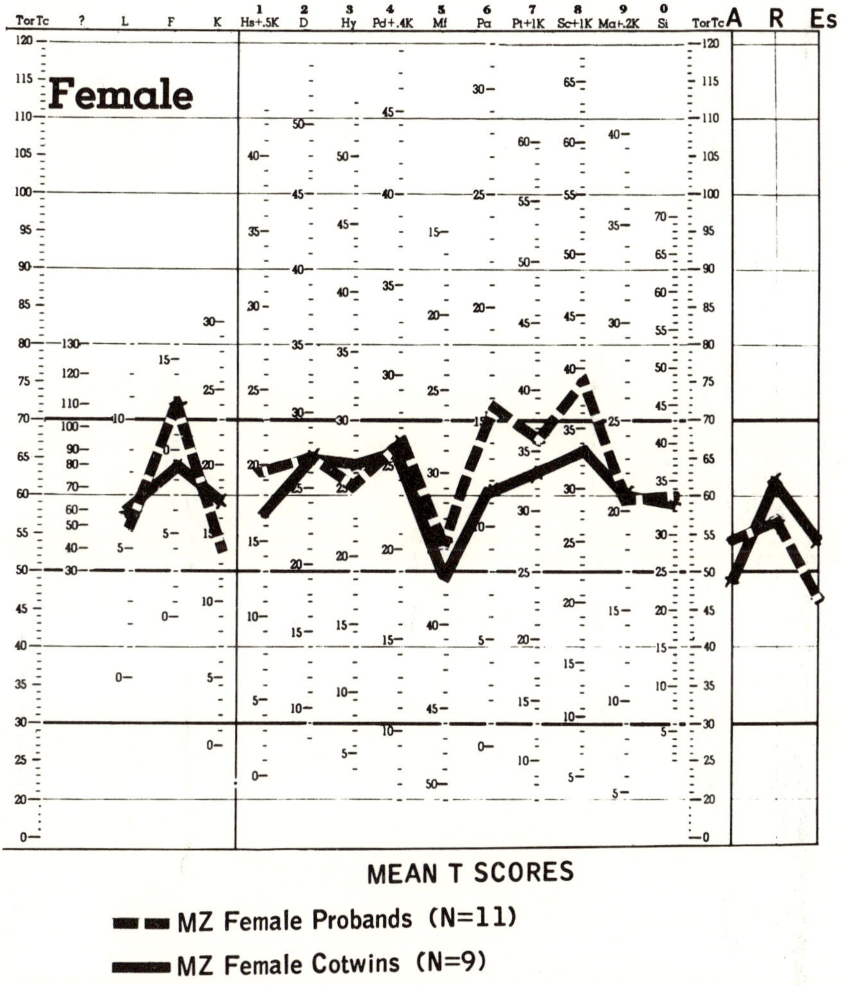

FIG. 6-3. MEAN MMPI PROFILES OF MZ FEMALE PROBANDS (N=11) AND CO-TWINS (N=9).

Individual pairs of MMPI profiles below illustrate some of the points we have raised above as well as some of the problems with the concept of discordance/concordance. In Figure 6-7 are the profiles of a pair of our Grade 1 MZ females, MZ 22, in their late twenties. Twin A was tested twice, once in a state of good remission while holding a job, maintaining a difficult marriage, and taking her medication when she felt the need. It had been almost three years since discharge from her last (fourth) hospitalization. Her first profile (Code 0′2) has F and the Social-introversion

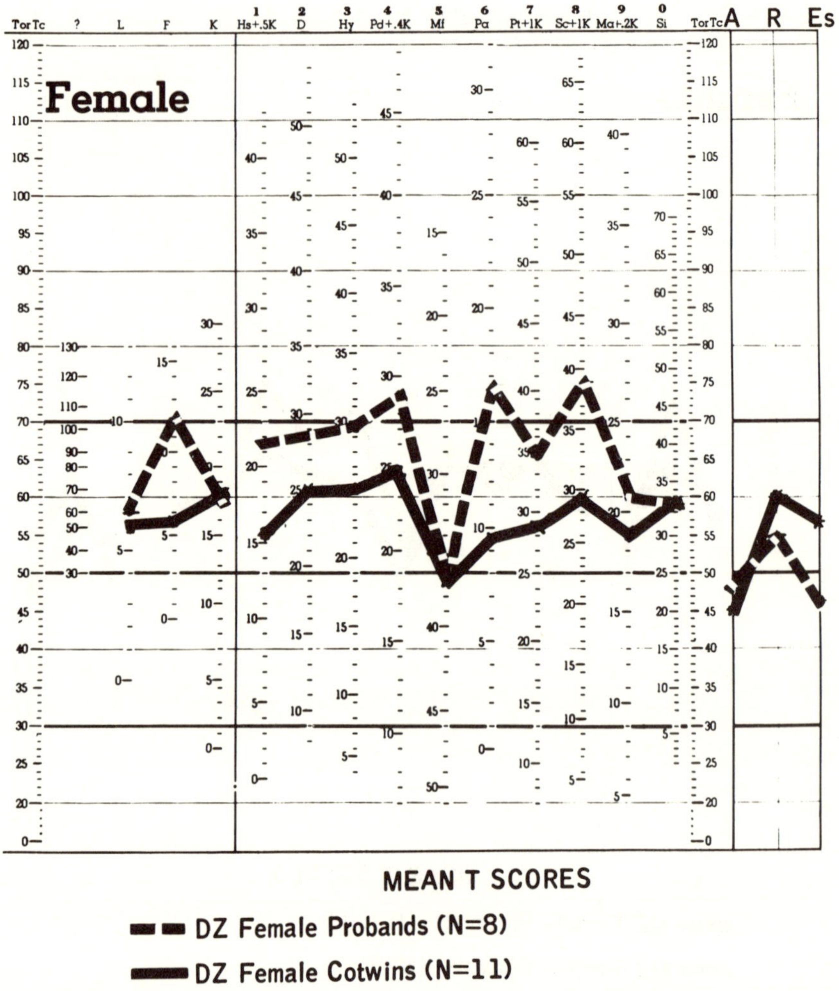

FIG. 6-4. MEAN MMPI PROFILES OF DZ FEMALE PROBANDS (N=8) AND CO-TWINS (N=11).

scale above a T score of 70. Her second profile (Code 86*742′) was obtained about seven weeks later, a week after admission to a day hospital with a recurrence of her schizophrenia. The second profile is a very clear example of a type associated with schizophrenia. Twin B was tested while in remission more than three years since discharge from her last (sixth) hospitalization. Her profile is within normal limits but remarkably similar to her twin's remission profile in that the three highest scales for each are Depression, Schizophrenia, and Social-introversion. The profiles illus-

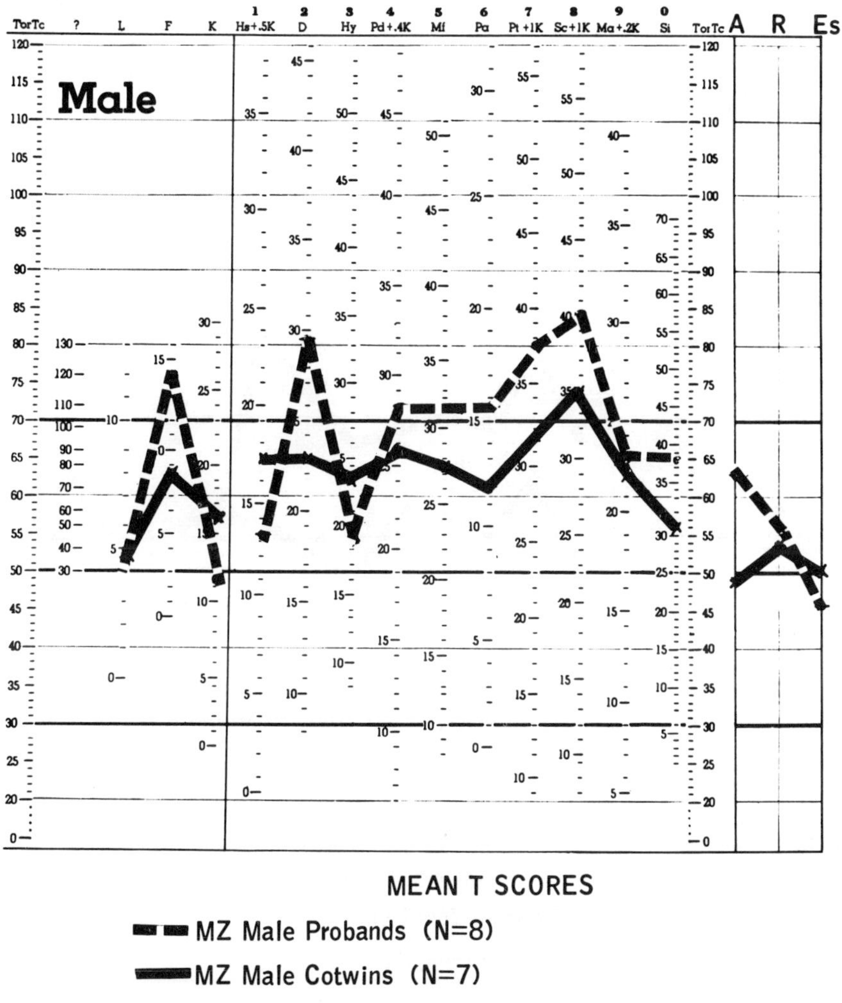

FIG. 6-5. MEAN MMPI PROFILES OF MZ MALE PROBANDS (N=8) AND CO-TWINS (N=7).

trate "psychometric concordance" as well as discordance as a function of clinical state. It is likely that the low MMPI profile with peaks on Social-introversion and Depression supplies a disproportionate number of future schizophrenics and is one of the cloaks worn by the compensated schizotype (cf. Sines, 1966; Marks and Seeman, 1963, p. 136).

Figure 6-8 illustrates the profiles of a pair of our Grade 3 MZ females, MZ 9. The proband was tested in her middle 40's some two years after discharge from her last hospitalization during which she had been diag-

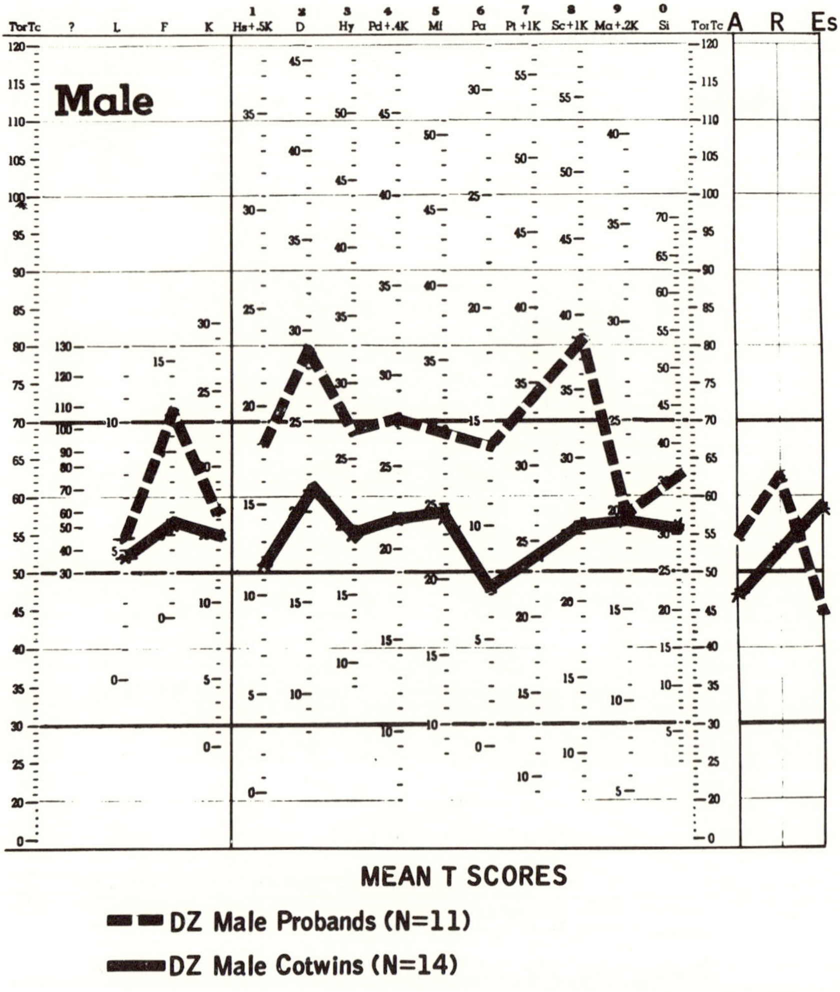

FIG. 6-6. MEAN MMPI PROFILES OF DZ MALE PROBANDS (N=11) AND CO-TWINS (N=14).

nosed as schizophrenic. Her first hospitalization did not occur until after 40 with a history of heavy drinking for the preceding 19 years. On interview she was in a gay mood but laughter merged quickly into crying and poetic sentimentality. She was regarded as having had an alcoholic psychosis (at times schizophrenic-like) by Dr. Slater from the case history. This MZ proband has probably been misdiagnosed as truly schizophrenic and we would not expect her sister to be of the schizophrenic genotype Twin B's profile is well within normal limits. She had a "nervous break-

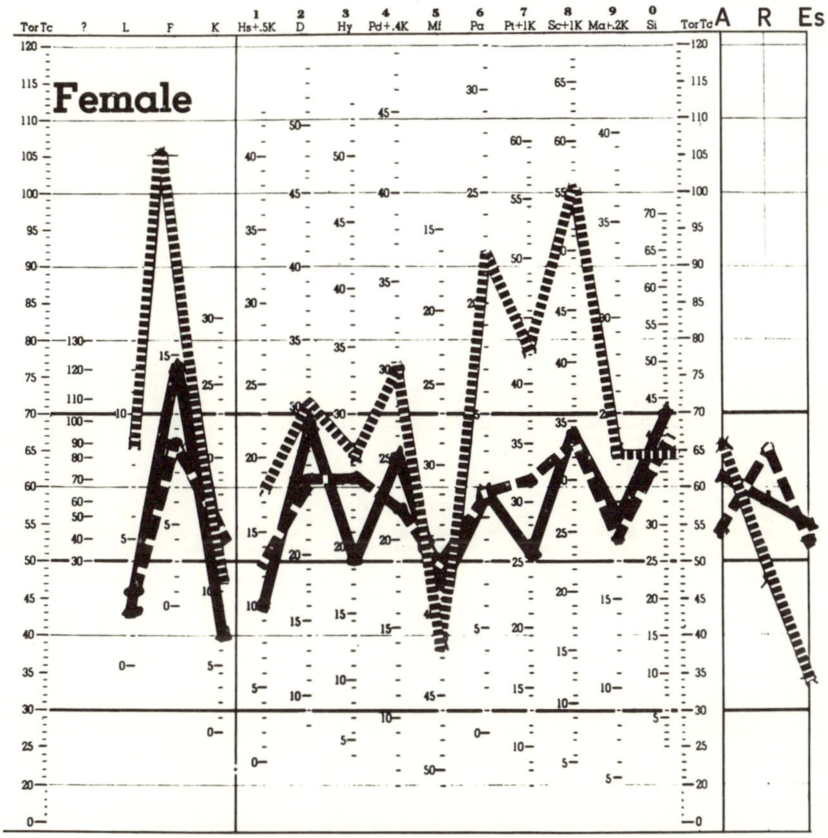

MZ 22-A_1 ▬▬▬ 0'284-6973/51:F'LK:

MZ 22-A_2 ▮▮▮▮▮ 86*742'9031-5#F*L-K:

MZ 22-B ▬ ▬ ▬ 80237-6491/5:F-K/L:

FIG. 6-7. MMPI PROFILES OF A PAIR OF MZ GRADE 1 FEMALES (MZ 22) WITH BOTH IN REMISSION (A_1 AND B) AND WHEN A HAD DECOMPENSATED (A_2).

down" five years prior to testing, quit work for three years, but has since taken up gainful employment. She was successfully treated with meprobamate. From the history she was considered by Dr. Slater to have been in an anxiety state.

The next pair of profiles in Figure 6-9 shows the kind of MMPI similarity we would expect to find when a pair of MZ twins are concordant for schizophrenia and in the same clinical state. The twins are male, aged 40 at testing. Both A and B in MZ 21 have been repeatedly diagnosed as

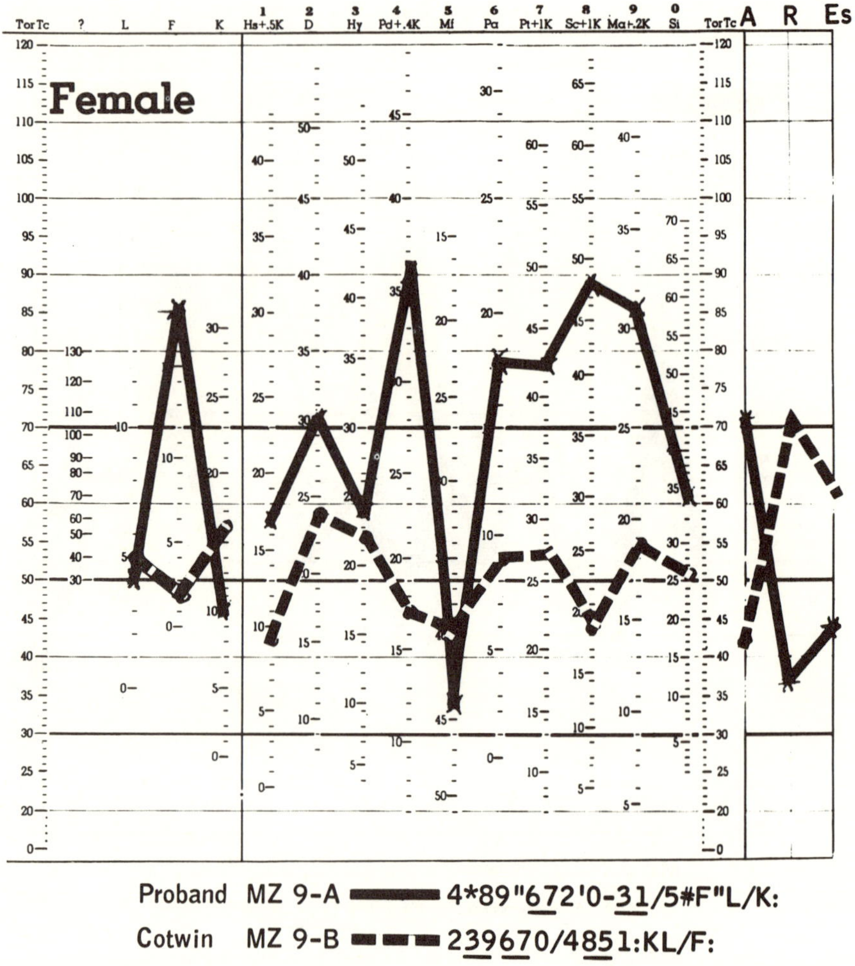

FIG. 6-8. MMPI PROFILES OF A PAIR OF MZ GRADE 3 FEMALES (MZ 9). PROBAND JUDGED TO BE PSYCHOTIC ALCOHOLIC RATHER THAN "TRUE" SCHIZOPHRENIC; CO-TWIN WITHIN NORMAL LIMITS ON TESTING.

schizophrenics. A has had 13 hospitalizations since his first break in late adolescence; B, 8, including prison terms for psychiatric "offenses." The twins were tested within a week of each other: A, 4 weeks after discharge and 2 weeks before readmission; B, 3 months before discharge. The degree of profile similarity is not unusual for our sample of MZ pairs when both twins are matched in clinical state.

Another pair of MMPI profiles for MZ twins who are phenotypically

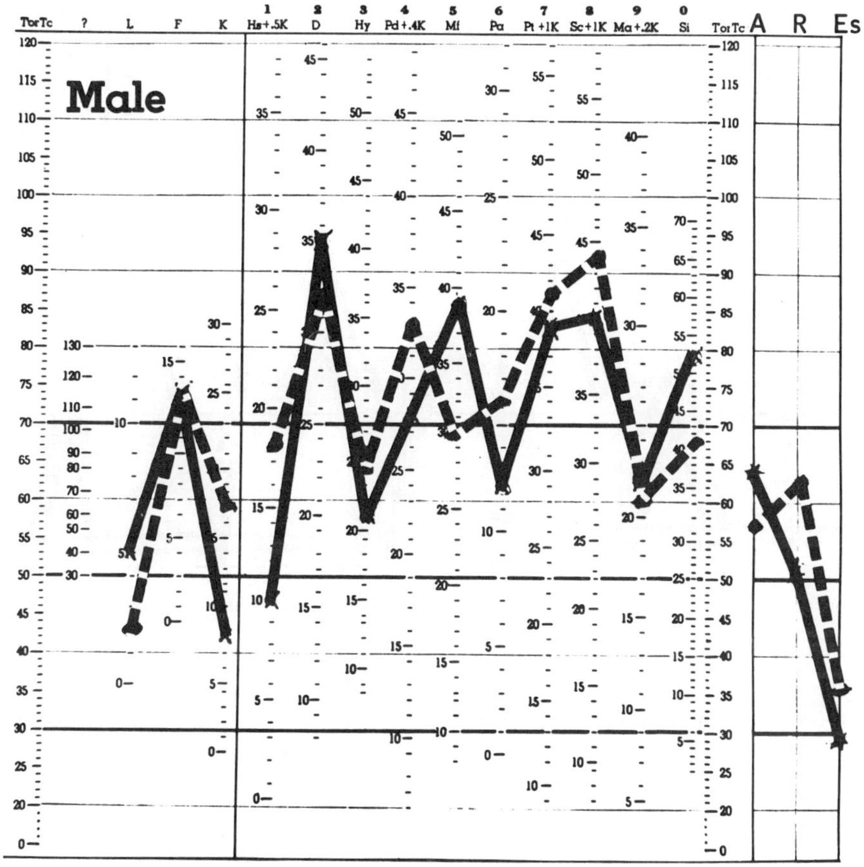

MZ 21-A ▬▬ 2*587"04'96-3/1:F'L/K:

MZ 21-B ▬ ▬ 8*274"6'50139-F'K/L:

FIG. 6-9. MMPI PROFILES OF A PAIR OF MZ GRADE 1 MALES (MZ 21) TESTED IN SIMILAR CLINICAL CONDITIONS.

discordant for a diagnosis of schizophrenia or psychiatric hospitalization is given in Figure 6-10. MZ 14 includes our youngest discordant co-twin, aged 20 at testing. Twin A had a hospital diagnosis of schizo-affective psychosis and was also so diagnosed blindly (from the history) by our expert judge. Twin B has had no contact with the world of psychiatry, but has hospitalized himself twice for physical complaints with no proved organic basis. He has put himself on an ulcer diet, has an unstable job history and lives with his overprotecting and rejecting mother. B was heavier at

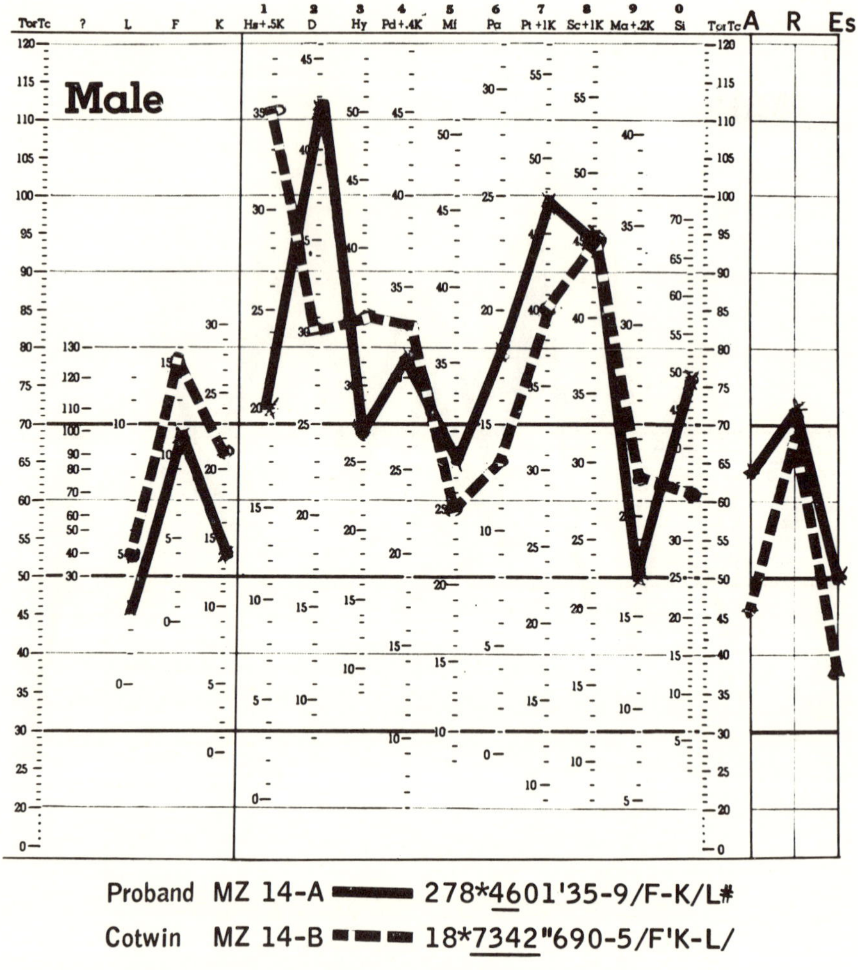

FIG. 6-10. MMPI PROFILES OF A PAIR OF MZ GRADE 3 MALES (MZ 14), CO-TWIN MARKEDLY HYPOCHONDRIACAL BUT NEVER BROUGHT TO PSYCHIATRIC ATTENTION.

birth, 4½ lbs. vs. 3¼ for A; A was delivered by forceps while B was not. In neither have neurological signs been recorded. From the history, B was judged to be diagnosable as an inadequate personality. Most blind readings of B's MMPI profile would place it in a schizophrenic category (Code 18*). The firm labeling of this pair as discordant may be premature in the light of B's youth and potential for decompensating into an overt psychosis.

One final example of MMPI profiles from a pair of discordant DZ

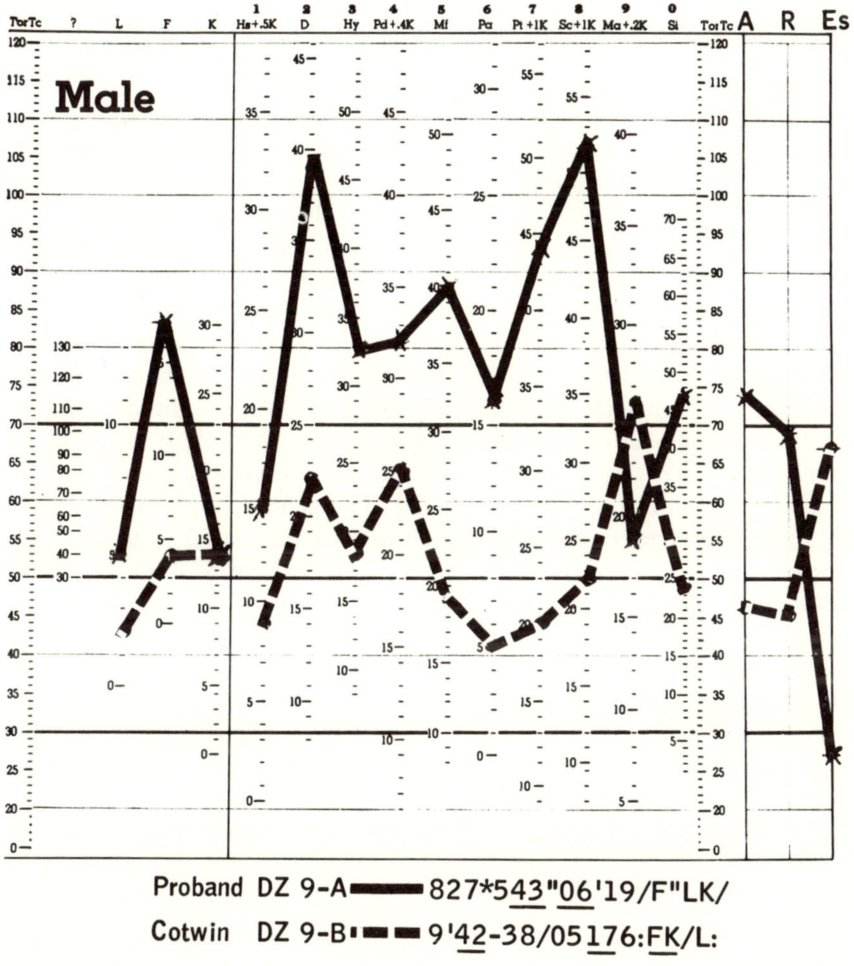

FIG. 6-11. MMPI PROFILES OF A PAIR OF DZ GRADE 4 MALES (DZ 9), CO-TWIN ENERGETIC EXTROVERT.

males, Grade 4, is given in Figure 6-11. A is a confirmed schizophrenic and despite the extreme elevation in his MMPI profile is being maintained at home on substantial amounts of phenothiazine due to a fortunate family situation that permitted this. It has been about four years since his last hospitalization but he has not been gainfully employed. His fraternal co-twin, on interview, was the picture of psychological health —gay, energetic, a sports car enthusiast, and a successful man-about-town. The dissimilarity of these fraternal twins is representative of our DZ sample. It has been 10 years since the onset of A's illness.

Time and space do not permit fuller histories of the twins or further examples from the collection of MMPI's we have gathered. Goldstein-Scheerer Object Sorting Tests were completed by 90 twins and parents. We hope to have these data analyzed blindly by Margaret Singer so that we may further characterize the co-twins on the dimension of thinking disorder or cognitive slippage.

In a preliminary search for MMPI scales other than the standard ones in a profile that might differentiate the MZ co-twins from the DZ co-twins, we have concentrated on those 5 scales derived by Rosen (1962) that attempted to cope with the difficult task of discriminating selected diagnostic groups from psychiatric patients in general, as contrasted with the original scale derivation that discriminated patients from normals. Only one of his five shows promise at this time and it is the one derived to discriminate schizophrenics from patients in general. It is called *Pz* and its discrimination power is helped by the addition of 1*K* as with the *Sc* and *Pt* scales in the standard MMPI. The general abnormal psychiatric population had a mean score of 50 with an S.D. of 10. The mean for our schizophrenic probands, N = 38, was 63 with an S.D. of 12. MZ co-twins, N = 16, obtained a mean of 60 with S.D. of 12 while the DZ co-twins, N = 25, obtained a mean of 50 with S.D. of 6. The mean for Rosen's cross validation group of 51 male overt paranoid schizophrenics was 60. We would like to see other workers make use of this scale in the hope that it may be a worthwhile indicator of a schizophrenic genotype, even if a fallible one.

INHERITANCE OF THE "LIABILITY" TO SCHIZOPHRENIA, ESTIMATED FROM INCIDENCES IN TWINS, SIBLINGS, PARENTS AND SECOND-DEGREE RELATIVES

Since Mendelian segregation ratios are not found with quantitative genetic differences, the methods of Mendelian analysis are inappropriate for continuously distributed variables. The theoretical structure for quantitative (biometrical) genetics was provided by Fisher, Haldane, and Wright starting in 1918. What use has been made of the techniques in animal and human behavior studies we largely owe to Cattell and the contributors to this symposium and Burt, Fraser Roberts, Broadhurst, and Jinks in England. Falconer (1960, 1965) has provided an important guide to the application of quantitative genetics to topics in agriculture and medicine. We have found this useful in our considerations of overt schizophrenia and the personality variation that may be associated with it. Crittenden (1961) has earlier derived a model for the interpretation of familial aggregation that was virtually identical with that of Falconer.

Especially worth exploring is the method recently developed (Falconer, 1965) for the handling of data on incidences of disease in the relatives of index cases for so-called threshold characters, i.e., diseases that appear to have an all-or-none manifestation but are in fact determined by an underlying graduation of some attribute *really* causing the disease. The latter attribute has been termed *liability* by Falconer, and is intended to convey not only the individual's innate tendency to develop the disease (i.e., susceptibility), but also the environmental milieu to which he is exposed that makes him likely to develop the disease. The point on the scale of liability above which all persons are overtly affected is called the threshold. It is the heritability of the liability to schizophrenia that is our chief concern in this section. Time and space permit only the briefest presentation of the theory involved and the reader is referred to Falconer (1965), Grüneberg (1952), and Carter (1965) for more complete details.

Heritability (h^2) expresses the degree to which phenotypes shown by parents are genetically transmitted to their children and is usually estimated from the degree of resemblance between relatives measured as a correlation (e.g., parent $\times$ offspring) or regression coefficient (cf. Gottesman, 1963a). Essentially Falconer's method converts incidences into regression coefficients which in turn lead to an estimate of the heritability of liability. Heritability is related to the *degree of genetic determination* in the following way: h^2 is the additive genetic variance as a proportion of the population phenotypic variance; degree of genetic determination is the total genetic variance (additive + non-additive) as a proportion of the total variance. The degree of genetic determination will equal h^2 in the absence of variance from dominance or gene interaction, or it may be greater, but it can never be less than h^2.

Imagine two normal distributions on a base representing a scale of liability, one for a reference population and one, displaced to the right, for the relatives of schizophrenics. We must introduce the following definitions after Falconer (1965):

G = mean liability of the general population
A = mean liability of schizophrenics in the general population
R = mean liability of relatives of schizophrenics
q = incidence of schizophrenia
x = distance of the threshold from the mean liability (normal curve deviate units)
z = height of ordinate at threshold
a = mean distance of schizophrenics in the general population ($A - G$) from mean liability of general population ($= z/q$).

Falconer provided tables of x and a which are entered with values of q, the incidence of schizophrenia in any degree of relative and the incidence in the general population. For the latter we have used the figure of one per cent but have also looked at the results if q should be two per cent. An important feature of the method is that it permits the calculation of the standard error of a regression coefficient which can then be converted into the standard error of h^2, a statistic that behavior genetics has lacked until now.

The regression, b, of relatives on probands, is given by

$$b = \frac{R - G}{A - G} \tag{1}$$

where the denominator is analogous to the *selection differential* and the numerator to the *response* in a selection-type of experiment (cf. Fuller and Thompson, 1960, p. 65). Equation (1) expressed in terms of normal curve statistics tabled by Falconer becomes

$$b = \frac{X_{\text{general pop.}} - X_{\text{relatives}}}{a} \tag{2}$$

The h^2 is readily derived from b as follows: Let P represent the liability of any person, R that of a proband's relative, and r the genetic coefficient of relationship. The regression of R on P is equal to the covariance RP divided by the total population phenotypic variance or, r multiplied by the additive variance divided by the phenotypic variance. Since h^2 was defined as the ratio of the latter two variances, we have $b = rh^2$, whence

$$h^2 = b/r \tag{3}$$

It will be recalled from the laws of Mendelian segregation that $r = 1$ for MZ twins, 1/2 for DZ twins, sibs, and parent-child, and 1/4 for second-degree relatives. Falconer has applied the technique to data from relatives other than twins [3] for renal stone disease, congenital pyloric stenosis, club foot, and peptic ulcer with h^2 values ranging from 37% ± 6% for ulcer to 79% ± 5% for pyloric stenosis. It is illustrative to note that even though the incidence of club foot in co-twins of index cases is only 32%, the h^2 is 77% ± 6%; the incidence in the general population is about .12%.

The application of Falconer's method to schizophrenia (Gottesman and Shields, 1967) gives h^2 values with their standard errors for $q = 1\%$

[3] We have applied the method to twin data on club foot (Idelberger, 1939), and have found the estimates to be within the range of values obtained from sibs and parents.

and q = 2% as shown in Table 6-6. Two percent is almost certainly too high for the incidence of schizophrenia generally, but some estimates from Norway (Ødegaard, 1946) and Sweden (Larsson and Sjögren, 1954) are nearer two percent than one percent. A lifetime incidence of two per cent would also make provision for a broader concept of schizophrenia than the one used by the earlier epidemiological workers and more in keeping with the standards used for diagnosing schizophrenia in unhospitalized relatives of probands.

The Falconer method requires that all independently ascertained index cases be counted and we have done this for those twin samples where the information was available as to whether both members of a pair were index cases. We have not, however, applied age corrections to the co-twin concordance rates. It is immediately apparent that the heritability of the liability to schizophrenia, however defined in the various studies, is quite substantial whether *q* is one or two per cent, and all considered, remarkably consistent when estimated from MZ twins, or DZ twins, or siblings, or

TABLE 6-6
THE HERITABILITY OF THE LIABILITY TO SCHIZOPHRENIA
(AFTER GOTTESMAN & SHIELDS, 1967)

	Investigator	Incidence		Falconer's h^2 q = 1%	q = 2%
	Slater (1953)				
	MZ Co-twins	28/41	68%	105% ± 8%	104% ± 8%
	DZ Co-twins	11/61	18%	106% ± 14%	94% ± 15%
Same-sex Twins	Gottesman & Shields (1966)				
	MZ Co-twins	14/28	50%	87% ± 9%	85% ± 3%
	DZ Co-twins	4/34	12%	86% ± 21%	72% ± 23%
	Kringlen (1966)				
	MZ Co-twins	28/64	44%	82% ± 6%	79% ± 7%
	DZ Co-twins	12/100	12%	86% ±12%	73% ± 13%
Parents, Sibs, and Children	Ødegaard (1963) Age Corrected	84/832	10%	79% ± 4%	64% ± 5%
	Erlenmeyer-Kimling et al. (1966)				
Sibs	Observed	131/2007	6.5%	61% ± 3%	45% ± 4%
	Age Corrected	131/1260.5	10%	80% ± 2%	66% ± 3%
Aunts, Uncles, etc. (Second Degree Relatives)	Ødegaard (1963) Age Corrected	81/1749	4.6%	96% ± 8%	61% ± 9%

aunts and uncles.[4] The standard errors, except for DZ twins, are reasonable enough to give confidence in the h^2 values obtained. It is necessary to caution that the heritability is a property not only of the trait but also of the population and its effective environmental milieu. The stability of the values in Table 6-6 seems impressive considering that they come from three different cultures, from mild and severe schizophrenics, and from relatives sharing a wide range of communality of environment with the probands.

We readily admit that our findings are far from monolithic. Values of h^2 greater than 100 per cent are obvious signs that the method is subject to errors, or that underlying assumptions have not been fully met. The sharing of trait-relevant environmental factors is a possible source of error that increases the regression computed from the incidence in relatives. This error is especially likely in siblings, whether they are twins or not. Another source of error in the estimates of h^2 arises only with DZ twins and ordinary sibs. Doubling this particular regression coefficient leads to an estimate of the additive genetic variance, together with one half of the non-additive variance arising from dominance, as a proportion of the total variance. Thus, the presence of appreciable non-additive variance leads to an exaggerated value of h^2 when the regression coefficient is doubled; but the value obtained is still less than the degree of genetic determination. An evaluation of the relative importance of non-additive genetic variance can be obtained, however, by subtracting the regression for DZ twins from that for MZ twins. The remainder estimates half of the additive genetic variance plus three-fourths of the dominance variance. Applying this procedure to our own data and using an independent estimate of h^2 (i.e., additive variance that could then be halved) from Ødegaard's second degree relatives showed that, at this stage of model development, non-additive variance was relatively unimportant for estimates of the heritability of the liability to schizophrenia. The error from sharing trait-relevant environment also appears to be unimportant as evidenced by the comparability of h^2 estimates obtained from aunts and uncles with those obtained from twins or sibs. Falconer[5] doubts that his method can be directly applied to MZ twins since the

[4] It should be noted that the affected relatives counted by Ødegaard were psychotic but not necessarily schizophrenic in the more limited sense in which it is used in Scandinavia; how many of the non-schizophrenic appearing psychoses were etiologically related to schizophrenia is problematical. Ødegaard (1967, personal communication) had commented that with a polygenic theory it is almost impossible to delimit a definition of "affected relative of a schizophrenic."

[5] Personal communication, 1966. We wish to thank Dr. D. S. Falconer for his comments and criticism of our application of his method to twin data.

liability distribution in MZ co-twins is skewed, not normal, thus leading to over-estimates of h^2.

Falconer heritability estimates will also be erroneous if the assumption of a continuous distribution of liability is untenable. Such a situation could arise from incidence data gathered from an unrepresentative sample of schizophrenics, overloaded with severe cases. In addition, a dominant gene with incomplete penetrance might cause a discontinuity in the distribution of liability to schizophrenia; its presence would be inferred if the estimated h^2 were very obviously too high to be credited (Falconer, 1965). The point can be made in another way. An abnormality known to be inherited as a dominant trait, like Huntington's chorea, with 50 percent of sibs affected and rare in the population will clearly have a heritability well beyond 100 percent if one were to evaluate it as for an additive polygenic trait with a manifestation threshold.[6] If a trait is continuous and occurs in the general population with a frequency of 0.1 per cent an incidence in first-degree relatives greater than eight per cent yields a value of h^2 greater than 100 per cent with the Falconer formulas; if the general population incidence is one per cent, incidences in the same relatives greater than 15 per cent have the same effect on h^2 values.

Up to now, estimates of the heritability of apparently discontinuous disorders such as schizophrenia have depended, for want of anything better, on a formula (Neel and Schull, 1954) which does not take the frequency of the condition into account and which simultaneously treats the concordance rates in MZ and DZ twins. It regards concordance rates as if they were the equivalent of correlation coefficients in respect of a continuous variable. We have questioned this last assumption (Gottesman and Shields, 1966b), and few have been pleased with the formula (cf. Report of Meeting of Investigators on Methodology of Twin Studies, WHO, 1966), which has in fact been little used. The formula would yield values of 44 per cent and 61 per cent respectively for our twin data and those of Slater in Table 6-6.

[6] A threshold character may itself be graded (Grüneberg, 1952). Suppose there are 5 pairs of genes (S+ and S−) in a polygenic system so that an individual may have from zero to a maximum of 10 S+ genes. If the critical value defining the threshold is 6 S+ genes, all individuals with 6 or more are affected, but what about those persons with more than six? A number of possibilities exist: each additional S+ gene may add an increment to the severity of the disorder so that 10 is worse than 9 and so forth; the character may be graded up to a point so that 7 is worse than 6, 8 worse than 7, but 9 and 10 are not worse than 8; and, the character may not be graded beyond the threshold so that 5 S+ genes and below are phenotypically normal and genotypes with 6, 7, . . . 10 are equally affected. It is perhaps worth noting that the first two possibilities provide for the finding that the children of dual mating schizophrenics are disturbed in various degrees as well as being phenotypically normal.

The present Falconer formula offers an alternative method, based, as we have seen, on the assumption of an underlying normally distributed attribute. Estimates can be based on MZ and DZ twin concordances indipendently, and it is gratifying to have these match. An advantage of the method is that it permits a quantitative evaluation of the general impression one has about the importance of genetic factors when confronted in schizophrenia by such seemingly low incidences as 6.5 per cent in sibs or 50 per cent in MZ co-twins. It takes into account the lifetime-expectation of schizophrenia in the general population of around one per cent. Even if we were to take as replicable our concordance rate of 17 per cent in the MZ co-twins of mild schizophrenics, the Falconer h^2 would be 51 per cent.

Some of the assumptions of the Falconer formula may be oversimplified so far as schizophrenia is concerned, but we believe it is of interest to see what values it gives before more complex models are put forward. Our values of the heritability of the "liability" to schizophrenia, as defined by Falconer, would appear to be as high as, if not higher than, those calculated for congenital abnormalities and physical disease of later onset that are commonly regarded as having a strong hereditary component. As we shall see later, they are similar to those found in diabetes. Indeed, there is uncomfortably little variance left to be accounted for by environmental stress but this provides no information about the potential for curing or preventing schizophrenia. A polygenic theory and a value of h^2 based on it are not an explanation one should be satisfied with for long. The task ahead is to identify some of the specific contributing genetic factors and to explore how they interact with other such factors and the environment. If one espouses a polygenic model it is in the hope that it will prove a springboard for further advances (Shields, 1968).

POSSIBLE LEADS TO SCHIZOPHRENIA FROM PARALLELS WITH DIABETES MELLITUS

The genetics of common diseases (i.e., those occuring with an incidence of one per cent or more) are complex and relatively unexplored (Penrose, 1953; Edwards, 1963).

In our opinion there are a number of striking similarities between the problems of carrying out research on diabetes mellitus and on schizophrenia. The incidences in the general population and among the relatives of diabetic index cases are much like those we have observed for schizophrenia. Table 6-7 presents our heuristic application of Falconer's method to data from the MZ and DZ co-twins of diabetics as well as from their sibs and parents (Falconer, 1967).

TABLE 6-7
THE HERITABILITY OF THE LIABILITY TO DIABETES MELLITUS

	Investigator	Incidence		Falconer's h^2 $q = 1\%$	Falconer's h^2 $q = 2\%$
	Then-Berg (1938)				
	MZ Co-twins	18/47	38%	75% ± 7%	71% ± 8%
Twins	DZ Co-twins	9/84	11%	83% ± 14%	69 % ± 15%
	MZ Co-twins (>age 43)	15/35	43%	81% ± 8%	78% ± 9%
	DZ Co-twins (>age 43)	9/64	14%	94% ± 15%	80% ± 16%
	Harvald and Hauge (1965)				
	MZ Co-twins	36/76	47%	84% ± 5%	82% ± 6%
	DZ Co-twins	22/239	9%	74% ± 4%	59% ± 5%
	MZ Co-twins (>age 40)	58/91	64%	101% ± 5%	100% ± 6%
	DZ Co-twins (>age 40)	12/135	9%	73% ± 11%	58% ± 13%
	White & Joslin (1959)				
Sibs and first-degree relatives	Sibs & Parents	300/4434	7%	63% ± 2%	47% ± 2%
	Working Party, College of General Practitioners (1965)				
	Sibs*	170/5683	3%	33% ± 1%	14% ± 1%
	Parents	157/2614	6%	58% ± 1%	41% ± 2%

* The h^2 values calculated from the sibling rate in the total sample (3%) may not accurately reflect the genetic loading in the Birmingham sample. If only probands under age 50 are used, the sib rate becomes 2.8%, but in age-matched controls it becomes 0.3%. These data lead to an h^2 value of 55% ± 2%.

Neel, Fajans, Conn, and Davidson (1965) called diabetes mellitus "a geneticist's nightmare." It seemed to present "almost every impediment to a proper genetic study which can be recognized." As with schizophrenia the nature of the basic defect is unknown so that heterogeneity could not be ruled out. A gamut of biochemical and physiological theories have been proposed and the genetic mechanisms suggested are mirror images of those proposed for schizophrenia. There is a juvenile form of diabetes that some consider to have a different etiology from the adult form (Simpson, 1962). We have not even broached the complications to our own field of research from such categories as childhood schizophrenia and infantile autism.

The power of the glucose tolerance test (GTT) in detecting latent diabetes is not yet matched by any psychological technique for detecting the compensated schizotype. Even greater refinements in detecting subclinical diabetes with the cortisone glucose tolerance test or by differential destruction rates of insulin by *insulinase* (Roy *et al.,* 1966) would correspond to the detection of the hypothesized schizotaxic individual (Meehl, 1965). Age correction of the data on diabetes presents problems since the incidence goes up sharply with age. Whatever the true incidence of diabetes might be, population surveys show that many mild cases will never be detected unless proper tests are made.

Two other problems are shared by both disorders. First, it is not clear whether diabetes is a qualitative departure from normal glucose metabolism or whether it represents the tail of a normal distribution. Thompson (1965) found glucose tolerance to be normally distributed in the relatives of diabetics, thereby suggesting polygenic control. On the other hand, Vallance-Owen (1964) thought that the presence or absence of anti-insulin antagonists was a critical factor that determined whether "essential" diabetes developed in an otherwise predisposed individual. He believed that excessive activity of the antagonist might be inherited as a simple dominant gene. If this proved to be true, we would then have a major gene whose effects would depend on a multifactorially determined and normally distributed background; this would parallel some theories about the etiology of schizophrenia. Secondly, the frequency with which diabetes is diagnosed is strongly influenced by trait-relevant environmental variables such as general nutritional level. Does the constriction of personality development by some parents of schizophrenics or the identity crises of some adolescents correspond to such trait-relevant environmental variables for schizophrenia?

The data on diabetes mellitus in twins are of most interest to us. Then-Bergh (1938) at the Kaiser Wilhelm Institute found concordance rates for clinical diabetes in her total sample of 47 MZ pairs and 84 DZ pairs to be 38 per cent and 11 per cent respectively. If only twins over the age of 43 were used so as to minimize the need for age corrections, the incidence of overt diabetes in MZ co-twins was 43 per cent and 14 per cent in the DZ. It is worth noting that if either overt diabetes or abnormal results of the GTT were taken as indicators of the diabetic genotype, the MZ concordance rose to 100 per cent, and the DZ to 40 per cent in the subsample over age 43. In the much larger sample of twins with diabetes gathered by Harvald and Hauge[7] (1965 and personal communica-

[7] We are grateful to Drs. B. Harvald and M. Hauge for providing us with unpublished data on the age distribution of their diabetic twins.

tion) from the Danish Twin Register, the incidence of probands of overt diabetes in the MZ and DZ co-twins after excluding pairs where the healthy co-twin had not reached age 40 was 64 per cent and 9 per cent, respectively. Excluding healthy co-twins under age 70 yielded concordances of 73 per cent and 34 per cent. An analysis of twin data paralleling our own with schizophrenics where probands are divided into severe and mild cases would be informative. Severity could be assessed by the amount of insulin required to treat the patient, age at onset, enzymatic breakdown rates, or by levels of blood insulin. Neel *et al.* (1965) recognized the relevance of severity when trying to predict the numbers of affected offspring from the matings of one or both affected parents. Grunnet (1957) found an increased risk of developing diabetes before age 50 among the relatives of severe cases as compared to those of mild diabetics, and the risk is higher in relatives of probands with earlier onsets (Working Party, 1965).

Other parallels between schizophrenia and diabetes might be drawn. Neel *et al.* said, ". . . we believe it to be genetic because we have no other explanation for the familial constellations of the disease." They then went on to favor complex genetic models "involving such possibilities as a 'principal gene' with modifiers, or genes at several different loci with approximately additive effect. . ." The experts also posed a question: "Why, you may wonder after this dreary recital, would any geneticist ever venture into this obviously genetically unprofitable arena?" Fascination and exploration of the unknown were sufficient answers for them as they are for us.

One advantage of a symposium such as this is that it permits the interchange of ideas with our specialist peers about ideas not yet in a form suitable for the generalists. In anticipation of the remaining papers in this section on schizophrenia, we may venture the thought that the interchange will not lead to inbreeding depression since a great deal of heterozygosity still remains in the symposium gene pool. Clearly, research into the etiology of schizophrenia is not going to fall by the wayside for want of ideas. While welcoming any fruitful new approach, we should not permit the careful work of the past to be ignored; psychiatric genetics rests on a solid foundation of heroic individual research efforts upon which modern teams of researchers can build a polished and augmented superstructure.

References

Bleuler, E. 1911, 1950. *Dementia Praecox oder Gruppe der Schizophrenien.* Leipzig: Deuticke. (English language edition, translated by Joseph Zinkin,

1961. *Dementia Praecox or the Group of Schizophrenias.* New York: Internat. Univ. Press.)

Carter, C. O. 1965. The inheritance of common congenital malformations. In *Progress in Medical Genetics,* ed. A. G. Steinberg and A. G. Bearn, *Vol. IV,* pp. 59-84. New York: Grune & Stratton.

Crittenden, L. B. 1961. An interpretation of familial aggregation based on multiple genetic and environmental factors. *Ann. N.Y. Acad. Sci.* **91:** 769-80.

Edwards, J. H. 1963. The genetic basis of common disease. *Amer. J. Med.* **34:** 627-38.

Erlenmeyer-Kimling, L.; Rainer, J. D.; and Kallmann, F. J. 1966. Current reproductive trends in schizophrenia. In *Psychopathology of Schizophrenia,* ed. P. H. Hoch and J. Zubin, pp. 252-76. New York: Grune & Stratton.

Essen-Möller, E. 1941. Psychiatrische Untersuchungen an einer Serie von Zwillingen. *Acta Psychiat. Scand.,* Suppl. 23.

Falconer, D. S. 1960. *An introduction to quantitative genetics.* New York: Ronald.

———. 1965. The inheritance of liability to certain diseases, estimated from the incidence among relatives. *Ann. Hum. Genet.* **29**: 51-76.

———. 1967. The inheritance of liability to diseases with variable age of onset, with particular reference to diabetes mellitus. *Ann. Hum. Genet.* **31**: 1-20.

Fuller, J. L. and Thompson, W. R. 1960. *Behavior genetics.* New York: Wiley.

Gilberstadt, H. and Duker, J. 1965. *A Handbook for Clinical and Actuarial MMPI Interpretation.* Philadelphia: Saunders.

Gottesman, I. I. 1963a. Genetic aspects of intelligent behavior. In *Handbook of mental deficiency: psychological theory and research,* ed. N. Ellis, pp. 253-96. New York: McGraw-Hill.

———. 1963b. Heritability of personality: a demonstration. *Psychol. Monog.* **77:** No. 9 (Whole No. 572).

———. 1965. Personality and natural selection. In *Methods and goals in human behavior genetics,* ed. S. G. Vandenberg, pp. 63-80. New York: Academic Press.

———. 1968. Severity/concordance and diagnostic refinement in the Maudsley-Bethlem schizophrenic twin study. In *The Transmission of Schizophrenia,* ed. S. Kety and D. Rosenthal. (In press.)

Gottesman, I. I. and Shields, J. 1966a. Schizophrenia in twins: 16 years' consecutive admissions to a psychiatric clinic. *Brit. J. Psychiat.* **112:** 809-18.

———. 1966b. Contributions of twin studies to perspectives on schizophrenia. In *Progress in Experimental Personality Research,* ed. B. A. Maher, Vol. 3, pp. 1-84. New York: Academic Press.

———. 1967. A polygenic theory of schizophrenia. *Proc. Nat. Acad. Sci.* **58:** 199-205.

Gregory, I. 1960. Genetic factors in schizophrenia. *Amer. J. Psychiat.* **116:** 961-72.

Grenell, R. G. 1962. Molecular biology and psychopathology. *Ann. N.Y. Acad. Sci.* **96:** 345-52.

Grüneberg, H. 1952. Genetic studies on the skeleton of the mouse. IV. Quasi-continuous variations. *J. Genet.* **51:** 95-114.

Grunnet, J. 1957. Heredity in diabetes mellitus. *Opera ex Domo Biologiae Hereditariae Humanae Universitatis Hafniensis,* No. 39. Copenhagen: Munksgaard.

Harvald, B. and Hauge, M. 1965. Hereditary factors elucidated by twin studies. In *Genetics and the Epidemiology of Chronic Diseases,* ed. J. V. Neel, M. W. Shaw and W. J. Schull, pp. 61-76. Washington, D.C.: Public Health Service Publ. No. 1163.

Idelberger, K. 1939. Die Zwillingspathologie des angeborenen Klumpfuss. *Beilageheft zur Z. Orthop.* **69.**

Inouye, E. 1963. Similarity and dissimilarity of schizophrenia in twins. *Proc. Third World Congress Psychiatry (1961).* **1:** 524-30. Montreal: University of Toronto Press.

Juel-Nielsen, N. 1965. Individual and environment, a psychiatric-psychological investigation of monozygotic twins reared apart. *Acta Psychiat. Scand.,* Suppl. 183.

Kaij, L. 1960. *Alcoholism in Twins.* Stockholm: Almqvist and Wiksell.

Kallmann, F. J. 1946. The genetic theory of schizophrenia: an analysis of 691 schizophrenic twin index families. *Amer. J. Psychiat.* **103:** 309-22.

———. and Roth, B. 1956. Genetic aspects of preadolescent schizophrenia. *Amer. J. Psychiat.* **112:** 599-606.

Kay, D. W. and Roth, M. 1961. Environmental and hereditary factors in the schizophrenias of old age ("late paraphrenia") and their bearing on the general problem of causation in schizophrenia. *J. Ment. Sci.* **107:** 649-86.

Kretschmer, E. 1948. *Körperbau und Charakter,* 19th ed. Berlin: J. Springer.

Kringlen, E. 1966. Schizophrenia in twins. An epidemological-clinical study. *Psychiatry* **29:** 172-84.

Larsson, T. and Sjögren, T. 1954. A methodological, psychiatric and statistical study of a large Swedish rural population. *Acta Psychiat. Scand.,* Suppl. 89.

Luxenburger, H. 1928. Vorläufiger Bericht über psychiatrische Serienuntersuchungen an Zwillingen. *Z. ges. Neurol. Psychiat.* **116:** 297-326.

Marks, P. A. and Seeman, W. 1963. *Actuarial Description of Abnormal Personality.* Baltimore: Williams and Wilkins.

Meehl, P. E. 1962. Schizotaxia, schizotypy, schizophrenia. *Amer. Psychol.* **17:** 827-38.

———. 1965. Detecting latent clinical taxa by fallible quantitative indicators lacking an accepted criterion. Unpublished ms. University of Minnesota Medical School. Minneapolis.

Neel, J. V., *et al.* 1965. Diabetes mellitus. In *Genetics and the Epidemiology of Chronic Diseases,* ed. J. V. Neel, M. W. Shaw, and W. J. Schull, pp. 105-32. Washington, D.C.: Public Health Service Publ. No. 1163.

Neel, J. V. and Schull, W. J. 1954. *Human heredity.* Chicago: University of Chicago Press.

Ødegaard, Ø. 1946. A statistical investigation of the incidence of mental disorder in Norway. *Psychiat. Quart.* **20:** 381-99.

———. 1963. The psychiatric disease entities in the light of a genetic investigation. *Acta Psychiat. Scand.,* Suppl. 169, pp. 94-104.

Penrose, L. S. 1953. The genetical background of common diseases. *Acta Genet.* **4:** 257-65.

Rosanoff, A. J.; Handy, L. M.; Plesset, I. R.; and Brush, S. 1934. The etiology of so-called schizophrenic psychoses with special reference to their occurrence in twins. *Amer. J. Psychiat.* **91:** 247-86.

Rosen, Albert. 1962. Development of MMPI scales based on a reference group of psychiatric patients. *Psychol. Monog.* **76**: No. 8 (Whole No. 527).

Rosenthal, D. (ed.) 1963. *The Genain Quadruplets.* New York: Basic Books.

Roy, C. C.; Elliott, R. B.; Shapcott, D. J.; and O'Brien, D. 1966. Resistance of insulin to insulinase. *Lancet* **ii**: 1433-35.

Rüdin, E. 1916. *Zur Vererbung und Neuenstehung der Dementia Praecox.* Berlin: J. Springer.

Shields, J. 1962. *Monozygotic twins brought up apart and brought up together.* London: Oxford University Press.

———. 1967. The genetics of schizophrenia in historical context. In *Recent Developments in Schizophrenia,* ed. A. Coppen and A. Walk, pp. 25-41, *Brit. J. Psychiat.* Special Publ. No. 1.

———. 1968. The transmission of schizophrenia: the genetic evidence. In *The Transmission of Schizophrenia,* ed. S. Kety and D. Rosenthal. (In press.)

———. Gottesman, I. I.; and Slater, E. 1968. Kallmann's 1946 twin study in the light of new information. *Acta Psychiat. Scand.* **43**: 385-96.

Simpson, N. E. 1962. The genetics of diabetes: a study of 233 families of juvenile diabetics. *Ann. Hum. Genet.* **26**: 1-21.

Sines, J. O. 1966. Actuarial methods in personality assessment. In *Progress in Experimental Personality Research,* Vol. 3, ed. B. A. Maher, pp. 133-93. New York: Academic Press.

Slater, E. (with the assistance of Shields, J.) 1953. Psychotic and neurotic illness in twins. *Med. Res. Counc. Spec. Rept. Ser. No. 278.* London: Her Majesty's Stationery Office.

———. 1958. The monogenic theory of schizophrenia. *Acta Genet. Stat. Med.* **8**: 50-56.

Stromgren, E. 1950. Statistical and genetical population studies within psychiatry. Methods and principal results. *Congrès International de Psychiatrie V.,* pp. 155-83. Paris: Herman.

Then-Bergh, Hildegard. 1938. Die Erbbiologie des Diabetes Mellitus. Vorläufiges Ergebnis der Zwillingsuntersuchungen. *Arch. f. Rass. u. Gesellsch. Biol.* **32**: 289-340.

Thompson, G. S. 1965. Genetic factors in diabetes mellitus studied by the oral glucose tolerance test. *J. Med. Genet.* **2:** 221-26.

Tienari, P. 1963. Psychiatric illnesses in identical twins. *Acta Psychiat. Scand.* Suppl. 171.

Tsuang, M. 1967. A study of pairs of sibs both hospitalized for mental disorder. *Brit. J. Psychiat.* **113:** 283-300.

Vallance-Owen, J. 1964. Insulin antagonists and inhibitors. In *Advances in Metabolic Disorders,* Vol. 1, ed. R. Levine and R. Luft, pp. 191-215. New York: Academic Press.

White, Priscilla and Joslin, E. P. 1959. The etiology and prevention of diabetes. In *The Treatment of Diabetes Mellitus,* 10th ed., E. P. Joslin *et al.,* pp. 47-98. Philadelphia: Lea & Febiger.

Working Party, College of General Practitioners. 1965. The family history of diabetes. *Brit. Med. J.* **1:** 960-62.

World Health Organization. 1966. The use of twins in epidemiological studies: Report of the WHO Meeting of Investigators on Methodology of Twin Studies. *Acta Genet. Med. Gemell.* **15:** 109-28.

7

L. ERLENMEYER-KIMLING
Department of Medical Genetics
New York State Psychiatric Institute

THE SIBSHIPS OF SCHIZOPHRENICS

So many family and twin studies of schizophrenia-in-retrospect already exist that one almost hesitates to introduce a new one. What we would really like to do, of course, would be to approach the problems of genotype-environment interaction prospectively, to trace forward from the earliest years the intrafamilial dynamics and developmental course in genotypically vulnerable individuals, and to attempt thereby to delineate critical variables that foster the expression of the disease. Studies of that type are now being approximated (Mednick and Schulsinger, 1965; Rosenthal, personal communication; Sobel, 1961). Meanwhile, our current stock of knowledge about heredito-environmental relationships in schizophrenia having been drawn from retrospective analyses, a few more bits of information may be wrung from an over-the-shoulder view.

At any rate, when a larger study at the Department of Medical Genetics afforded an opportunity for looking into 214 sibships with two or more schizophrenic members, we began a backward trek to cull data about the individual and family histories of these siblings. The potential of such data for studying early experience factors was evidently limited. On the other hand, the detailed analyses of hospitalization patterns that could be carried out permitted us to evaluate certain aspects of intrapair similarity and dissimilarity with respect to the course of the illness. This report describes the sample and briefly considers three issues of theoretical importance in relation to the sibling data.

Subjects and Methods

Since subjects for the sib study derived from a larger survey of marriage and reproductive trends in schizophrenic patients, it is necessary to

---

This research was supported by grants from the National Institute of Mental Health (Public Health Service Research Grant MH–03532) and from the Scottish Rite Committee on Research in Schizophrenia.

Grateful acknowledgment is made to members of our research staff for their assistance in collecting and analyzing the data. In particular, I wish to thank Miss Susan Nicol and Miss Patricia Palmer, Mr. Bruce Denham, Mr. George Nowacek, and Miss Elyse Van den Bosch.

outline briefly the procedures used in the main investigation. More detailed descriptions appear elsewhere (Deming, 1962; Erlenmeyer-Kimling *et al.*, 1966).

The purpose of the main investigation was to compare marriage and reproductive rates of schizophrenics admitted to New York State hospitals during the two periods—1934-6, prior to the introduction of current treatment methods, and 1954-6, after such measures had become prevalent. Random samples were drawn of consecutive admissions to eleven New York State hospitals during each of the two periods. To ensure comparability of diagnostic criteria for both periods, all of the patient admissions (approximately 9,800) drawn into the initial samples were reviewed by our staff psychiatrists. Patients receiving a verified diagnosis of schizophrenia became *index cases* for our study of marriage and reproductive trends. Patients receiving other diagnoses or a questionable status were discarded. Changes of the hospital diagnosis (usually manic-depressive psychosis) *to* schizophrenia occurred in 7.3 per cent of all 1934-6 admissions and 5.0 per cent of all 1954-6 admissions; changes from schizophrenia to another diagnosis occurred in 0.5 per cent of the 1934-6 admissions, 0.4 per cent of the 1954-6 admissions.

A total of 3,337 verified schizophrenic index cases was obtained [1] of which 1,922 belonged to the 1934-6 admission period and 1,415 to the 1954-6 period. Preliminary data were obtained on each index case from hospital records. Follow-up procedures were instituted for the majority of the white index cases in an attempt to bring their hospitalization, marital, and family histories up to date. Follow-up consisted of: (a) tracing forward all subsequent hospitalizations of the patient; (b) checking known relatives for admission to mental hospitals; (c) completing personal and family histories through interviews with the index case and/or relatives.

Family information was sufficient to warrant investigation of the possibility of mental illness in the siblings of 1,610 index cases. All known siblings were routinely checked with the New York State Department of Mental Hygiene where a record is maintained of all persons admitted to State hospitals and to the majority of private institutions. Further inquiries were made directly with veterans', private, and state hospitals in New York, as well as in any other state in which a sibling was known to have resided. Hospitalized siblings were diagnostically evaluated according to the same criteria used in drawing the main samples.

The sib study reported here includes only those siblings who had at

[1] The total was previously reported as 3,354 index cases. Follow-up revealed that 17 cases had not been residents of New York State at the time of admission and they were therefore discarded from the study.

least one admission to a mental hospital and had a verifiable diagnosis of schizophrenia based on the clinical picture presented during hospitalization.

Basic Description of the Sample

Of the 1,610 index cases available for study, 211 (13.1 per cent) were found to have at least one sibling—not an index case—who met the foregoing criteria. In addition, there were three instances in which two index cases came from the same family. Thus, there were 217 index cases and 258 siblings (secondary cases), from 214 families. As shown in Table 7-1:

Table 7-1
Distribution of Sib Study Cases, by Number of Schizophrenic Siblings in Family (double index case families listed separately)

Schizophrenic siblings in family (excluding index case)	Number of families=number of index cases	Number of schizophrenic siblings (not index case)	Number of index-sib pairs
1	174	174	174
2	32	64	64
3	3	9	9
4	0	0	0
5	1	5	5
6	1	6	6
Subtotal	211	258	258
+3 double index families	3 families, 6 index cases	0	6
Total	214 (families) 217 (index cases)	258	264

174 of the families yielded one index case and one (secondary) affected sibling each, giving 174 pairings; 37 families yielded one index case each with two or more (secondary) affected siblings, giving 84 pairings;[2] and three families yielded two index cases each with no secondary cases, giving six pairings. In total, 264 pairs have been derived from the 475 subjects. Among the 264 pairs are included seven twin pairs of undetermined zygosity, although the latter have been omitted from some of the analyses of hospitalization patterns.

It was possible to follow the histories of most of the index cases and siblings to the present or to the date of death. The modal year of last

[2] Had the index case not been used as a pivot in pairing, each schizophrenic individual would have paired once with each other schizophrenic individual within the sibship, giving a total of 330 sibling pairs.

information was 1963 for the index cases, 1965 for the siblings; the median year was 1963 for all subjects. One hundred of the 475 subjects (43 index cases and 57 siblings) had died.

In addition to tracing admissions of the siblings, a search was made for hospitalization of the parents in the 214 families. Twenty mothers and 13 fathers had been hospitalized with verifiable diagnoses of schizophrenia (Table 7-2).

The family in which both parents were schizophrenic is of some interest and may be described briefly. There were ten children from this marriage, of whom seven (index case + six siblings) were affected. The seven included a pair of opposite-sexed twins. The children's ages at onset ranged from 14 to the mid-twenties; three were hospitalized prior to the admission of either parent. Both parents entered hospital in their mid-forties and had several hospitalizations of short duration. The father became inaccessible for follow-up when he escaped from the hospital. At last information, the mother was discharged and making a marginal adjustment at home, separated from her husband. The seven children were hospitalized for relatively brief periods—no more than two years in any case—but were acutely ill during those periods. At last information, however, all had married and appeared to be making marginal to good adjustments in their own homes.

Besides the 33 known schizophrenic parents, 31 (7.2 per cent) other parents had been in mental institutions. Diagnoses could be verified for all except one of these parents and were established as other than schiz-

TABLE 7-2
DISTRIBUTION OF SIB STUDY CASES, BY SCHIZOPHRENIA IN PARENTS

Parents:		Number single pair families	Number multiple pair families	Number index cases	Number schizophrenic siblings	Total ever born in sibship
Neither schizophrenic	(85.1%)	150 [a, b]	32	184	215	1110 [b]
One schizophrenic						
mother	(8.9%)	17 [c]	2	20	23	116
father	(5.5%)	10	2	12	14	49
Both schizophrenic	(0.5%)	0	1	1	6	10
Total	(100.0%)	177 [a,b,c]	37	217	258	1285 [b]

[a] Two families with two index cases each—each counted as one family.
[b] Total sibship size unknown in one family.
[c] One family with two index cases—counted as one family.

ophrenia (see Table 7-3). Another six parents classified as "doubtful" in Table 7-3 had histories suggesting schizophrenia although they had never been hospitalized. The tally of schizophrenia in the parents may thus be an underestimate.

The sibships included, in addition to the index cases and their known schizophrenic siblings, a total of 810 other members. Of these, 143 (82 of known sex and 61 of unknown sex) died prior to the age of fifteen. The remaining 667 who survived to the age of fifteen or older were included in the hospital checks. As shown in Table 7-3, 19 siblings were hospitalized with confirmed diagnoses other than schizophrenia, nine were in mental institutions but could not be diagnostically verified, ten were mentally retarded, four were suicides, one was alcoholic, and two were in the "doubtful" class described above. Again, it is possible that our conservative approach resulted in an underestimation of schizophrenia in these sibships. The siblings hospitalized with unverified diagnoses, the

TABLE 7-3
CLASSIFICATION OF KNOWN "ABNORMALITIES"
IN PARENTS AND "NONSCHIZOPHRENIC" SIBLINGS IN THE 214 SIB STUDY FAMILIES

	Parents		"Nonschizophrenic" siblings	
Total	428		810	
Information totally lost	—		61	
Died prior to age 15	—		82	
Subtotal for study	428		667	
Known "abnormalities":				
Schizophrenia	33	(7.7%)	—	
Other psychiatric diagnoses [a]	30	(7.0%)	19	(2.8%)
"Mental hospital," diagnosis unknown	1	(0.2%)	9	(1.3%)
Alcoholism	25	(5.8%)	0	—
Suicide	4	(0.9%)	4	(0.6%)
Mental retardation	5	(1.2%)	10	(1.5%)
Doubtful, not hospitalized	6	(1.4%)	2	(0.3%)
Total with "abnormalities" [b]	104	(24.2%)	44	(6.5%)

[a] Other psychiatric diagnoses: alcoholic psychosis (1 sib); epilepsy (6 sibs); general paresis (2 parents); involutional psychosis (8 parents, 2 sibs); manic-depressive psychosis (2 parents, 3 sibs); personality disorders (5 parents, 2 sibs); psychoneurosis (1 parent, 1 sib); psychoses of aging (9 parents); psychosis with encephalitis (1 parent); psychosis with mental deficiency (1 parent, 1 sib); psychosis due to metabolic disorder (1 parent); reactive depression (3 sibs).

[b] Four parents and two sibs with more than one "abnormality" are listed only once each.

suicides and the "doubtfuls" could conceivably have been schizophrenic. They would have added another 15 pairs to the study.

The possibility remains that extended follow-up of the "nonschizophrenic" siblings would reveal further cases of schizophrenia in these families. At final observation, 497 "nonschizophrenic" siblings were still alive and were not included among the various "other abnormality" categories of Table 7-3. Approximately 56 per cent (281) were below the age of 45 and thus, technically, still in the schizophrenia-risk period. Since the sibships under study represent a specially selected group, calculation of schizophrenia-risk figures in the usual manner would not be appropriate. It would be reasonable to assume, however, that latent cases of schizophrenia would be found largely among the younger members of the sibships—those who had passed through the relatively smaller portion of the risk period. By the same token, those who had already demonstrated the disorder should be found to be clustered among the older members of the families. The fact that the older and younger siblings of the index cases contain an approximately equal frequency of persons already known to be affected suggests that there is little remaining risk of schizophrenia in the families under study.

In studying the sibling pairs, we were interested particularly in considering certain variables that are sometimes thought to be of significance in the etiology of schizophrenia. The present report briefly reviews the findings with respect to birth order, and sex role identification, examining also their influence on intrapair differences in severity of the disease.

Birth Order

Both parental attitudes and the role of the child vis-à-vis his siblings may be markedly different for later born than for earlier born children, although it is unclear which extreme should, theoretically, be subjected to the greater degree of stress. Search for a connection between birth order and the incidence of schizophrenia has been reported in a number of studies (cf. Farina *et al.,* 1963; Gregory, 1959; Schooler, 1961; Smith and McIntyre, 1963). Results of such investigations are contradictory, with some showing a preponderance of later-born persons among schizophrenic patients, some showing a preponderance of earlier-borns, others showing no relationship between birth order and the disease.

The sib study families provide an unusual opportunity for considering the question of birth order. A commonly used method, devised by Greenwood and Yule (1914), for calculating expected frequencies of different birth orders has been applied (a) to the index cases alone, (b) to their affected siblings alone, and (c) to index cases and schizophrenic siblings combined. A similar series of computations was made under a method suggested by Slater (1962). Although the Greenwood-Yule formulation shows no significant deviations from the expected birth order distributions, the Slater method shows shifts from the theoretical mean for the index cases as well as for the affected siblings. Index cases, with a mean birth order of 0.55 ± .03 (opposed to the expected mean of 0.50), exhibited a slight tendency to be born late in the sibships. The affected siblings, on the other hand, had a mean of 0.44 ± .03 and thus tended toward earlier birth orders. The shifts observed for the separate computations offset each other, so that the combined material for schizophrenic members of these families shows no deviation from the expected mean.

Although the sib study families show no apparent association between schizophrenia and birth order, according to the standard methods of calculation, we were interested in a second possibility: namely that there might be individual family patterns, with environmental pressures exerted upon the older children in some sibships, and the younger ones in others. The importance of birth order would thus be family-specific, while examination of absolute birth order data across families would show no effect. Within a family, however, the birth orders of two affected members would be likely to be adjacent rather than randomly distributed throughout the sibship. If, on the other hand, the affected siblings are as frequently separated by intervening births of nonschizophrenic siblings, as would be expected on a probability basis, the effect of birth order as such would not seem to be particularly strong. For each sibship size of $n > 2$, it is possible to determine the expected proportion of families in which at least one of the nonschizophrenic siblings would be born between the two affected siblings. As seen in Table 7-4, the observed distribution of intervening births agrees remarkably well with theoretical expectations. Thus, even in a family-specific sense, a birth order effect fails to appear in this sample of schizophrenic sibling pairs.

Hospitalization History in Relation to Birth Order

Within the affected pairs, the earlier-born members may be compared with their later-born siblings with respect to hospitalization history in an attempt to evaluate a possible association between birth order and severity of the disease. In 62.3 per cent of 244 pairs,[3] the earlier-born member was admitted first. On the average, the earlier-born was admitted at an older age than was the later-born (mean difference = three years)—a

[3] Base for analyzing intrapair comparisons of hospitalization history through Table 7-7 is 244 pairs. The 20 excluded pairs are: 7 twin pairs, 3 double-index case pairs and 10 pairs whose hospitalization histories were inadequate for detailed analysis.

TABLE 7-4
THEORETICAL AND OBSERVED DISTRIBUTION OF NONSCHIZOPHRENIC SIBLINGS WHOSE BIRTH ORDERS INTERVENE BETWEEN THE BIRTHS OF THE TWO SCHIZOPHRENIC SIBLINGS (INCLUDES ONLY SINGLE-PAIR FAMILIES, SIZE N>2, FOR WHICH BIRTH ORDER OF ALL MEMBERS IS KNOWN)

Sibship size (n)	Number of families (f)	Probability that at least one nonschizophrenic sibling intervenes (p = 1—2/n)	Expected distribution (E = fp)	Observed distribution (O)	χ^2 values $[(O - E)^2/E]$
3	14	.333	4.67	5	0.02
4	23	.500	11.50	15	1.06
5	22	.600	13.20	12	.11
6	17	.667	11.33	9	.48
7	8	.714	5.71	7	.29
8	9	.750	6.75	6	.08
9	5	.778	3.89	2	.92
10	5	.800	4.00	3	.25
11	4	.818	3.27	4	.16
12	3	.833	2.50	3	.10
13	1	.846	.85	1	.03
Total	111		67.67	67	3.50

$\chi^2 = 3.50$, df = 10, not significant

finding similarly reported by Tsuang (1965) in his study of 71 sibling pairs hospitalized for mental illness. On the average, also, the earlier-born member appeared to have a poorer hospitalization history, with longer duration of first hospitalization, and more readmissions, than his younger sibling.

For Tsuang's cases, the mean difference in calendar year of admission was 1.3 years, indicating a tendency for the second sibling to follow the first into hospital in fairly rapid succession. Such a tendency is not observed in our study; the mean intrapair difference in calendar year of admission was 7.7 years. Nevertheless, we had initially thought that the later-born sibling might in some way be influenced by the hospitalization of the earlier-born. The generally younger age at admission for the later-born, for instance, might be attributable to earlier recognition of symptoms when the second member of the family becomes ill. Similarly, the somewhat better subsequent history of the later-born might reflect the fact that he had received more prompt treatment. On the other hand, such findings might indicate that the later-born was actually the less severely ill of the two siblings—which would imply an association between birth order and severity of schizophrenia.

Further analysis demonstrates that the foregoing dichotomy is not appropriate. There are actually three classifications to be made among the sibling pairs: A. earlier-born admitted first, later-born subsequently admitted at a younger age—74 pairs; B. earlier-born admitted first, later-born subsequently admitted at an older age—78 pairs; C. later-born admitted first, earlier-born subsequently admitted, as a matter of course, at an older age—92 pairs.

A birth order effect with respect to age at first admission is simulated by considering the combined pairs of groups A and C (166 pairs total) in which the earlier-born sibling is admitted at an older age than the later-born. Clearly, however, these two groups are disimilar in other ways—for example, the later-born members in group C are not admitted as a consequence of some influence exerted by the admission of their earlier-born siblings. A birth order effect is also simulated with respect to prognosis when groups A and B are combined (152 pairs total). Earlier-born siblings in these groups have poorer hospitalization histories than do later-born members of the pairs. But in group C, it is the later-born who have the poorer histories. Combinations of groups A and B show no effect with respect to age at admission; combinations of A and C show no effect with respect to prognosis; combinations of B and C show no birth order effect at all. There is no evidence, therefore, that severity of illness covaries with birth order.

First- versus Second-admitted Siblings

Although the course of hospitalization is not correlated with order of birth, it does appear to be associated with order of admission. As illustrated in Table 7-5, prognosis was on the average less favorable for the first-admitted members of the 244 pairs. They entered hospital at a younger age (mean intrapair difference = 5.5 years) and remained in hospital significantly longer for the initial residence ($p < .04$). They were less often discharged from the initial hospitalization and, of those discharged, significantly more returned to hospital than was true for the second-admitted sibling ($p < .005$). It follows that the first-admitted siblings spent more *total* time in hospital ($p < .00003$, Wilcoxon matched-pairs signed-ranks test).

It is difficult to account for the sizable intrapair differences between first- and second-admitted siblings. Several hypotheses might be advanced, none of them easily testable. One might speculate that there is no real

TABLE 7-5
COMPARISONS OF HOSPITALIZATION HISTORIES OF FIRST- AND SECOND-ADMITTED MEMBERS OF THE SCHIZOPHRENIC SIBLING PAIRS (N = 244 PAIRS)

Hospitalization history	First-admitted	Second-admitted	Intrapair difference
First admission:			
Mean age (in years)	24.7 ± 0.5	30.2 ± 0.6	5.7 ± 0.6
Per cent younger than sib	69.7%	24.6%	—
Mean duration (in years)	6.8 ± 0.7	5.3 ± 0.6	1.5 ± 0.8
Per cent in longer than sib	43.0%	37.7%	—
Subsequent history:			
Per cent ever discharged	74.6%	77.0%	—
Per cent readmitted (of discharged)	77.5%	63.8%	—
Mean number admissions	2.5 ± 0.1	2.1 ± 0.1	0.3 ± 0.2
Per cent with more admissions than sib	41.4%	27.9%	
Mean number years hospitalized	13.1 ± 0.8	8.1 ± 0.6	5.0 ± 0.9
Per cent in longer than sib	61.5%	31.6%	—

difference between the two siblings in severity of illness and that the more favorable pattern observed for the second-admitted sibling reflects increased support and perhaps better management by the family in the face of its second affliction. Nearly 50 per cent of the individuals (first- and second-admitted) were married at the time of hospitalization, however, while another 15-20 per cent were otherwise living away from the nuclear family, so that it would be difficult to assign a large role to family cohesion.

On the other hand, the first-admitted individual may actually be more severely affected than his sibling. The difference could stem from unequal experiential factors during development or could be accounted for by genotypic differences. According to Kallmann's hypothesis (1953), for instance, the predisposition to schizophrenia is mediated by a recessive allele, with modifying factors being important in determining resistance to the disease. Two affected siblings could thus be homozygous for the main recessive allele without necessarily sharing the same combination of modifiers. The probability of sib-sib similarity for all modifying factors, and for severity of illness, would depend upon the number of factors hypothesized.

The concept of modifying factors quickly leads one to consider a polygenic model of schizophrenia under which, as Rosenthal (1963) has pointed out, both the degree of inherited predisposition and the intensity of potential environmental stresses would be continuously distributed—and severity of the disorder would be dependent upon the degree of each.

Prospective, rather than retrospective, studies may be especially useful in detecting psychological influences which mark the child who will become the first-admitted sibling more sharply than the second-admitted, and the second-admitted more than the siblings who will remain free of the disorder.

Hospitalization History in Relation to Sex of the Pairs

Hypotheses regarding environmental factors in schizophrenia would predict greater similarity in severity of illness between like-sexed siblings than between opposite-sexed siblings, since the latter less often share quantitatively, or even qualitatively, similar experiences. Moreover, female pairs, being more restricted to within-home contacts, might be expected to undergo the same detrimental influences more frequently than male pairs. Intrapair differences in hospitalization patterns should, therefore, be least for female pairs, greatest for opposite-sexed pairs.

Table 7-6 presents intrapair comparisons for each of the three sex pairings for various measures of hospitalization. Curiously, the *males* showed the smallest intrapair differences on all measures, although there are no statistically significant differences among the three groups. Opposite-sexed siblings were less similar than like-sexed with respect to age at first admission. This appears to be due to the general sex differential in age at first admission—male schizophrenics are, on the average, three to five years younger than females at first admission—rather than to special intrafamilial differences between opposite-sexed siblings. On other meas-

Table 7-6
Intrapair Differences in Hospitalization History, by Sex-Type [a]

Average intrapair difference		Male-male (59 pairs)	Female-female (70 pairs)	Opposite-sexed (115 pairs)
Age at 1st admission				
(in	Mean	6.98 ± 0.88	7.16 ± 0.75	9.27 ± 0.66
years)	Median	5.0	6.0	9.0
Duration of 1st admission				
(in	Mean	6.07 ± 1.03	8.54 ± 1.40	6.76 ± 0.90
years)	Median	2.0	3.0	2.0
Total number of admissions				
	Mean	1.29 ± 0.18	2.34 ± 0.36	1.43 ± 0.18
	Median	1.0	1.0	1.0
Total length, all admissions				
(in	Mean	8.90 ± 1.22	12.01 ± 1.33	10.65 ± 0.89
years)	Median	5.0	8.0	7.0

[a] Excludes 7 twin pairs and 13 pairs with incomplete hospitalization histories on siblings.

ures of hospitalization history (length of first admission, number of readmissions, total length of all hospitalizations) the intrapair differences for the opposite-sexed siblings are strikingly similar to those for like-sexed pairs. Opposite-sexed siblings are just as apt also to be concordant for discharge from the first admission and for readmission as are like-sexed siblings (Table 7-7).

TABLE 7-7
CONCORDANCE OF SIBLINGS FOR DISCHARGE FROM FIRST ADMISSION AND FOR READMISSION, BY SEX TYPE [a]

Concordant-discordant	Male-male (59 pairs)	Female-female (70 pairs)	Opposite-sexed (115 pairs)
Discharge from 1st admission			
% both discharged	54.2	61.4	65.2
% neither discharged	11.9	5.7	11.3
% discordant	33.9	32.9	23.5
Readmitted (based on pairs with both members discharged)			
% both readmitted	46.9	55.8	48.0
% neither readmitted	15.6	4.7	10.7
% discordant	37.5	39.5	41.3

[a] Excludes 7 pairs of twins and 13 pairs with incomplete hospitalization histories on siblings.

In short, the data fail to support the expectations. Sex of the paired siblings is not an important factor in determining whether they will display a more or a less similar course with respect to severity of the disease.

SEX-ROLE IDENTIFICATION

In several previous studies of mentally ill relatives, a disproportionate number of like-sexed, particularly female, pairs has been observed. Rosenthal (1962), in reviewing the earlier work, has suggested that like-sexed relatives may be more apt to be concordant for mental illness because they share a stronger role identification than do opposite-sexed family members. Although data confined to schizophrenia alone do not exhibit as marked a discrepancy as those dealing with "mental illness" generally,[4] the issue is theoretically important from the point of view of psychological interpretations of schizophrenia.

[4] Data on "mentally ill" sibling pairs from Mott (1910), Myerson (1925), Penrose (1942, 1945), Tsuang (1965) combined give the following: 451 (24.3%) male pairs, 625 (33.7%) female pairs, 780 (42.0%) opposite-sexed pairs. Data on schizophrenic pairs only from Myerson (1925), Penrose (1945), Rüdin (1916), Schulz (1932), Tsuang (1965), Zehnder (1940) combined give the following: 200 (25.9%) male pairs, 219 (28.4%) female pairs, 353 (45.7%) opposite-sexed pairs.

Of the 264 schizophrenic sibling pairs in the present study, 25 per cent (66) were male-male, 28.4 per cent (75) were female-female, and 46.6 per cent (123) were opposite sexed. The expected percentage distribution would be 25:25:50, assuming equal probability of each type of pairing. χ^2 tests have been computed for the 177 families that contributed only one pair each, for the 37 families yielding more than one pair each, and for the combined material (Table 7-8). A second set of calculations groups the pairs according to the admission period of the index case (1934-6 admissions in one group, 1954-6 admissions in the other). Whichever way the data are viewed, there is no significant deviation of the sex pairings from the expected proportions.

Nevertheless, the directional tendency observed in some of the earlier investigations is noted in the present material. There is a slight preponderance of female pairs, a slight deficit of opposite-sexed pairs. Further analyses indicate that this may be the result of two types of sampling difficulties which apparently have a residual effect on our data but which may have substantially influenced some of the previous studies. The first of these is concerned with the problem of heavier migration of male relatives out of the research area; the second is concerned with the sex ratios of the surviving siblings of index cases.

In the present study, an attempt was made to trace hospitalizations of siblings throughout the United States. Use of a broad geographical base for ascertainment of secondary cases thus reduced possible losses due to migration, in contrast to previous investigations which were confined to a single hospital or a relatively limited area. Male members of the families migrated more frequently than females: 9.5 per cent of the schizophrenic brothers of our New York State index cases were never hospitalized within the state, as opposed to 5 per cent of the schizophrenic sisters. The difference, while small, would have contributed to a distorted sex distribution had investigations not been carried out in other states.

As seen in Figure 7-1, female-female pairs tended to be admitted to the same hospital (63.5 per cent), while this was less often true for male-male pairs (50.0 per cent). Curiously, opposite-sexed pairs showed a slightly greater tendency than male pairs to enter the same hospital (59.5 per cent), though not as frequently as female pairs. The sex distribution of pairs ever in the same hospital does not differ significantly from the expected 25:25:50 (per cents), but there is a significant divergence for pairs in the same hospital at the *same time* ($p < .025$). Female pairs are thus easier to locate than are pairs involving one or two males. Apparently the more restricted the sampling frame, in time and space, the greater will be the loading of females. The foregoing analyses suggest

TABLE 7-8
DISTRIBUTION OF MALE, FEMALE, AND OPPOSITE-SEXED PAIRS, BY TYPE OF FAMILY

Type of pair	Single pair families (177 families, 3 with 2 index cases each)	Multiple pair families: 11 families with like-sex pairs only	Multiple pair families: 26 families with mixed pairings	Total number of pairs
Male-male	39	16	11	66 (25.0%)
Female-female	58	6	11	75 (28.4%)
Like-sex	97	22	22	141 (53.4%)
Opposite-sex	83	—	40	123 (46.6%)
Total	180	22	62	264 (100.0%)
Like: opposite	Obs. 97:83 Exp. 90:90 $\chi^2 = 1.09$ Not signif. at .05 level	Obs. 44:40 Exp. 42:42	$\chi^2 < 1.00$	Obs. 141:123 Exp. 132:132 $\chi^2 = 1.23$ Not signif. at .05 level
M-M:F-F	Obs. 39:58 Exp. 48.5:48.5 $\chi^2 = 3.72$ Not signif. at .05 level	Obs. 27:17 Exp. 22:22	$\chi^2 = 2.27$ Not signif. at .05 level	Obs. 66:75 Exp. 70.5:70.5 $\chi^2 < 1.00$

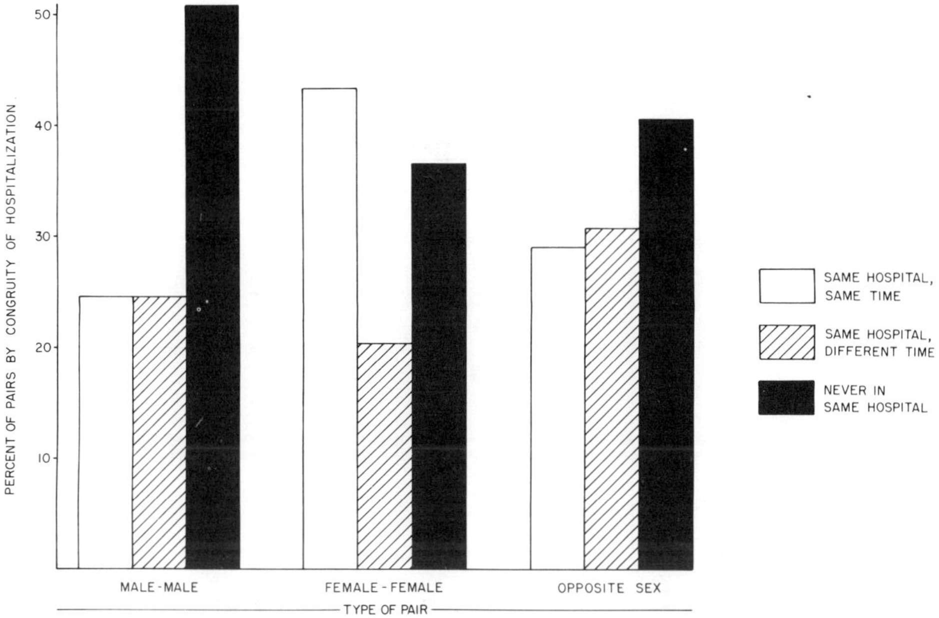

FIG. 7-1. CONGRUITY OF HOSPITALIZATION (SAME HOSPITAL, SAME TIME; SAME HOSPITAL, NON-OVERLAPPING RESIDENCE; NEVER IN SAME HOSPITAL), BY TYPE OF PAIR.

that at least part of the female excess found in previous studies may stem from the difficulty of locating and ascertaining the hospitalizations of male siblings with the sampling techniques used.

The second question to be asked about the probability of observing each of the three types of pairs concerns the sex ratio among the total sibships of the index cases. As noted by Penrose (1942, p. 314), ". . . it might be true that sibs of males were altogether more likely to be brothers than to be sisters." The expected proportions of the three types of pairs should be adjusted, therefore, to account for the sex ratio of the *total* surviving siblings of index cases in a given sample.

Table 7-9 shows the total number of brothers and sisters surviving to age 15 or older. Male index cases had 224 brothers and 216 sisters. The schizophrenic pairs formed between male index cases and their brothers represent 29.5 per cent of all potential male-male pairs, while those between male index cases and their sisters represent 30.1 per cent of all po-

TABLE 7-9
SEX DISTRIBUTION OF ALL SIBLINGS[a]
(SCHIZOPHRENIC AND NONSCHIZOPHRENIC) OF SCHIZOPHRENIC INDEX CASES

		Male index cases (N = 106)			Female index cases (N = 110[b])		
Siblings[a,b,c]		Total	Schizophrenic	Nonschizophrenic	Total	Schizophrenic[b]	Nonschizophrenic
All siblings	N	440	131	309	500	132	368
	%	100.0	29.8	70.2	100.0	26.4	73.6
Brothers	N	224	66	158	239	58	181
	% →	100.0	29.5	70.5	100.0	24.3	75.7
	% ↓	50.9	50.4	51.1	47.8	43.9	49.2
Sisters	N	216	65	151	261	74	187
	% →	100.0	30.1	69.9	100.0	28.4	71.6
	% ↓	49.1	49.6	48.9	52.2	56.1	50.8

[a] Siblings surviving to age 15 and of known sex. Excluded: 143 siblings dead prior to age 15, of which 61 were sex unknown.
[b] One female index case and schizophrenic sister excluded because total sibship size unknown.
[c] Three families with two index cases each counted twice here.

tential male-female pairs. Female index cases, with 239 brothers and 261 sisters, formed 24.3 per cent of all potential female-male pairs, 28.4 per cent of all potential female-female pairs.

(*Note*: Although it is likely that very few of the "nonschizophrenic" siblings will ultimately manifest the disease, a slightly greater probability might be assigned to brothers of female index cases than to the other siblings. At last information, 54 per cent of the living brothers of females were still in the schizophrenia-risk period, contrasted with 50 per cent of the living sisters. Among the living brothers of male index cases, 48 per cent were still in the risk period; among the living sisters, 46 per cent.)

Adjusting the expected sex distribution according to the actual sex ratios of the sibships would give corrected theoretical proportions for this group of index cases of : 23.8 per cent male pairs, 27.8 per cent female pairs, and 48.4 per cent opposite-sexed—which coincides closely with the observed figures (25 to 28.4 to 46.6 per cent) .

How the previous studies are to be evaluated remains questionable. Rosenthal (1962) has pointed out that studies restricted to a single hospital (e.g., Mott, 1910; Myerson, 1925; Zehnder, 1940) may show the influence of differential migration rates for males and females. The number of male-male and of opposite-sexed pairs located would thus be lower than the number of such pairs actually formed. Other investigations with searches confined to the hospitals of limited geographical areas (e.g., a given city or county) are subject to the same problem in lesser degree.

Only one study (Penrose, 1942) presents data that permit appropriate calculations with respect to possible distortions of the sex ratios of total surviving siblings. In that material (based on 500 male and 500 female index cases) , 7.8 per cent of all surviving brothers of male index cases, 7.5 per cent of all surviving sisters of female index cases, and 6.5 per cent of all opposite-sexed siblings—of male and female index cases combined —were counted as mentally ill. A slight, though far from significant, excess of like to opposite-sexed pairs was found, while the proportion of female pairs relative to male pairs was approximately equal. Unfortunately, the methods of tracing the relatives are not specified in detail, and the data were based on all types of mental illness combined (as were the Mott and Myerson data) , so that it is difficult to disentangle the picture for schizophrenia alone.

Data from the present study suggest that the higher incidence of like-sexed, especially female-female, sibling pairs observed elsewhere is more artifactual than real. To clarify the sex concordance issue, however, it would be necessary to carry out replications specifically designed to take into account: (1) estimated probabilities of drawing index cases of either sex for whom complete family information is available; (2) the number of male and female siblings ever born in the families; (3) differential survival and migration rates for the two sexes.

The uncertainties regarding ascertainment of sibling pairs apply also to parent-child pairings. With respect to twins, on the other hand, ascertainment difficulties are less apparent. Rosenthal (1962), discussing the various problems involved in studies of concordance among relatives, has pointed out that co-twins are more readily located because they are more likely to remain in geographic proximity or at least to remain in contact with each other, and because they represent a clearly defined group, with only one individual to be investigated per twin-index case. Accordingly twin studies should be less troubled by the sampling problems just discussed in connection with data on siblings. Migration, therefore, cannot be so readily invoked to account for the finding in several investigations (see review by Rosenthal, 1961) of a higher schizophrenia concordance rate among female, than among male, twin partners.

Data, especially from small samples, on siblings and other relatives may be obscured by distortions of the overall sex ratio of the group studied. At first glance, it would seem that the data on twins could not be similarly affected. It is important to note, however, that the investigations showing higher concordance for female twins generally contain more female pairs altogether (concordant and discordant). Undoubtedly the surplus of female cases reflects sampling biases of various types (Rosenthal, 1961), but it might also be partially due to a surplus of intact female pairs (with both twins surviving at least to adolescence) compared to intact male pairs in the general population. While male births exceed female births among twins, just as they do among single-born individuals, there is a progressive loss of jointly surviving male twins (Barr and Stevenson, 1961; Donnelly, 1956; Wilson and Jones, 1931). A male twin who survives to be hospitalized for schizophrenia may thus be less apt to have a living male co-twin than is a schizophrenic female twin to have a living female co-twin.

Critical data would be those reporting the sex ratio of hospitalized twins who were discarded for study because their co-twins had died early. For example, Slater (1953) noted that of 364 twins located among a resident hospital population, 167 were not useful because the co-twin failed to survive childhood. The sexes of the discarded cases were not given. It might be estimated that the larger proportion of the deceased twins were male and that the larger proportion of the patient twins discarded were also male. How many of these missing co-twins might have developed schizophrenia, had they lived, cannot be estimated. It seems unlikely, however, that all of the ones who died at an earlier age would have been mentally healthier than the survivors. Evaluation of the reputed higher concordance rate among female twins is thus handicapped by missing information.

CONCLUDING REMARKS

The importance of many issues touching upon gene-environment interaction had not yet been recognized when the earlier family and twin investigations were conducted. To some extent, it is possible to pull out of the existing data answers to the questions that we now ask. Largely, this is a frustrating business, and a variety of key analyses that might have been derived from the data will remain permanently lost to the literature.

This is perhaps one reason for new studies—still dealing with a retrospective approach to schizophrenia—to be undertaken. Their chief value will not lie in further computations of risk rates for the relatives of schizophrenic patients, but in the attempt to probe specific hypotheses concerning intrafamilial dynamics. Indeed, much of the newer work along these lines will be a "mopping up" process and should be designed primarily for the purpose of filling in some of the pieces that have been mislaid in the pioneering research. In other words, such new studies will give us new insights into the old material, but, on the whole will contribute relatively little in the way of new kinds of data.

The prospective approach which follows the suspected pre-schizophrenic individual forward from his early years is a different story. Such studies may be expected to yield completely new types of data, and to provide completely new opportunities for analyzing interaction patterns. They will have their own problems, of course, and will generate a further series of questions. Hopefully, however, research in the forward direction will be so formulated that it benefits from the accumulated retrospective work in terms of both the solid information already acquired and an awareness of the fine points that must be considered.

References

Barr, A. and Stevenson, C. 1961. Stillbirths and infant mortality in twins. *Ann. Hum. Genet.* **25:** 131-40.

Deming, W. E. 1962. Some statistical principles for efficient design of surveys and experiments. In *Expanding Goals of Genetics in Psychiatry,* ed. F. J. Kallmann, L. Erlenmeyer-Kimling, E. V. Glanville and J. D. Rainer, pp. 32-41. New York: Grune & Stratton.

Donnelly, M. M. 1956. The influence of multiple births on perinatal loss. *Amer. J. Obst. Gyn.* **72:** 998-1006.

Erlenmeyer-Kimling, L., and Rainer, J. D., and Kallmann, F. J. 1966. Current reproductive trends in schizophrenia. In *Psychopathology of Schizophrenia,* ed. P. H. Hoch and J. Zubin, pp. 252-76 New York: Grune & Stratton.

Farina, A.; Barry, H.; and Garmezy, N. 1963. Birth order of recovered and nonrecovered schizophrenics. *Arch. Gen. Psychiat.* **9:** 224-28.

Greenwood, M., Jr. and Yule, G. U. 1914. On the determination of size of family and of the distribution of characters in order of birth from samples taken through members of the sibships. *J. Roy. Stat. Soc.* **77**: 179-99.

Gregory, I. 1959. An analysis of family data on 1000 patients admitted to a Canadian mental hospital. *Acta Genet. Stat. Med.* **9:** 54-96.

Kallmann, F. J. 1953. *Heredity in Health and Mental Disorder.* New York: W. W. Norton.

Mednick, S. A. and Schulsinger, F. 1965. A longitudinal study of children with a high risk for schizophrenia: A preliminary report. In *Methods and Goals in Human Behavior Genetics,* ed. S. G. Vandenberg, pp. 255-95. New York: Academic Press.

Mott, F. W. 1910. The Huxley Lecture on hereditary aspects of nervous and mental disease. *Brit. Med. J.* **2:** 1013-20.

Myerson, A. 1925. *The Inheritance of Mental Disease.* Baltimore: Williams & Wilkins.

Penrose, L. S. 1942. Auxiliary genes for determining sex as contributory causes of mental illness. *J. Ment. Sci.* **88:** 308-16.

———. 1945. Survey of cases of familial mental illness. *Dig. Neurol. Psychiat.* **13:** 644 (unpublished data cited by D. Rosenthal, 1962, *Psychol. Bull.* **5:** 401-21).

Rosenthal, D. 1961. Sex distribution and the severity of illness among samples of schizophrenic twins. *J. Psychiat. Res.* **1:** 26-36.

———. 1962. Familial concordance by sex with respect to schizophrenia. *Psychol. Bull.* **5:** 401-21.

———. 1963. Theoretical overview: A suggested conceptual framework. In *The Genain Quadruplets,* ed. D. Rosenthal, pp. 505-11. New York: Basic Books.

Rüdin, E. 1916. *Zur Verebung und Neuentstehung der Dementia Praecox.* Berlin: Springer.

Schooler, C. 1961. Birth order and schizophrenia. *Arch. Gen. Psychiat.* **4:** 91-7.

Schulz, B. 1932. Zur Erbpathologie der Schizophrenie. *Z. Ges. Neurol. Psychiat.* **143:** 175-293.

Slater, E. (with the assistance of J. Shields). 1953. *Psychotic and Neurotic Illnesses in Twins.* London: H.M.S.O.

———. 1962. Birth order and maternal age of homosexuals. *Lancet* **1:** 69-71.

Smith, C. M. and McIntyre, S. 1963. Family size, birth rank and ordinal position in psychiatric illness. *Canad. Psychiat. Assn. J.* **8:** 244-48.

Sobel, D. 1961. Infant mortality and malformations in children of schizophrenic women. *Psychiat. Quart.* **35:** 60-5.

Tsuang, M. 1965. A study of pairs of sibs both hospitalized for mental disorder. Unpub. doctoral diss. Univ. London.

Wilson, P. I. and Jones, H. E. 1931. A study of like-sexed twins. I. The vital statistics and familial data of the sample. *Hum. Biol.* **3:** 107-32.

Zehnder, M. 1940. Über Krankheitsbild und Krankheitsverlauf bei schizophrenen Geschwistern. *Monatschr. Psychiat. Neur.* **103:** 231-77.

8

EINAR KRINGLEN
Institute of Psychiatry,
University of Bergen,
Bergen, Norway

CLINICAL VARIABILITY IN SCHIZOPHRENIC TWIN PARTNERS

By and large, investigations of twins have shown much higher concordance figures for monozygotic than for dizygotic twins with respect to schizophrenia and manic depressive psychosis. However, in a previous study I found no significant difference in concordance rates between monozygotic and dizygotic schizophrenic males (Kringlen, 1964). As the sample in that study was small, a more comprehensive study has been conducted in which all twins recorded in the Norwegian birth register from 1901 to 1930 have been checked against the central register of psychosis. This has provided a relatively large and, what is more important, an unselected sample of psychotic twins. In this paper I am going to focus my attention especially on the non-schizophrenic co-twin's personality.

SELECTION OF THE SAMPLE

A twin register was established for the period 1901 to 1930 on the basis of the Norwegian birth records. The twins born in this period now would be in the thirty-five to sixty-four age group, so that most of them would have passed the risk period of schizophrenia. Thus a sample of approximately 50,000 twins, or 25,000 pairs of twins, both monozygotic and dizygotic, was obtained. According to the Weinberg differential method, about 28 per cent of these twins should be monozygotic.

The next step was to check the twin register against the national register of psychosis. All psychiatric institutions in Norway must report annually all patients diagnosed as psychotic.

The checking of the twin register against the register of psychosis resulted in a sample of 342 pairs of twins, in which one or both had been hospitalized at some time because of "functional" psychosis—schizophrenia, manic depressive psychosis, or reactive psychosis. Pairs where the co-twin had died before the age of 15 had been excluded.

The next step was to get the case histories of the patients involved, and trace the twins. Although all the pairs of the same sex were then asked to visit their private physicians for blood samples to be taken, in most cases I had to visit the twins myself to get the blood samples. I succeeded in getting blood samples for 71 per cent of the group. However, rather reliable information was obtained about the similarities and dissimilarities existing in the doubtful group. The following blood and serum systems were used: ABO, MNSs, P, Rh, Le, Fy, Kell, Gm and Gc. About 99 per cent of all dizygotic twins can be classified by means of these systems (Juel-Nielsen, Nielsen, and Hauge, 1958).

PSYCHIATRIC INVESTIGATION OF THE FAMILY

My aim was to investigate as many as possible of the families of the 75 monozygotic twin pairs. Sixty-three families of monozygotics were investigated personally, including about 170 siblings.

Forty-two pairs of dizygotic twins of the same sex, but not their families, were also studied personally. The rest of the dizygotics were investigated on the basis of hospital records.

Information about the families was obtained through semi-structured interviews. In studying the identical twins, I was especially interested in the life history differences from early childhood on. As far as the siblings were concerned, attention was focused on the adult life and mental health status. Each twin was asked to describe his co-twin; and each sibling was asked to describe the other siblings and the twins. In this way the data were constantly monitored and supplemented.

CLASSIFICATION

Three types of classification were used principally: the classical type based on psychiatric diagnosis—for example, schizophrenia, reactive psychosis, and manic depressive illness; another one based on syndromes—for example depression, excitement, paranoid ideas, obsessions, and so on; and a global evaluation based mainly on symptomatology, personality, and social functioning in the form of a mental health rating scale ranging from 1 for "normality" to 7 for severely deteriorated schizophrenia. This mental health rating scale will be illustrated by clinical examples in another publication (Kringlen, 1967).

RESULTS

Concordance Figures

Table 8-1 gives an overall picture of the concordance rates, with the three main diagnostic groups—schizophrenia, reactive psychosis, and

TABLE 8-1
CONCORDANCE FOR ALL TYPES OF "FUNCTIONAL PSYCHOSES"*

	Number of pairs	Concordant pairs	Discordant pairs	Percent concordance
MZ	75	18	57	24
DZ, same sex	131	8	123	6
Unknown zygosity, same sex	10	0	10	0
DZ, opposite sex	126	8	118	6
Total pairs	342	34	308	10

* These figures are arrived at by checking the twin register against the register of psychosis. The figures are without age correction.

TABLE 8-2
CONCORDANCE FOR SCHIZOPHRENIA AND SCHIZOPHRENIFORM PSYCHOSES, BASED ON HOSPITAL RECORDS

	Number of pairs	Concordant pairs	Discordant pairs	Percent concordance
MZ	55	14	41	25
DZ, same sex	90	6	84	7
Unknown zygosity, same sex	6	0	6	0
DZ, opposite sex	82	8	74	10
Total pairs	233	28	205	8.3

manic depressive psychoses, combined. There is a significant difference between the concordance rates of monozygotics and dizygotics, but this difference is smaller than it is usually reported.

Table 8-2 gives the concordance rates for typical schizophrenia plus the more benign "schizophreniform" psychoses. Patients belonging to this last group will in most countries, including the United States and England, be classified as schizophrenics.

Table 8-3 gives the concordance for typical schizophrenia, with the exclusion of so called schizophreniform psychoses. Regardless of their orientation, most clinicians should be able to agree on a diagnosis of schizophrenia in these cases. The findings are based on hospital records. The results of my personal diagnoses would not introduce significant changes in these figures.

On the whole, the difference between the concordance rates—in regard to schizophrenia—of monozygotic and dizygotic twins is statistically significant, and therefore speaks in behalf of a genetic factor in the etiology of schizophrenia. However, this genetic factor appears to play a less im-

TABLE 8-3
CONCORDANCE FOR "STRICT" SCHIZOPHRENIA, BASED ON HOSPITALIZED REGISTERED CASES

	Number of pairs	Concordant pairs	Discordant pairs	Percent concordance
MZ	45	12	33	27
DZ, same sex	69	3	66	4
Unknown zygosity, same sex	2	0	2	0
DZ, opposite sex	64	3	61	5
Total pairs	180	18	162	10

TABLE 8-4
CLASSIFICATION OF THE CO-TWINS OF INDEX CASES WITH TYPICAL SCHIZOPHRENIA

Type of pathology	Number	Percentage
Typical schizophrenia	14	31
Reactive psychosis	1	2
Borderline state	3	7
Neurosis	13	29
Normalcy	14	31
Total co-twins	45	100

portant role than it is usually assumed. (For a more detailed discussion of concordance rates and sources of errors, see Kringlen, 1966, 1967.)

THE CO-TWIN OF THE SCHIZOPHRENIC TWIN

As can be seen from Table 8-4, the co-twins of typical chronic schizophrenics display a broad range of psychopathological conditions. Only 31 per cent of the co-twins show the same type of psychopathology as the index case. Our three border-line cases have never presented a clear-cut picture of psychosis; thus none of them has ever been hospitalized. However, they show certain personality traits that are of a psychotic rather than neurotic nature. One of them, for instance, is rather suspicious and withdrawn with slight thought disturbance, and the two others are extremely schizoid. Some would probably classify these three borderline cases as ambulatory or pseudo-neurotic schizophrenics.

In the group of neurotics, we find a broad spectrum of clinical symptoms, namely character disorders, anxiety states, depressive neuroses, "somatic" neuroses, and one case where alcohol was the main problem.

To show the personality types found among the so-called normal co-twins, I shall present briefly some case histories. I am deliberately selecting the co-twins of severely deteriorated index cases.

Case 1: Monozygotic male twins, aged thirty-eight.

The schizophrenic who had been of an introverted nature even at an early age, has become progressively worse, especially after the age of 19. At the age of 28 he developed neurotic symptoms and the following year, one evening he had been suddenly seized with a catatonic attack in which he had dashed naked into the street. Since then, he has remained either hospitalized or in private care. He has gone through recurrent periods of stupor, of agitation, displaying severely paranoid ideas of influence. A follow-up visit made a few years ago found him an incoherent patient in a state of total mental confusion. The only thing he was able to communicate was that he had arrived on earth in a space ship. To-day, he can be described only as an extremely advanced case of catatonic paranoid schizophrenia.

His twin brother has never been neurotic and has never manifested any psychosomatic disorders. Socially, he is well adjusted; he has steadily advanced in his work. He was married at the age of twenty-two, he has children and a nice home, and he is relatively affluent. I had several interviews with him in which, at the beginning, he was rather reserved, but later became more expansive. Although he may be slightly introverted, he is definitely not schizoid and he is free of any other deviant personality traits. Intellectually he appears to be well above average.

It is particularly interesting to note that from the age of twenty-three months to the time they were sixteen years old, the twins had been separated, and that they had been raised in two completely different worlds. (For a more detailed description see Kringlen, 1964.)

Case 2: Female monozygotic twins, aged thirty-eight.

The index case developed severe catatonic schizophrenia when she was eighteen years old and she has remained in a hospital since then. During most of this time she has been psychotic, with such symptoms as stupor, negativism, flashes of temper, lack of personal cleanliness, mutism, and withdrawal with total loss of interest in her surroundings. Last year she started showing some signs of improvement.

Her twin sister experienced at the age of twenty-two some mild depressive symptoms, which were accompanied by headaches. These episodes were precipitated by definitely stressful situations. Since then she has been practically free of symptoms being only occasionally troubled by some mild pain in her neck and shoulders. She has never been treated by a psychiatrist. Her husband, too, has a schizophrenic sibling. During the many hours I spent with her in the course of my follow-up study, I was favorably impressed by the various positive aspects of her personality. I could detect in her no marked nervous symptoms and she appeared to be quite relaxed. She seemed to be of an independent and cooperative nature and she was quite willing to speak frankly about herself.

Case 3: Female monozygotic twins, aged fifty-four.

Our index case had developed, at the age of twenty-six, a progressive paranoid catatonic schizophrenia. During the years that followed she remained with her parents who were living on a remote farm. Finally, when she was forty-six years old, she became aggressive and had to be taken to a hospital; she has been living ever since, as a schizophrenic displaying marked deterioration, either in hospitals or in homes that board mental patients. I saw her when she was fifty-four years old. She was prattling incoherently, she was incapable of taking care of her personal needs, and quite unmanageable.

Her twin sister is unmarried, and works as a house maid. At the age of thirty-five she experienced some emotional anxiety and sleeping difficulties, which probably had been precipitated by overwork. She was treated by a general practitioner, who prescribed sleeping pills. She had recovered within a few weeks. Since then, she has had no other nervous symptoms. During my interview with her she was rather reserved and quiet, but otherwise apparently normal. (For a more detailed description see Kringlen, 1967.)

The data show that the clinical pictures presented by individuals with the same herditary endowment as the schizophrenic partner, embrace a graduated series of personality patterns that range from a duplication of the psychosis, to total normality.

The relatively low concordance rates for schizophrenia in monozygotic twins prove that environmental factors play a significant role in the etiology of schizophrenia. One could argue that it is not schizophrenia, as such, that is inherited, but a certain personality structure, which predisposes to schizophrenia. However, the great variability of the clinical pictures encountered in the non-schizophrenic co-twins does not provide adequate support for this hypothesis. If most of the non-schizophrenic co-twins were more or less schizoid, this would suggest a relationship between the schizoid and the schizophrenic personality. However, it is clearly brought out by our data that many of the co-twins are quite normal, and furthermore, that many of those who are neurotic are often not strikingly introverted.

Do these findings agree with the literature? Actually, this field is still *terra incognita.* Thus Kallmann (1946, 1950) collected an extremely impressive volume of data on twins, but unfortunately, he did not publish his case histories.

Rather few monozygotic pairs of twins have been described where one twin is schizophrenic and the partner quite normal. The figures in Table 8-5 are based upon data from four comprehensive studies documented by case material, and I have been cautious in classifying any co-twin as "normal." Slater (1941) had apparently seven normal co-twins, but only one of these could be included in the table, because information on the other six was incomplete. Tienari (1963) classified ten of his co-twins as nor-

TABLE 8-5
NUMBER OF NORMAL MZ CO-TWINS OF SCHIZOPHRENICS IN VARIOUS STUDIES

Investigator		Number of normals	Total sample	Percentage
Essen-Møller	1941	1	7	14
Slater	1953	1	37	3
Tienari	1963	3	15	20
Kringlen	1967	14 (17) *	45 (55) *	31 (31) *

* Schizophreniform included.

mal, but since seven of them displayed schizoid traits, only three of them were included in the table.

Inouye's (1961) findings were published in a statistical form, which does not lend itself to the breakdown required by this table. However, they confirm to some extent the results mentioned above.

As to personality traits, the term schizoid is used to designate in Essen-Møller's sample two co-twins, in Slater's sample two or three, and in Tienari's seven, who were thought to have schizoid or introverted traits.

Inouye mentions also the high incidence of schizoid personality traits in the non-schizophrenic co-twins. Pollin and his co-workers (1966), whose main aim was to study discordant schizophrenic pairs, failed to find a single co-twin displaying outstanding schizoid or introverted traits.

Our next question is: Are all of the so-called normal or neurotic co-twins more or less randomly paired with various subtypes of schizophrenia and with the various degrees of severity this disease presents? Would a case of malignant schizophrenia—believed to be of a genetic origin—be more apt to be paired with a psychotic who shows either the same type of psychosis or a borderline type with marked schizoid traits, while, on the other hand, a case of benign schizophrenia—of a nongenetic type—would be paired with a merely neurotic co-twin, or even with a normal co-twin?

Table 8-6 shows that the normal partner may be paired with any type

TABLE 8-6
SUBTYPES OF SCHIZOPHRENIA PAIRED WITH
MZ CO-TWINS (SCHIZOPHRENIFORM EXCLUDED)

Subtype of MZ index cases	Total number	Number of clinically normal co-twins
Hebephrenia	6	1
Catatonia	7	4
Mixed syndromes	19	7
Paranoid type	13	2
Total	45	14

TABLE 8-7
SEVERITY OF ILLNESS IN MZ INDEX CASE IN RELATION TO NORMALCY IN THE CO-TWIN

Degree of severity in MZ index-case	Normal co-twins number
Extremely severe	4
Moderately severe	10 (8)
Slightly severe	0 (2)
Total	14

of schizophrenia. Among my own cases, there is only one hebephrenic with a clinically normal co-twin, but some of the schizophrenics classified as mixed types display hebephrenic traits.

Table 8-7 shows that the normal co-twins may be paired not only with moderately severe cases of schizophrenia, but even with extremely severe cases. Such instances are few, but the very fact that even the most severe case of schizophrenia may be paired with a normal twin is startling.

CONCLUSIONS

The principal findings and conclusions reached in this study are that:

The concordance figures for monozygotic twins in regard to schizophrenia are lower than it is usually reported, hence they de-emphasize the role of the genetic factor in schizophrenia.

The clinical picture found in the non-schizophrenic co-twins is very variable since it ranges from a duplication of the schizophrenic psychosis, to neurosis and even normalcy.

The normal co-twin may be paried with any of the Kraepelinian subtypes of schizophrenia. The normal co-twin may be paired not only with a milder case of schizophrenia, but even with a very severely affected partner.

The findings may suggest that schizophrenia is more apt to be determined by experiential than genetic factors, and that this applies also to the most extreme "nuclear" groups. Consequently, classifications based on genetic principles are not very promising.

REFERENCES

Essen-Møller, E. 1941. Psychiatrische Untersuchungen an einer Serie von Zwillingen. *Acta Psychiat. Scand.* Suppl. 23.

Inouye, E. 1961. Similarity and dissimilarity of schizophrenia in twins. *Third World Congr. Psychiat., Proc.* 1: 524-30.

Juel-Nielsen, N.; Nielsen, A.; and Hauge, M. 1958. On the diagnosis of zygosity in twins and the value of blood groups. *Acta genet. (Basel)* **8:** 256-73.

Kallmann, F. J. 1946. The genetic theory of schizophrenia. An analysis of 691 schizophrenic twin index families. *Amer. J. Psychiat.* **103:** 309-22.

———. 1950. The genetics of psychoses. An analysis of 1232 twin index families. *Congr. Internat. Psychiat.* Paris VI, 1-40.

Kringlen, E. 1964. Schizophrenia in male monozygotic twins. *Acta Psychiat. Scand.* Suppl. 178, and Oslo: Universitetsforlaget.

———. 1966. Schizophrenia in twins. An epidemiological-clinical study. *Psychiatry* **29:** 172-84.

———. 1967. *Heredity and Environment in the Functional Psychoses.* Oslo: Universitetsforlaget and London: Heinemann.

Pollin, W.; Stabenau, J. R.; Mosher, L.; and Tupin, J. 1966. Life history differences in identical twins discordant for schizophrenia. *Amer. J. Orthopsychiat.* **36:** 492-509.

Slater, E. 1953. *Psychotic and neurotic illnesses in twins.* London: Her Majesty's Stationery Office.

Tienari, P. 1963. Psychiatric illness in identical twins. *Acta Psychiat. Scand.* Suppl. 171.

9

WILLIAM POLLIN, JAMES R. STABENAU, AXEL HOFFER, LOREN R. MOSHER, AND BARBARA SPILLMAN
National Institute of Mental Health

THE *NIMH* STUDY OF A SERIES OF MONOZYGOTIC TWINS DISCORDANT FOR SCHIZOPHRENIA

This is perhaps an unusual paper to be included in a volume on Human Behavior Genetics, for it is an interim report of a study designed to use monozygotic twins in such a manner as to *exclude,* rather than to study, the role of genetics in behavior and the development of psychopathology. This does not imply that we believe that genetic factors do not play a role in the etiology of schizophrenia. We do believe that there are hereditary factors which play an important role in helping to shape personality and that genetic factors are involved in schizophrenia, either by this or by some more direct route. We have chosen to study discordant identical twins because we believe they represent the optimal study group for achieving the smallest number of variables that require examination in a study of the pathogenesis of schizophrenia.

Most twin studies have been designed to help elucidate the role of genetic factors in various biological and behavioral phenomena by measuring the difference in concordance rate between monozygotic and dizygotic twins for a given illness or subject of interest. In contrast, our group at NIMH initially set out to focus on the non-genetic, interpersonal, psychological factors related to schizophrenia and personality formation, and decided that the optimal group in which to study these phenomena would be a series of identical twins discordant for schizophrenia. We thereupon set about to collect a series of such pairs of adult twins, stipulating at the same time that we would only study them as part of an intact family unit in which both parents, as well as both twins, were willing and able to come to Bethesda and participate in an intensive, multidisciplinary evaluation. It seems appropriate to include our interim results in a collection such as this for several reasons. One of these reasons is that the results of our intensive workup of such discordant pairs may help genetically oriented investigators to visualize some of the kinds of factors which explain, or at least contribute, to discordance for schizophrenia in genetically identical individuals. Another is that our finding concerning the role played by non-genetic constitutional differences may carry significant theoretical implications for the classical twin method. In this series, such differences have played an unexpectedly consistent role in differentiating the schizophrenic from the non-schizophrenic twin's early history, both directly, and also by helping to determine psychological issues of major importance, such as intrafamilial roles and relationships.

This NIH twin study has been underway for a number of years and is a continuing effort. It was planned originally in collaboration with a number of investigators including Drs. David Rosenthal and Lyman Wynne. It very much represents a group effort in which, at the present time, many collaborators are involved. We felt that in view of the highly selective nature of the same being accumulated, a major multidisciplinary effort in studying these carefully screened families was in order, and at the present time data are being accumulated on each of the families by the following investigators and procedures:

Collaborating investigators are separately analyzing data in their own fields of competence and will individually report these in appropriate settings. Eventually we plan to bring these separate reports into an integrated monograph. This report deals primarily with the results obtained by the NIMH Section on Twin and Sibling Studies, and will also report some of the relevant biochemical and neurological findings obtained in conjunction with Dr. C. Frohman, R. Cohn, R. Paine and F. Guggenheim.

The current study differs from other twin studies not only with respect

TABLE 9-1
TWIN STUDY PROCEDURES AND INVESTIGATORS

Procedures	Investigators/Consultant
I. *Zygosity*	
1. Blood Typing	Human Genetics Branch, NIDR Mr. Webster Leyshon
2. Finger printing	Section on Twin & Sibling Studies
3. Body photos	Research Assistants†
4. Anthropometric characteristics	
5. Taste test	
II. *Psychiatric Evaluation* — Individual and group interviews	
1. Personal and family histories; formulation	Drs. William Pollin, James Stabenau, Axel Hoffer* and Miss Barbara Spillman

TABLE 9-1 (CONTINUED)

Procedures	Investigators/Consultant
2. Home and community visits	Miss Barbara Spillman
3. Diagnostic Consultation, Inpatient Multidimensional Psychiatric Scale	Drs. Donald Burnham, Marvin Adland and Earle Silber
4. Hospital Ward Observations	Nursing staff
III. *Psychological Evaluation*	
1. Rorschach, Spouse Rorschach, Family Rorschach	Psychology Unit, APB, (Dr. Winfield Scott**)
2. Sentence Completion, Draw-a-person, Draw-a-house, Draw-a-family, Bender, Proverbs, Hooper, Shipley-Hartford, TAT, Color naming, Binet Vocabulary, Writing and Recall, Harris Lateral-Dominance, Ishihara	Section on Twin and Sibling Studies Research Assistants†
3. Thought Process and Intelligence, Tell-me Thinking Test, Object Sorting, Raven Progressive Matrices	Drs. Edward Jerome, Marguerite Young and Mr. John Van Dyke
4. WAIS	
5. Leary, Developmental History, Family History, PARI, PBI, CARI, CBI, Embedded figures (Gottschalk), interviews, Q-sort, Moral Values, Cornell Medical Index, Rosenberg Ego Scales, Self Esteem Scale, Hand Writing Sample	Section on Twin and Sibling Studies Research Assistants†
IV. *Polygraph, Conditioning and Reaction Time Studies*	Dr. Theodore Zahn
V. *Biological*	
1. Physical workup, X-ray, Blood, and Clinical Chemistries	Unit on Psychosomatics, LCS (Drs. P. Cardon, H. Mirsky***)
2. Neurological evaluation and EEG	Drs. Richmond Paine and Robert Cohn
3. Serum Protein Factors	Dr. Charles Fohrman, William Turner
4. Chromosomal Study	Dr. Cecil Jacobson
5. Corticosteroids	Dr. Morton Lipsett
6. Catecholamines	Dr. I. Kopin
3,4-dimethoxyphenylethylamine	Dr. A. Friedhoff

* Formerly Drs. Joe Tupin, Loren Mosher
** Formerly Dr. Nathene Loveland
*** Formerly Drs. F. Guggenheim, L. Baer
† Mrs. Martha Werner, Miss Christine Walter

to the difference in goal, described above, but also, we believe, with regard to the extensiveness of the historical material which is available, and to which we have turned in the attempt to understand the sources of the discordance in these identical twin pairs. We realized that our selection criterion which required that both parents be admitted along with their twins to the Clinical Center, and that both participate fully in the study, would sharply reduce the size of the sample available to us. However, we felt it was essential to have such parental involvement in order to obtain the type of material having to do with pregnancy, birth, and the neonatal and infant period which seemed essential for optimal reconstruction of early life experience. During our major initial two-to-three-week period of intensive family evaluation, we average over 100 hours of psychiatric interview time with each family in individual and group interviews. In addition, through extensive use of family photo albums and movies; letters and diaries; an extensive home visit, which involves interviews with other relatives, neighbors, family physicians, teachers, and the like; and in some few instances, an extended period of intensive therapy with the index twin, we have been able to obtain a body of historical material, interview data, and current observations which gives us a compelling and convincing case study in depth for each of the families we have worked with. These same techniques also appear to make it possible to correct, to a considerable extent, the type of distortions which are inherent in the retrospective method.

Thus far, 19 families have been admitted to the Clinical Center for purposes of study.* Thirteen of these included a pair of twins discordant for psychosis. In nine, the index twin showed clear-cut, hard-core schizophrenia; the other four were patients who had previously been diagnosed as schizophrenic in various psychiatric hospitals, but about whom one or several of our five-man diagnostic panel raised some question concerning the diagnosis of schizophrenia at the time of our work-up. We have previously described this latter group as our "borderline" cases, though we now believe that this diagnostic material can best be described in terms of a continuum of psychotic pathology. Five other families represent matched control families—in which neither twin had ever previously required psychiatric treatment—or families with monozygotic twins concordant for schizophrenia. In one pair the index case had a brief psychotic episode which was initially noted and treated at NIH, and which will be reported separately. This report will concern itself almost exclusively with the 15 pairs discordant for psychosis. Table 9-2 summarizes the diagnostic history, ratings, and consensus at NIMH for these twins.

Our sample was obtained by sending letters to mental hospitals and clinics throughout the country describing the study and the type of patients we were interested in and asking for referral of any such pair which might meet our criteria. Dr. Frederick Guggenheim analyzed the

* Six more have been admitted since this presentation.

Table 9-2
Diagnostic Summary

Family Number	Index Age at Hospitalization	Diagnostic Index	Diagnoses of Index	Length of Discordance*	Diagnosis of Control Twin
			Group A: "Schizophrenic"		
(1)	1. 17	7.0	Acute schizophrenic reaction	7.5 yrs.	Obsessive compulsive character disorder (consensus)
	2. 17 (transfer)		Catatonic reaction		
	3. 18		Schizophrenic reaction, hebephrenic type		
	4. 20		Schizophrenic reaction, hebephrenic type		
	5. 23—NIH		Chronic schizophrenic reaction, undifferentiated type		
(2)	1. 16	7.0	Schizophrenic reaction, catatonic type	6.5 yrs.	"Obsessional with excessive use of denial and rationalization."
	2. 18	7.0	Schizophrenic reaction, catatonic type		"Passive dependent character with obsessive and depressive features."
	3. 21—NIH	7.0	Schizophrenic reaction, catatonic type		
(3)	1. 17	7.0	Acute schizophrenic reaction, catatonic type	2.5 yrs.	"I found little that signified any serious emotional difficulty for this young lady."
	2. 18		Chronic undifferentiated schizophrenic reaction		"No evidence of psychopathology."
	3. 18—NIH		Chronic undifferentiated schizophrenic reaction		"No evidence of psychopathology."
(5)	1. 26	7.0	Schizophrenic reaction, paranoid type	3.5 yrs.	"Compulsive personality."
	2. 26	7.0	Schizophrenic reaction, paranoid type		"Epileptic personality type."
	3. 27	7.0	Schizophrenic reaction, paranoid type		"Obsessive compulsive character."
	4. 28—NIH	7.0	Schizophrenic reaction, paranoid type		

* As of 1965 follow-up.

TABLE 9-2 (CONTINUED)

Family Number		Index Age at Hospitalization	Diagnostic Index	Diagnoses of Index	Length of Discordance*	Diagnosis of Control Twin
(6)	1.	33	7.0	Schizophrenic reaction, chronic undifferentiated type	8 yrs.	"I regard this woman as a somewhat depressed person."
	2.	35		Schizophrenic reaction, paranoid type		"Passive dependent character with depressive and obsessive features."
	3.	39		Schizophrenic reaction, paranoid type		
	4.	40		Schizophrenic reaction, paranoid type		
	5.	40—NIH		Schizophrenic reaction, residual type		
(8)	1.	22	7.0	Schizophrenic reaction, paranoid type	1.5 yrs.	"Obsessive compulsive personality with depressive trends."
	2.	23		Schizophrenic reaction, paranoid type		"Passive dependent character disorder with obsessive compulsive features."
	3.	23—NIH		Schizophrenic reaction, undifferentiated type		
				Group A: "Schizophrenic"		
(17)	1.	26	6.5*	Paranoid Schizophrenia	6 years	**Hysterical character disorder
	2.	(Information not available at this time)				
	3.	31—NIH		Paranoid Schizophrenia		
(18)	1.	42	6.5*	Paranoid Schizophrenia	3 years	**Present psychiatric diagnosis; history of depressive reaction with passive, dissociative episode, mild, transient.
	2.	(Information not available at this time)				
	3.	45—NIH		Paranoid Schizophrenia		Obsessive-compulsive character structure; with marked dependency needs.
				Group B: "Borderline"		
(7)	1.	25	5.5	Schizophrenic reaction, hebephrenic type	1.5 yrs.	"No diagnosis; imprecision, vagueness as in sister."

	2.	25		Acute undifferentiated schizophrenic reaction		"No gross psychopathology; some hysterical features."
	3.	26—NIH		Residual schizophrenic reaction (?) schizo-affective type; hysterical character disorder		
(10)	1.	33	4.5	Schizophrenic reaction, paranoid type	1 yr.	"Passive dependent with depressive potential."
	2.	33		Psychoneurosis, mixed type		"No evidence of psychosis; guarded, hypomanic."
	3.	34—NIH		Residual schizophrenic reaction (?) undifferentiated type; passive dependent character disorder with depression		"Passive aggressive personality disorder."
(9)	1.	22	3.0	Psychoneurosis, obsessive compulsive reaction	1.5 yrs.	"No psychiatric illness."
	2.	22		Schizophrenic reaction acute undifferentiated type, or decompensated obsessive compulsive neurosis		"Evidence of potential instability." "No gross psychotic process." "Obsessive compulsive personality."
	3.	22		(as above)		
	4.	23		Obsessive compulsive reaction		
	5.	23—NIH		Psychoneurotic reaction, obsessive compulsive type, severe, decompensated; many borderline schizophrenic features		
(4)	1.	13	1.5	Adjustment reaction of adolescence, passive aggressive personality	2 yrs.	"Depressive and phobic symptoms."
	2.	14		Schizoid personality		"Depressive features and some phobic thoughts."
	3.	15		Schizophrenic process with depression		"No psychiatric diagnosis, but depressive and phobic features present."
	4.	16—NIH		Passive aggressive personality disorder with depressive reaction and marked schizoid features		

* As of 1965 follow-up.
** Preliminary diagnoses and ratings; all diagnostic ratings have not yet been integrated.

first 850 referrals we obtained (until Jan. 1, 1966) to determine what sample bias, if any, might be influencing our results (Guggenheim *et al.* 1966). This initial analysis indicated that 143 of these referrals were probably monozygotic, and that 45 of these presented good presumptive evidence of discordance for schizophrenia. In 23 of these, both parents were available for study. Of these 23, eight rejected our invitations to participate in the study and 15 accepted. The major overt characteristic distinguishing those who accepted from those who rejected was that 13 of the 15 who accepted were families in which the index twin continued to require psychiatric treatment and/or hospitalization, whereas only one of the eight who rejected our invitation required such further treatment. We offer these families free travel to and from Bethesda for the family unit, a daily stipend for each non-patient member of the family, and free evaluation and intensive therapy, either inpatient or outpatient, as required. Thus, such need for continuing therapy, in particular where prior therapy has been unsuccessful, serves as an important inducement for involvement in the study. The results of this preliminary analysis do not thus far suggest any sample bias other than severity of illness, and continued consequent need for therapy.

Zygosity determination was based initially on a history which included the fact that the twins were of the same sex and were often misidentified, both in childhood and in adulthood. This was subsequently confirmed by determination of 28 blood group factors. In addition, fingerprints and anatomical features and parental blood types have been taken into account. Diagnostically, a five-man panel including two senior consultants not members of the Section and who might therefore be considered free of investigational bias, were required to agree unanimously on the presence of schizophrenia in the index and the absence of any psychosis in the control in order for any family to be considered part of the hard core schizophrenia group. In this group of 13 families, the parents range in age from forty-two to seventy, with a mean of fifty-seven years; 9 pairs of twins are female and 4 male, the twins ranging in age from sixteen to forty-five, with a mean of twenty-six years; the age of first hospitalization averaged twenty-two years and the average length of discordance thus far is 5 years.

The procedures which we have employed and additional detailed information concerning the nature of the sample, the criteria for diagnosis, and examples of the clinical pictures involved have been previously reported (Pollin *et al.*, 1965, 1966; Stabenau and Pollin, 1967; and Pollin and Stabenau, 1968).

To date we have found a consistent pattern of differences tending to distinguish the index from the control twins. We will group these differences into two categories, for purposes of this review, dealing first with constitutional factors, and secondly with differences in family roles and relationships. We will then describe our present view of the relationship between these two groups of phenomena.

The non-genetic constitutional differences distinguishing the index from the control twins are noted in both historical data and current observations. (a) In 12 of 13 of the pairs of twins the index twin was the lightest in birthweight. The birthweights for the index range between 2 lb. 11 oz. and 6 lb. 2 oz., with a mean of 4 lb. 12 oz., and for their control non-schizophrenic twins were 3 lb. 11 oz. to 7 lb. 8 oz. with a mean of 5 lb. 6 oz. (b) There was a marked preponderance of evidences for neonatal physiologic disequilibrium in the index as compared to his control twin. Many more episodes of neonatal or infantile cyanosis, feeding difficulties, colic, and sleep disturbances were described for the indexes than the controls. (c) Careful neurological evaluation undertaken by Drs. Robert Cohn and Richmond Paine, focusing on soft signs such as difficulties in 2 point discriminations, unsustained clonus, praxis disturbances, and the like, reveal that in 11 of the 13 pairs the index shows a preponderance of soft signs as compared to the controls; when such signs were quantified by an independent observer on a 0 to 4 point scale, the mean value for the indexes was 3.07 and for the controls was 1.61. (d) Current measures of protein-bound iodine show that 11 out of 12 of the indexes have a lower PBI than do their corresponding non-schizophrenic control twins. These PBI values are all within normal limits and the extent of the difference is often minor, but the consistency is nonetheless impressive. (e) The lactate/pyruvate ratios, determined by Dr. Charles Frohman and his associates from Lafayette Clinic in a blind analysis, show that 9 out of 11 of the indexes have a higher ratio than the controls. (Two control twins could not be tested simultaneously.)

The second major group of differences between the index and control twins we wish to describe has to do with family roles and relationships. The index twins were not only the smaller of the two at birth, but in addition they were also seen characteristically as the more vulnerable of the two twins. They were thus selected, by the circumstance of this early difference, as the twin with whom one of the parents, usually the mother, developed a more intense involvement, an involvement which was also usually more ambivalent and colored by a greater degree of anxiety and concern. In those cases where one of the twins came home from the hospital with the mother and another twin was forced to remain behind in the incubator, it would be the index who remained behind. In some instances the concern about the smaller index-twin-to-be reached high levels of anxiety about his continued survival. These concerns led to significant differences in feeding patterns. A frequent description was the story of the mothers attempting to make the smaller twin increase his early food intake, so that he would match that of the larger one. In view of their smaller size these twins would frequently show a tendency to take less, and then fall off to sleep more quickly. A number of the mothers described slapping, pinching, and using cold water in an effort to keep the smaller ones awake, to see that both twins had equal food intake.

With respect to both smoothness and effectiveness in biological function as evidenced by the number of episodes of eating and sleep disturbance, colic, respiratory difficulties and the like, and also with regard to later psychological behavioral competence and degree of relatively comfortable independence in school and peer activities, the control twins characteristically exceeded the indexes from the earliest years on. There was a persistent predominance of anxiety symptoms, such as phobias, free anxiety, compulsive rituals, and the like, in the childhood of the index as compared to the control twins. In no instance did these reach or even approach disabling levels in childhood, and our impression has been that without the constant comparison of the more effective controls, the index twins as a group would have been seen as not significantly troubled or psychologically unhealthy children.

An additional notable difference between the indexes and the controls was the greater degree to which the controls were able to demonstrate independent, sustained, goal-directed activity. Characteristically, with regard to school tasks, we heard of the controls sitting down and going about their business while the indexes would delay and at the last moment appeal for help from someone in the family. In later years and in a manner similar to this, the indexes showed a strong tendency to be more uncertain and less successful in defining life and career goals, in contrast to the controls. With few exceptions, there was a tendency for the indexes overall to show less marked and less successful social outgoingness, and less intimacy and success in their peer relationships.

We have not thus far systematically evaluated parental personalities and interparental communication patterns in these families in comparison with control, nonschizophrenic families. It is our impression, however, that there are interesting and possibly consistent differences. The most vivid of these impressions to date have to do with: (1) a relative deficiency and lack of involvement, often superficially hidden, on the part of the father, particularly with regard to the extent to which he adequately fulfills his male parental and authority role within the family;

and (2) a reciprocally matching tense, tight, overassertive, aggressive or controlling role on the part of the mother.

In passing, an unexpected finding has arisen from a careful investigation carried out by Dr. Frederick Guggenheim as part of the earlier mentioned study (Guggenheim *et al.*, 1966) of physical illness in the parents of the first 12 sets of twins. An unexpectedly high prevalence of documented thyroid disease was found in these mothers. Seven of the 12 mothers had had thyroid disease, 6 of the 7 cases developing at least several years prior to the onset of the twins' psychosis. The particular form of thyroid pathology varied, but all of them were well documented. The increased prevalence rate of thyroid disease (58 per cent) is statistically highly significant when compared to all other available prevalence studies in age and area matched samples of women.

The data, though consistent thus far, do not as yet permit extensive conclusions to be drawn. It is still not clear whether there is a relation between these data and schizophrenia per se, or to a susceptibility to various forms of psychopathology in general; or whether they are derived in part or in whole from our selection methods, which may have produced a skewed sample of schizophrenic subjects and their families. It certainly must be noted that our requirement that both parents jointly participate with both twins in an inpatient study has resulted in our obtaining a sample which excludes disrupted families, or families in which intense, overt rejection or hostility has led to an overt cleavage between parents and patient.

Our current tentative formulation of these data is that, in the group of families thus far studied, there existed initial non-genetic, constitutional differences between the twins as a result of intrauterine circulatory and mechanical differences. These initial constitutional differences involved, most importantly, a different level of biological maturity or competence, which resulted in less smooth and effective operation of various adaptational and internal environmental regulators in the smaller twin. These sometimes slight biologic differences contributed to or determined the very early establishment of role differences within the family relationship patterns. The smaller twin, as a result of these relationship and role differences, experienced a sequence of reinforcing events in childhood years which *in toto* accentuated rather than mitigated the initial minimal disparity of coping potential. The fact that there was such reinforcement rather than attenuation of these initial disparities results from certain characteristics of the personality and/or of the pattern of relationships between the parents, which are yet to be convincingly documented in our group, and thus far provide therefore only an impressionistic hypothesis.

Increasing inter-twin differences in personality, particularly with regard to adequacy of ego function and type of ego defenses, led with passing years to an increasingly unfavorable stress-coping ratio in the index twins. That is to say, there was increasing tendency to generate stress in dealing with ongoing developmental events and transitions, and a relatively decreased ability to cope with them.

These differences and theoretical formulations have been detailed in greater length in earlier publications (Pollin and Stabenau, 1968; Stabenau and Pollin, 1967a, b).

If we continue to find similar patterns of constitutional and life history differences as we continue with the series, it would become especially interesting and important to compare our cases with matched concordant identical and fraternal controls. Only in that way will it be possible to evaluate the possibility suggested by our data thus far, which brings into question some of the basic hypotheses of the classical twin study methods currently in use in behavioral genetics. Such studies, of course, assume that the difference in concordance rates between identical and dizygotic twins are a useful, if only approximate, measure of the genetic contribution to the psychopathology or personality trait in question. Our data suggest that, to some significant degree, such differences in concordance rates may result not from genetic differences, but instead, from differences in non-genetic constitutional variables. Such constitutional variables show greater variability in dizygotic than in monozygotic twins, and thus would be likely to lead to more marked role and relationship differences between fraternal than between identical twins. Such a non-genetic, though constitutionally precipitated sequence, could be expected to bring about differences in concordance rates.

Another question of considerable interest is the extent to which our findings may be relevant to non-twin schizophrenics. We are interested in the possibility that in the development of singletons there is a psychological image of the ideal child within the parents' unconscious, or some similar pattern embodied in a sibling or other relative, which plays a similar though less clear role in terms of comparative phenomena than does the identical twin in these discordant pairs. The finding of similar pre-psychosis life patterns characterizing the history of non-twin schizophrenics, when compared with non-schizophrenic siblings or other well-matched control groups, suggest that this may well be the case (Lane and Albee, 1965, 1966; Fleming, 1968; Pollack *et al.*, 1966).

[*Note:* Since presenting this paper, we have studied two additional pairs of discordant identical twins. There has been no basic change in the pattern of findings described above. However, two additional points have begun to stand out. The first is that in those pairs in which the twin who was biologically more competent and larger at birth becomes schizophrenic, we have up to this point in each instance found some specific major early event explaining the subsequent switch of developmental pathways. In one case, this was an episode of severe life-threatening cyanosis during the first month of life; and in another a severe case of Rocky Mountain spotted fever at age three. Second, the relationship between birth weight and protein bound iodine described above has subsequently been found to be independent of the presence or absence of schizophrenia. Possible implications of this relationship for central nervous system and body-wide endocrine functioning are now being further explored (Stabenau and Pollin, 1967).]

References

Fleming, P. 1967. Emotional antecedents of schizophrenia: Inner experiences of children and adolescents who were later hospitalized for schizophrenia. Presented at Conference of Life History Research in Psychopathology. New York.

Guggenheim, F. G.; Pollin, W.; Stabenau, J.; and Mosher, L. 1968. Prevalence of physical illness in the parents of monozygotic twins discordant for schizophrenia. Presented at Annual Meeting of American Psychomatic Society, Chicago, Illinois, March 1966. (To be published.)

Lane, E. and Albee, G. 1966. Childhood intellectual differences between schizophrenic adults and their siblings. *Amer. J. Orthopsychiat.* **35** (4): 747-53.

—— and ——. 1966. Comparative birth weights of schizophrenics and their siblings. *J. Psychol.* **64:** 227-31.

Pollack, M.; Woerner, M.; Goodman, W.; and Greenburg, I. 1966. Childhood development patterns of hospitalized adult schizophrenic and non-schizophrenic patients and their siblings. *Amer. J. Orthopsychiat.* **36:** 510-17.

Pollin, W. and Stabenau, J. R. 1968. Biological, psychological and historical differences in a series of monozygotic twins discordant for schizophrenia. In *The Transmission of Schizophrenia,* ed. S. Kety and D. Rosenthal. London: Pergamon Press.

——; ——; and Tupin, J. 1965. Family studies with identical twins discordant for schizophrenia. *Psychiat.* **28:** 60-78.

——; ——; Mosher, L.; and Tupin. 1966. Life history differences in identical twins discordant for schizophrenia. *Amer. J. Orthopsychiat.* **36:** 492-509.

Stabenau, J. R. and Pollin, W. 1967a. Early characteristics of monozygotic twins discordant for schizophrenia. *Arch. Gen. Psychiat.* **17:** 723-34.

——. 1967b. Maturity at birth and adult protein bound iodine. *Nature* **215:** 996-97.

PART II
Twin Studies

Introduction

Recent twin studies have shown a marked increase in sophistication in research methodology. Vandenberg has elsewhere (1966, 1968) and at the beginning of Part III of this volume reviewed some of the criticisms of older twin studies and suggested that some of these criticisms are no longer valid, while other objections have lost much of their importance through the introduction in later twin research of objective zygosity criteria, better statistical methods, the use of larger samples and, more adequate consideration of the nature of the twin sample and of the psychological measures used.

In the following four chapters one or another of these points contributed in important ways to the design of the study described. In chapter 10 the Louisville Twin Study is described. This is perhaps the largest study of twins ever undertaken in terms of the number of subjects, the number of variables studied, and in length (more than eight years). In the Louisville study, efforts were made to deal more or less adequately with all these criticisms of older studies of twins. In chapter 11 the effect of varying parental influences on twin differences is examined to see whether this variation may be a function of the zygosity of the twin pair. The evidence is negative. In chapter 12 it is reported that earlier separation of identical twins raised apart does not produce greater pair differences in intelligence, another finding which shows that the importance of environmental causation of twin differences may sometimes have been exaggerated. In chapter 13 it is shown that, while nearly all subtests of the Wechsler Intelligence Scale have a significant hereditary component, there are nevertheless interesting differences in the degree of hereditary determination between these subtests.

I do not wish to deny the usefulness of an initial demonstration by a twin study of the presence of an hereditary component in some human trait in which it had not been suspected before, but it is my fond hope that future twin studies will take adequate account of the innovations reported here so that it will be possible to move on to a comparison of the importance of heredity in a variety of behavioral traits and to the study of the more rewarding ones with other methods of behavior genetics, rather than merely to continue adding variables for which in some study a suggestion was found of some degree of hereditary control. The four chapters in Part II thus, in a sense, clear the ground for the final part of this volume where methods are described which attempt to show how one can isolate, within a given body of data, those variables most strongly determined by hereditary components. Minor modifications of these methods may be used to select from psychological tests those items which best measure a hereditary component, as a first step in the design of tests specifically constructed for use in genetic studies.

A minor complication for twin studies has recently come to light. On very rare occasions twins arising from a single fertilized egg may differ chromosomally, so that they have, for instance, different sexes. In view of the very infrequent occurrence of this phenomenon, this is not likely to affect the statistical findings of the conclusions from twin studies, provided that the studies are based on adequate numbers of twin pairs. Nevertheless, the possibility of such a complication should be kept in mind whenever unusually large differences are found in a pair of twins who are concordant for all blood tests or whenever extreme similarity is found in twins of different sexes. In such cases it may be worth while to investigate the karyotype of the parents, the twins, and all siblings of the twins.

References

Vandenberg, S. G. 1966. Contributions of twin research to psychology. *Psychol. Bull.* **66**: 327-52.

———. 1968. In defense of twins and the twin method. In *Essays in Honor of Professor Essen Möller*, ed. E. Nyman, *Acta Psychiat. Scand.*, suppl.

10

STEVEN G. VANDENBERG [2]
RICHARD E. STAFFORD [3]
and ANNE M. BROWN
University of Louisville School of Medicine

THE LOUISVILLE TWIN STUDY [1]

INTRODUCTION

Since the classic twin study of Newman, Freeman, and Holzinger in 1937, there have been several twin studies of psychological variables—for instance, by the Thurstones and Strandskov (1953), by Gottesman (1963), and by Vandenberg (1962). These were all cross-sectional studies, with variables observed in only one instant of time, and until they are replicated, they permit only limited conclusions. A longitudinal twin study, on the other hand, provides for regular replication. Longitudinal work has the further advantage of allowing the consideration of hereditary factors in rates of growth and in age-related behavioral changes.

To those of our readers who may feel that we are beating the heredity-environment controversy into its final demise, we would say: "We hope we are, and a good thing, too!" Perhaps we can even lay the controversy to its final rest by changing the approach from a forced choice between two extremes to an approach which integrates the effects of heredity and environment. It is time for psychologists to cease ignoring either source of variation and proceed with full recognition that the two are highly interdependent. Progress can only be made by collecting the kind of material which will permit analysis of this interaction of heredity and envi-

[1] The Louisville Twin Study was started in 1957 by Dr. Frank Falkner, a pediatrician, to take advantage of recent methodological advances in zygosity diagnosis and statistical evaluation. Some time thereafter Falkner assumed the chairmanship of the Department of Pediatrics, which left him little time for personal research. In 1960, Steven G. Vandenberg joined the study as a research psychologist, and later became director. Primary emphasis was placed on the development of cognitive and personality variables. A second emphasis was on the nature of the twin situation itself. Richard E. Stafford joined the study in February, 1965, as associate director. In September, 1965, Ronald S. Wilson took charge of psychophysiological work for the study, replacing Roy Griffiths. On September 1, 1967, Wilson assumed direction of the study as a result of the departure of Vandenberg for the University of Colorado and of Stafford for the Ohio State University.

[2] Now at the University of Colorado.

[3] Now at Ohio State University.

ronment. The twin method, albeit imperfect, is one way—and possibly the most economical way—of obtaining such material.

Zygosity Determination

When conclusions about the role of heredity are to be drawn from a comparison of the relative similarity of fraternal and identical twins, it is essential that the zygosity determination be made independent from the variables to be studied. If similarity in anthropometric variables is used in this determination, heritability estimates of those variables should not be included in that study. We have, therefore, based our work entirely on blood groups.

Whether or not a twin pair is regarded as identical (MZ) or fraternal (DZ) in our study depends exclusively on the results of an extensive battery of serological tests. The antisera used are shown below.

Genetic markers used in this study to determine zygosity:

Always used	Not always used
A_1A_2BO	Mt^a Martin
MNSs	Mi^a Miltenberger
Rhesus tests CcDEe (Rh factor)	P_1
Lutheran a and b	C^w
Lewis a and b	Wr^a Wright
Kell K	Vw Verweyst
Cellano k	Yt Cartwright
Kidd (Jk^a)	Do^a Dombrock
Duffy (Fy^a)	

If there is a discordance on any of these tests the twins are classified as fraternal, otherwise they are regarded as identical. This procedure will, on the average, lead to an accuracy of 95 per cent. For the type of analysis we are doing this is adequate. Most of our psychological variables probably have a considerably lower precision anyway.

Statistical Methods

A number of different statistical methods for use in twin studies have been suggested by various investigators. We have chosen the following:

For single variables: To evaluate the importance of the increased differences between fraternal twins when compared to the differences between identical twins, we test the statistical significance of the F ratio of the fraternal (DZ) and identical (MZ) within-pair variances:

$$F = \frac{\sigma^2_{wDZ}}{\sigma^2_{wMZ}} \qquad (1)$$

with degrees of freedom N_{DZ} and N_{MZ}, where the within-pair variances

$$\sigma^2_w = \frac{1}{2N} \Sigma (x_A - x_B)^2 \qquad (2)$$

and where N is the number of pairs, and x is a score observed for twin A or twin B.

In addition, we frequently calculate Holzinger's index of heritability according to the formula

$$H = \frac{\sigma^2_{DZ} - \sigma^2_{MZ}}{\sigma^2_{DZ}} \qquad (3)$$

to allow comparison with older publications which frequently used this index. For further discussion of these methods see Vandenberg (1966a).

Multivariate Analysis of Twin Differences

For a determination whether different variables are in large measure controlled by the same genetic components a measure is used based on the comparison of fraternal and identical within-pair covariances. The method and some results have been described by Vandenberg (1965a, 1965b).

Longitudinal Analysis

Where possible, we attempt to fit our data to growth curves of the form

$$y = a + bx + cx^2. \qquad (4)$$

Other formulas may later be considered, especially when we analyze the learning tasks recently presented to the high school age twins.

Eventually we hope to investigate the question whether parameters of such curves, such as the rate of growth, *for different variables* are correlated. Where different measures have been obtained in different years correlations between them will be calculated and twin differences on different tests will be compared and correlated.

DEVELOPMENTAL STUDY

Recruitment of "Neonate" Twins

1. *Original Method.* In the beginning of the study, arrangements were made with the hospitals of the Louisville area to report all multiple births to the Child Development Unit. Placentas were saved at the hospitals, collected by a worker from the Child Development Unit and taken to a laboratory in the Louisville General Hospital for study. Meanwhile the parents of the twins were contacted to see if they would be interested in joining the study, and if so, an appointment was made to bring the twins in for their first visit, hopefully, at one month of age. No families were recruited unless a placenta report had been completed.

This procedure required the interaction of many people and rather close timing in order that the placenta would be in condition for study. While the cooperation of the hospitals was in most cases good, some hospitals were better equipped than others to follow through with the procedure and as a result a higher percentage of reports came in from some hospitals than from others.

When the placenta study was discontinued, a new approach to recruitment of twin families was sought and a new procedure developed which has proved to be inexpensive, efficient, and quite successful.

2. *Present Method.* In March, 1965, arrangements were made with the Health Department of Louisville and Jefferson County to receive Xerox copies of birth certificates for all multiple births in Louisville and Jefferson County. A letter is then sent to the parents of the twins, introducing the study to them and asking them to consider whether or not they would be willing to participate. A follow-up contact is made by phone or by letter, and if the parents are interested a visit is scheduled when the twins are three months old. Approximately 40 per cent of the twins born in Louisville and Jefferson County this year have been recruited for the Twin Study by this procedure. One of the unexpected benefits of the recruiting from birth certificate data has been the elimination of the socioeconomic selectivity which apparently was operating in the recruitment from hospital reports, which for some reason resulted in a bias toward the upper socioeconomic levels.

Present Sample

Zygosity of twins in the study is determined by blood typing. The blood sample is taken when the twins are about three years old. This delay in blood typing is unavoidable for technical and psychological reasons, but it is emphasized here because the number of twins enrolled in the study appears somewhat misleading unless one realizes that about one-half of this total are children who have not yet been blood typed and therefore cannot be included in current data analysis.

1. *Same Sex Twins.* One hundred and sixty sets of same sex twins have been started in the "neonate" study. These presently range in age from three months to seven years. As mentioned before, while a considerable amount of data has been collected on all these children, only about half of them can be included in current concordance studies because the rest have not yet been blood typed.

2. *Boy-Girl Twins.* Eighteen pairs of infant boy-girl twins have recently been recruited for a study of the emergence of sex differences in abilities and interests. Plans are to build this sample up to about 30 pairs.

Characteristics of the Present "Neonate" Sample, Based on 140 Interviews

The socioeconomic status of the families, as judged from the occupation of the fathers, was rated according to the socioeconomic scale reported in Reiss *et al.* (1961). This scale is based on a survey of the status value of jobs and occupations by the National Opinion Research Center (NORC). This survey was performed in 1947 and the few small shifts that may have occurred in the relative status of some occupations would not affect the present work significantly. One of the great advantages of the scale is that it makes it possible to obtain a rating when only a minimum of information is available, and hence makes it possible to compare the Louisville Twin Study sample with other populations in which information about the occupation of the father is available.

Table 10-1 shows the percentage distribution for the occupations of the fathers of twins in the neonate sample of the Louisville Twin Study, classified by 10-point intervals of the NORC socioeconomic scale. Examples of occupations typical of each interval are listed. Since the Louisville Twin Study neonate sample is recruited primarily from Louisville and Jefferson County, the percentage distribution is also shown for occupations of fathers of all twins born in Louisville and Jefferson County for a nine month period (the time for which these data have been available to us).

TABLE 10-1
A COMPARISON OF THE NUMBER OF FATHERS OF TWINS IN VARIOUS SOCIO-ECONOMIC GROUPS FOR THE LOUISVILLE TWIN STUDY, AND FOR ALL TWINS BORN DURING A NINE MONTHS PERIOD IN JEFFERSON COUNTY, KENTUCKY

NORC Rating	Examples of Occupations	Twin Study n	Twin Study %	Jefferson Co. Sample n	Jefferson Co. Sample %
0–9	Unskilled laborer, janitor	10	7.1	10	12.1
10–19	Carpenter, painter, enlisted serviceman	10	7.1	27	32.5
20–29	Shipping clerk, bus driver, cabinet maker, machine operator	12	8.6	19	23.0
30–39	Gas station mgr., retail sales, policeman, plumber	16	11.4	5	6.0
40–49	Surveyor, electrician, business machine operator	10	7.2	4	4.8
50–59	Mgr. retail trade, foreman, technician	19	13.6	4	4.8
60–69	Insurance agent, draftsman, wholesale salesman	24	17.14	6	7.2
70–79	Buyer, credit mgr., public school teacher, large corp. middle mgt.	17	12.1	4	4.8
80–89	College professor, banking mgt., engineer, natural and social scientist	13	9.3	2	2.4
90–99	Architect, dentist, physician, lawyer	9	6.4	2	2.4
	Totals	140	100%	88	100%

Figures 10-1a and 10-1b show the percentage distributions of these two samples along with that of the male experienced civilian labor force population used in the construction of the NORC scales.

Of the twins born in Louisville and Jefferson County during the period studied, approximately 68 per cent were born to families in the lowest one-third of the socioeconomic scale. In the present Louisville Twin Study sample, 23 per cent of the twin families fall into this category. Recruitment of these 140 families was dependent on voluntary reports to the Twin Study of twin births by the hospitals of the Louisville area. However, as explained earlier, as of April 1965, the method of recruiting twins for the study was changed when an arrangement was made with the Jefferson County Health Department for obtaining Xerox copies of birth-certificates of all multiple births in Louisville and Jefferson County. Figure 10-1 shows that this method has resulted in a distribution which reflects more nearly the general population of the Louisville-Jefferson

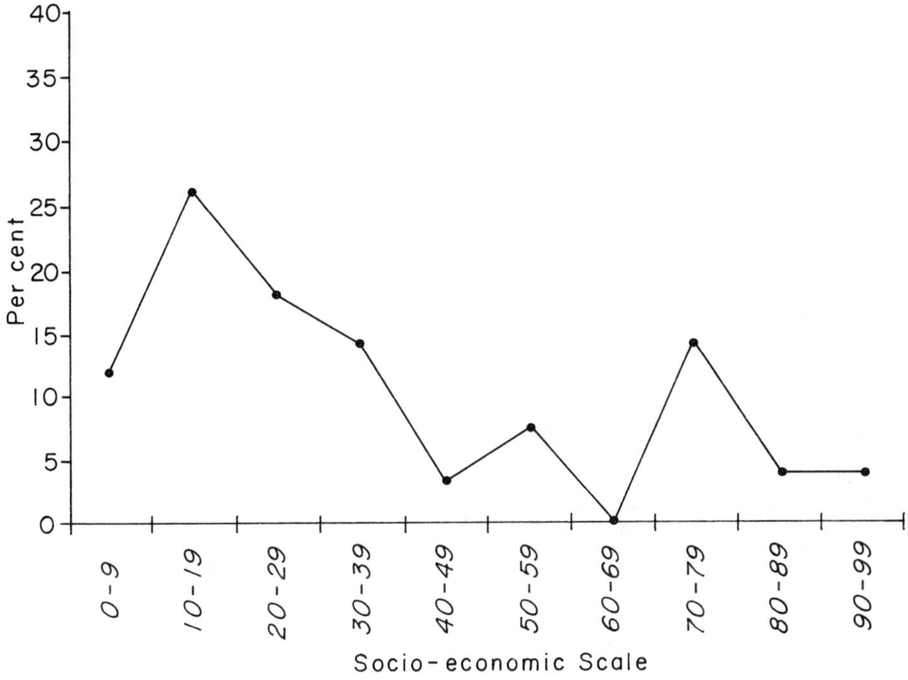

FIG. 10-1A. SOCIOECONOMIC STATUS OF TWIN FAMILIES RECRUITED FROM BIRTH CERTIFICATE DATA.

County area. For example, 52 per cent of the new recruitments are in the lowest one-third of the socioeconomic scale. (It should be noted, however, that otherwise an analysis of information from these 27 recently recruited families is not included in the present report.)

In addition, it may be expected that this longitudinal research sample will be somewhat above average in cooperativeness and intelligence since rewards for participation are largely non-material ones related to such values as interest in growth and development of children, contributing to education and research, and possibly a certain amount of social prestige in being connected—however tenuously—with the University. Families in the study are reimbursed for the actual expenses of transportation, but the mother still must have leisure time to spend a half day at the Child Development Unit, frequently must make arrangements for other children in the family, and be able to commit herself to a definite appointment on a certain day—none of which are characteristic ways of operating for families in the lowest socioeconomic categories.

The educational levels of the twins' fathers lend support to the socio-

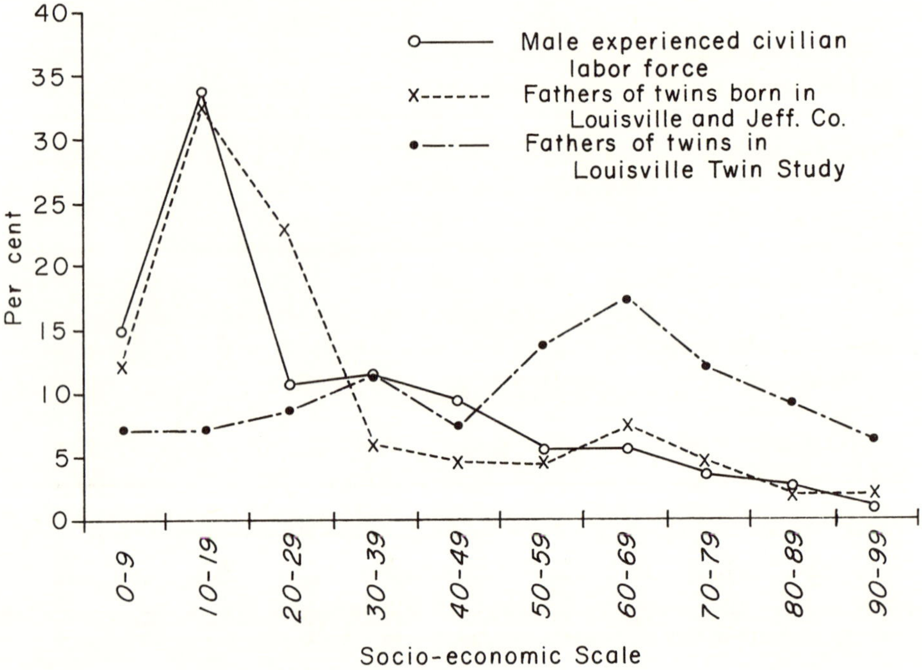

FIG. 10-1B. PERCENTAGE DISTRIBUTION OF SOCIOECONOMIC INDEX FOR MALE EXPERIENCED CIVILIAN LABOR FORCE, FATHERS OF ALL TWINS BORN IN LOUISVILLE AND JEFFERSON COUNTY, AND FATHERS OF TWINS IN LOUISVILLE TWIN STUDY.

economic ratings. Mean education level for fathers was 13.4 years, distributed as follows:

Years formal education:	No. of cases:
less than high school (6-11 yrs.)	23
high school completed	48
1-3 years college	25
bachelor's degree	20
graduate school (17-22 yrs.)	20
Total	136

The educational level of mothers was also somewhat above average, with a mean of 12.5 years. (The 1960 census reports median educational levels of 9.3 years for Louisville and 9.9 years for Jefferson County.)

Ninety per cent of the mothers were full time homemakers at the time of these initial interviews; 6.5 per cent worked outside the home on a part-time basis, and only 3.5 per cent worked full time.

The age of mothers at the birth of the twins is shown in Table 10-2. The mean age was 28.6 years for all the mothers. Mean ages differed very little for mothers of MZ and DZ twins—28.9 and 29.1 years respectively. However, the mean is lower (28.2 years) for mothers of untyped twins. It has already been noted that the twins who have not been bloodtyped are less than three years old and of lower average socioeconomic levels. They would, of course, include both MZ and DZ twins. In the sample of all twin births in Louisville and Jefferson County for the last nine months (n = 83) the average age of mothers at the birth of twins was found to be 27.5 years.

The twins' position in relation to other children in the family and to maternal age is shown in Table 10-3. For comparative purposes, ages of mothers and ordinal position of the twins within the sibship are shown for the general population sample along with the Louisville Twin Study data.

In the Louisville Twin Study, twins are firstborn children in about 14 per cent of the cases. In two-thirds of the cases, the twins are second, third or fourth born children. There appears to be no significant difference in the parity of mothers of monozygotic and dizygotic twins in this sample. (Table 10-4)

Within-pair birth weight differences for the twins are shown in Figure 10-2 The mean within-pair difference was 11.6 ounces. No significant correlations were found between birth weight and birth order of the twins, either for the entire group, or for the MZ or the DZ twins separately.

Procedures Used in the Neonatal Study

The twins in the longitudinal or "neonatal" study are brought to the Child Development Unit at three months, six months, nine months,

Table 10-2
Maternal Age at Birth of Twins for MZ, DZ and Untyped Pairs, Louisville Twin Study Sample

	Mother's age at birth of twins (n = 140)						
Zygosity	20 or less	21–25	26–30	31–35	36–40	41 or more	Totals
MZ pairs	3	10	6	9	6	1	35
DZ pairs	2	13	5	12	6	1	39
Untyped	2	25	20	9	9	1	66
Totals	7	48	31	30	21	3	140

TABLE 10-3
MATERNAL AGE AT BIRTH OF TWINS AND PARITY—PERCENTAGES FOR LOUISVILLE TWIN STUDY (LTS), AND IN SAMPLE OF ALL TWIN BIRTHS IN JEFFERSON COUNTY (ALL)

	Age of mothers at birth of twins													
	20 or less		21–25		26–30		31–35		36–40		41 or more		Totals (%)	
Number living children older than twins	LTS	All	LTS	All	LTS	All	LTS	All	LTS	All	LTS	All	LTS	All
0	2.1	8.4	10.0	7.2	.7	—	.7				.7		14.2	15.7
1	2.1	2.4	12.9	12.1	6.4	8.4	1.4		1.4				24.2	22.9
2	.7	1.2	5.7	4.8	7.1	4.8	5.7	2.4	3.6	1.2			22.8	14.5
3		1.2	4.3	4.8	3.6	7.2	5.7	2.4	4.3	1.2	.7	1.2	18.6	18.1
4			.7	2.4	2.1	1.2	4.3	3.6	.7	2.4			7.8	9.6
5					.7	1.2	2.1		1.4			1.2	4.2	2.4
6			.7		1.4	1.2	.7	2.4	2.1	2.4	.7		4.9	6.0
7							.7	1.2	.7	2.4			1.4	3.6
8									.7	1.2			.7	4.8
9														
10														
11										1.2				1.2
12								1.2						1.2
Total %	5.1	13.2	33.3	21.3	22.5	24.1	21.7	16.8	15.2	12.0	2.1	2.4	100%	100%

TABLE 10-4
PARITY AND ZYGOSITY OF TWINS—LOUISVILLE TWIN STUDY

	Number of siblings older than twins									
	0	1	2	3	4	5	6	7	8	Totals
MZ	5	9	9	6	1	1	3	1	—	35
DZ	5	8	8	10	4	3	—	1	—	39
Untyped	9	17	16	10	6	2	5	—	1	66
Totals	19	34	33	26	11	6	8	2	1	140

twelve months, eighteen months, two years, two and one half years, three years and yearly thereafter. (Boy-girl twins have an additional visit at fifteen months.) At every visit, each twin is given a psychological test, physical measurements are taken, and the mother is interviewed about the twins as she sees them.

1. *Psychological Tests.* The Bayley Infant Scales of Mental and Motor

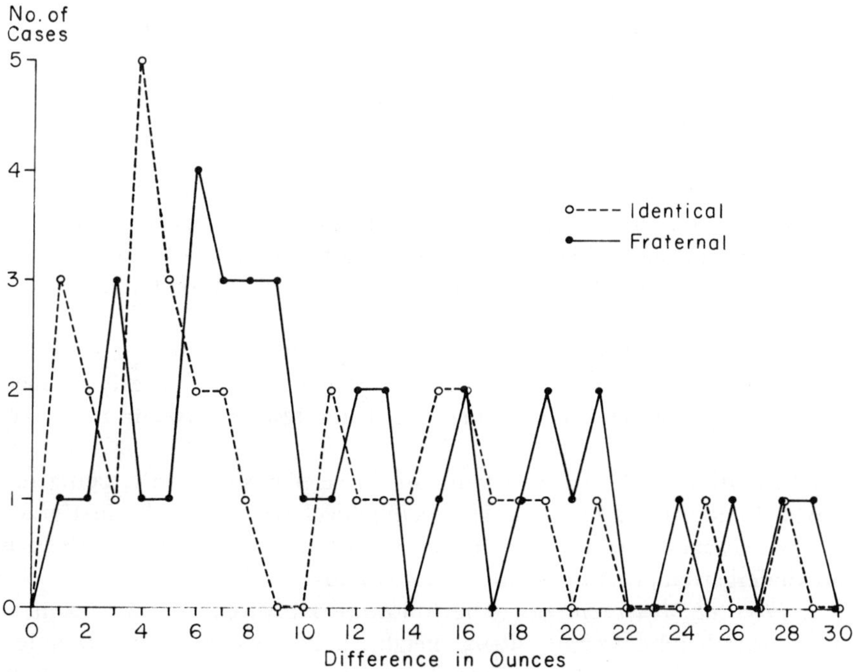

FIG. 10-2. WITHIN-PAIR DIFFERENCES IN BIRTH WEIGHT.

Development are administered from three months through two years. These scales need no description, other than that we are using an "in-between" version which has been somewhat modified by Dr. Bayley since our study started.

From two and one-half through seven years of age the Pacific Multifactor Test designed by Meyers, Dingman and others is used. This test is designed to measure the same five separate abilities at three age levels, beginning in the second year. In addition we used a number ability scale developed by Shotwell *et al.* (1956). The difficulty of the tasks employed to measure these abilities increases, but the nature of the tasks are as similar from one level to the next as the rapid expansion of the developing child's behavior repertoire permits. Currently we administer Level 2 at two and one half and three years; Level 4, at four and five years and Level 6 at six and seven years. Originally the Level 2 battery was administered at age two, but it was found that some of the subtests were too difficult for our subjects at this age. The materials used in these tests are described in detail in Meyers *et al.* (1962, 1965). We followed their instructions carefully in constructing our two sets from masonite and plywood. To facilitate the rapid and smooth administration of these tasks to infants with short attention span, a special cabinet was constructed which houses each subtest in a most convenient way. (We are grateful to Mr. Warren Gliessner for the design of this cabinet.) Figure 10-3 shows the cabinet itself; Figure 10-4 shows the cabinet and some of the materials spread out on a table.

Experimental work has been started using a scale developed by Drs. Uzgiris and McVickers Hunt to assess mental development with procedures based on Piaget's theories. Schemas of "object permanence" and "means of causing desired results" are being tested at nine months and at one and two years. The schema "concept of space" will be tested at later ages.

The mother is with the twins during the psychological testing until the twins are about three years old. After this time they are usually satisfied to be tested alone and mother may watch the testing through a one-way window.

Observation of the boy-girl twins in a free play situation is being initiated. In addition a rating scale is being developed for use by first grade teachers after the twins enter elementary school. Unfortunately, there is no kindergarten program in our school systems.

2. *Physical Measurements.* The following measurements are taken at each visit: height, weight, sitting height, pelvic width, chest circumference, head circumference, and maximum upper arm circumference (Vandenberg & Falkner, 1965).

Fig. 10-3. Cabinet used to house Pacific Multifactor Test (Meyers et al., 1962, 1965).

Fig. 10-4. Cabinet and some of the test material for the Pacific Multifactor Test.

3. *Interviews with Mothers.* When the physical and psychological measures are completed, a staff member takes over the care of the children while the mother talks with the interviewer about the twins' simi-

larities and differences as she sees them. In the initial interview with a mother, basic socioeconomic information about the family is also taken and this is updated in subsequent interviews if necessary.

Maintaining Parental Interest in the Twin Study

Continued interest in and support of the project is, of course, crucial to the success of any longitudinal research. No substitute has been found for genuinely appreciating the real contribution these families make by their faithful participation and letting them know by the attitude of the entire staff that their efforts are appreciated. Over and above this, however, some tools have been found to be helpful in strengthening the ties between twin families and the twin study.

1. *Twin Study Newsletter.* Often the mothers say that it is "interesting" to be a part of the twin study. The Twin Study Newsletter was inaugurated to report to them some of the interesting findings of the study. Newsletters are sent three or four times a year. These newsletters are written in a popular style and care is taken to avoid presenting data which might bias future observations of the twins by the mothers.

2. *Twin Study Brochure.* A 14 page illustrated booklet was sent to all participating families. In the booklet the general purposes of the study are explained, the need for such research is stressed, and some of the findings of the study are presented. The local press has used this booklet to prepare newspaper articles about the Twin Study.

3. *Physical Growth Charts.* The height and weight of each twin are plotted on growth charts which show norms for boys and girls separately. At each visit the mother is give an updated copy of our chart showing the growth pattern of her twins in relation to each other and in relation to the norms.

Results

1. *Analysis of Initial Interviews with Mothers of Young Twins.* Although the Louisville Twin Study is longitudinal in nature, the data reported here represents a cross section of twins whose ages ranged from one month to six years. This cross sectional analysis was undertaken as a first step in the development of an interviewing procedure which would be appropriate for twins throughout early childhood.

At every visit of the twins to the Child Development Unit, each twin is given a psychological test and several physical measures are taken. The mother stays with the twins during the psychological testing usually until the twins are three years old, after which she observes the testing through a one-way window. The mother is always with the twins during the physical measurements. Then a staff member takes over the care of the children while the mother talks with the interviewer about the twins. The interviewing situation is kept as informal and non-threatening as possible. Some specific questions are asked, such as, "Which of the twins has more sleeping problems?" The mother is encouraged to talk about the twins, and she is assured that questions about differences in the twins are not directed toward criticism of either twin, but rather toward finding the ways in which the children's individuality is expressed in very early life. She is instructed not to strain for differences that are not fairly obvious, but to answer "no difference" if that is the case.

In the eight variables reported upon here, the mother was asked which twin exhibited certain traits more than the other. No attempt was made to evaluate "how much" more. The responses were analyzed for relationships of the variables to zygosity, birth weight, birth order within the twinship, and were correlated with each other.

The eight variables under consideration are: which twin (1) laughs and smiles more readily, (2) has more feeding problems, (3) shows temper more often, (4) has more sleeping problems, (5) cries more, (6) generally succeeds in taking toys from his twin, (7) has more tantrums, and (8) is more like the mother herself in personality.

Perhaps it is worth noting again that within-pair differences, as seen by the mother of the twins, are under consideration here, rather than comparisons against any outside criteria or norms. For this reason a discussion of the relations of our data to those of investigators such as Birch or Macfarlane is not attempted.

Since several factors are simultaneously involved here, it was decided to sort out portions of the data for separate analysis. The questions to be answered were these:

Are the number of within-pair differences reported by the mothers—or conversely, the number of "no difference" responses given by the mothers—related to zygosity of the twins? If so, this could be directly interpreted as evidence of hereditary components in the behavior.

Disregarding zygosity, can the differences between the twins be attributed to birth order or to birth weight differences?

Is there evidence of interaction between zygosity, birth weight, birth order, and these particular variables?

When there is a difference reported, which traits tend to occur in the same child?

Hereditary Factors in Eight Variables. Only twins whose zygosity was determined by blood typing are considered here. Table 10-5 shows χ^2 and

TABLE 10-5
PRESENCE OR ABSENCE OF DIFFERENCES SEEN BY MOTHERS
RELATED TO ZYGOSITY OF TWINS

	χ^2	p
1. Laughs, smiles more readily	1.543	.22
2. More feeding problems	9.586	.002
3. Shows temper more often	1.389	.23
4. More sleeping problems	5.685	.017
5. Cries more	2.448	.12
6. Takes toys from twin	0.218	.65
7. More tantrums	0.0	1.00
8. More like mother	6.441	.01

(Variance was in direction of more "no difference" responses than expected for MZ twins, and more differences than expected for DZ twins.)

p values for the eight variables when only the presence or absence of concordance and the zygosity of the twins were examined. Variance in all cases was in the direction of fewer observed differences than expected for MZ twins and more than expected for DZ twins. Discrepancies were significant for "feeding problems" ($p = .002$), "sleeping problems" ($p = .017$), and for the mothers' seeing one twin as more like herself in personality ($p = .01$). The results for temper tantrums ($\chi^2 = 0$) might almost be considered significant in the opposite direction—i.e., there is no evidence here for even the slightest hereditary component operating in the tendency toward temper tantrums, because the concordance rates are exactly equal.

Relationship of Birth Weight and Birth Order to Reported Differences. Disregarding zygosity for the moment, percentage frequencies were calculated to see if the presence or absence of these variables was related to birth weight or to birth order, per se. Table 10-6 shows that there is a tendency for the child who is small at birth to be the one who laughs and smiles more, has more feeding problems, shows temper more often, has more sleeping problems, and succeeds in taking toys from his twin. It tends to be the heavier twin who cries more, has more tantrums, and is seen by the mother as more like herself.

However, in χ^2 tests of the extent to which these frequencies differed from chance expectation, only the results for feeding problems ($p = .003$) reached an acceptable level of significance, with the twin who is lighter at birth also being the one with more feeding problems over this cross section of ages.

TABLE 10-6
OCCURRENCE OF TRAITS IN RELATION TO BIRTH ORDER AND WEIGHT

Traits	Trait characteristic of twin who was: Heavier at birth %	Lighter at birth %	1st born %	2nd born %
1. Laughs, smiles more readily	42.7	57.3	44.7	55.3
2. More feeding problems	31.0	69.0*	41.4	58.6
3. Shows temper more often	45.5	54.5	47.3	52.7
4. More sleeping problems	42.1	57.9	60.5	39.5
5. Cries more	51.9	48.1	45.7	54.3
6. Takes toys from twin	44.2	55.8	50.0	50.0
7. More tantrums	51.6	49.4	38.7	61.3
8. More like mother	55.3	44.7	42.1	57.9

(*Difference significant at .003 level.)

In regard to birth order within the twinship, there was a tendency for the first born twin to be the one with more sleeping problems. All the other variables were more typical of second born twins, with the exception of "taking toys from twin" which was divided evenly.

χ^2 tests hovered about a p value of .20 for the probability of more feeding, temper, sleeping problems, and tantrums occurring in the second born twin, but none reached significance. Perhaps it should be mentioned again here that no significant correlation was found between birth order and birth weight for MZ, DZ or untyped twins in the Louisville Twin Study population.

Relationships between Zygosity, Birth Order and Birth Weight. Some trends can be seen in the data which indicate that being a first born MZ twin might be quite different from being a first born DZ twin, and that a weight advantage at birth might have different correlates for MZ and for DZ twins. However, with so many categories, the number of cases is not sufficient to permit reliable statistical conclusions. When the younger babies are old enough to be blood typed, more dependable conclusions may be reached. Only cell frequencies are presented here, with some discussion of trends which may bear watching.

	Heavier at birth	Lighter at birth	Zygosity	1st born	2nd born
The twin who laughs and smiles more is:	6	10	MZ	7	9
	18	24	Untyped	23	22
	11	13	DZ	8	16

These frequencies would indicate that birth order here is more important for DZ twins than for MZ, and it does not seem to be as important among the untyped infant twins as among the older twins. The twin who is lighter at birth is rather consistently more likely to be the one who smiles and laughs more.

	Heavier at birth	Lighter at birth	Zygosity	1st born	2nd born
The twin with more feeding problems is:	2	7	MZ	6	3
	7	18	Untyped	8	17
	9	15	DZ	10	14

Lower birth weight is significantly related to feeding problems, as has already been mentioned. In regard to birth order, there is a tendency for the *first* born MZ, but the *second* born DZ twin to have this problem. Since there are fewer MZ cases (more of them being in the "no difference" category which is not being considered here) no firm conclusions can be drawn. It does suggest a possible relationship which is even more clearly demonstrated in regard to sleeping problems.

	Heavier at birth	Lighter at birth	Zygosity	1st born	2nd born
The twin with more sleeping problems is:	1	5	MZ	6	0
	6	9	Untyped	8	7
	9	8	DZ	9	8

Although the number of MZ cases with a difference in sleeping problems is small, the data suggests that when there is such a difference in the MZ twins, and when the first born is also the smallest in birth weight, there may exist a strong relationship to sleeping difficulties. No similar trend can be detected for the DZ twins.

	Heavier at birth	Lighter at birth	Zygosity	1st born	2nd born
The twin who shows temper more is:	15	13	MZ	15	13
	24	25	Untyped	19	30
	12	23	DZ	19	16

The relatively high frequencies here show that this is one variable for which there is usually a difference between the twins which is readily recognized by the mothers. For MZ twins, it appears to have little to do with either birth weight or birth order. For DZ twins, the one who was smaller at birth tends to be the one who has more frequent outbursts of temper, and among the untyped babies it is the second born who shows temper more often. Temper tantrums appear to be quite a different matter as the next analysis shows:

	Heavier at birth	Lighter at birth	Zygosity	1st born	2nd born
The twin with more temper tantrums is:	7	3	MZ	2	8
	3	6	Untyped	4	5
	6	6	DZ	6	6

Again the *n* is considerably reduced and the variability is greater in the MZ twins, with the heavier, second born, identical twin showing more of a tendency towards tantrums, in contrast to the situation with feeding and sleeping problems where it was the lighter and first born who had the problems.

	Heavier at birth	Lighter at birth	Zygosity	1st born	2nd born
The twin who cries more is:	9	9	MZ	9	9
	15	21	Untyped	15	21
	18	9	DZ	13	14

Age differences seem to be operating here along with other factors, with the twin who was smaller at birth and the second born more frequently reported as crying more. Among the DZ twins there seems to be a trend for the twin who was heavier at birth to be the one who cries more.

	Heavier at birth	Lighter at birth	Zygosity	1st born	2nd born
The twin who is more likely to be taking toys from the other is:	8	8	MZ	8	8
	7	8	Untyped	7	8
	8	13	DZ	11	10

Probably the most striking thing here is the lack of variation from what might be expected on the basis of chance. If one can take one twin's success getting things away from his twin as any indication of dominance within the twinship, the low frequencies are also of interest since they would indicate that there was frequently no noticeably dominant twin.

	Heavier at birth	Lighter at birth	Zygosity	1st born	2nd born
The twin who is more					
like mother in	4	1	MZ	2	3
personality is:	8	9	Untyped	4	13
	9	7	DZ	10	6

This particular question unfortunately was included only in about one-half of the interviews. The mothers of DZ twins were more likely to see one or the other of the twins as more like herself, and among the younger untyped twins there was a tendency for the second born to be more often reported to have a personality similar to mother's own.

Intercorrelations for the Eight Variables. When a difference is reported by the mothers, do the same traits tend to occur in the same child? Is the twin who has more feeding problems the same one who has more sleeping problems, shows more temper, does more crying, etc.? Phi coefficients (corrected for unequal means by phi/phi max.) for eight traits are shown in Table 10-7.

TABLE 10-7
PHI COEFFICIENTS, CORRECTED BY PHI MAX, FOR THE TRAITS REPORTED BY MOTHERS AS OCCURRING IN SAME TWIN

	Laughs more	Feed. prob.	More temper	Sleep prob.	Cries more	Takes toys	Tantrums	Like mother
Laughs, smiles more		−.026	−.028	.120	−.049	−.116	.077	−.257
More feeding problems	−.026		−.022	.684[a]	−.058	.256	−.250	−.076
Shows temper more often	−.028	−.022		.510[b]	.196	.379[c]	.497[e]	.086
More sleeping problems	.120	.684[a]	.510[b]		.194	.650[d]	.099	.083
Cries more	−.049	−.058	.196	.194		−.050	.222	.081
Takes toys from twin	−.116	.256	.379[c]	.650[d]	−.050		.499	−.259
More tantrums	.077	−.250	.497[e]	.099	.222	.499		−.314
More like mother	−.257	−.076	.086	.083	.081	−.259	−.314	

(Lower case letters indicate significance of correlation coefficients by x^2 at following levels of probability: a) $p = .066$; b) $p = .088$; c) $p = .01$; d) $p = .077$; e) $p = .04$.)

2. *Results on Bayley Scales of Infant Development.* The results for 91 pairs who were blood typed were available for analysis. These pairs fell in the following categories:

Identical (MZ) pairs, boys	17
Identical (MZ) pairs, girls	25
Fraternal (DZ) pairs, boys	23
Fraternal (DZ) pairs, girls	26

The Bayley scales were administered six times: at three, six, nine, twelve, eighteen and twenty-four months of age. (A visit at one month of age was discontinued because of the high incidence of broken appointments or inability to make an appointment.) Many twins missed one or more visits so that the actual number of cases which could be analyzed for a given age varies considerably. Total scores were assigned to the performance of these infants at each visit. To obtain a total score, one point credit was given for each item passed of the ones administered, beginning with the item occurring two months below the child's chronological age unless perusal of the previous records indicated that the child had been doing unusually poorly or well, such cases being rare in our sample. "Basal level" credit was given for all the items which came before the first item administered.

The means and sigmas of the scores on the Bayley Mental and Motor Development Scales at the six age levels, as well as the number of cases available, are shown in Table 10-8. It may be noted there is not an equal increase in the mean score between visits: the increases are 43.30, 26.84, 16.55, 26.28 and 27.78 for the Mental Scale, and 11.65, 13.67, 6.89, 4.41 and 4.57 for the Motor Scale.

The unequal increases are mainly due to the fact that there are an unequal number of items available at different age levels. This fact is a function of the extreme limitations placed on the test developer by the infant's very restricted range of capabilities and very limited attention span. The number of items available in the scale of motor development is especially small at the higher age levels. For example, the month levels 14 through 18 are grouped and only three items are presented.

Because of the small number of items at some age levels, failure on a single item carries a relatively heavy load in determining performance at that level. On the other hand, the variability in total score at that level is quite limited due to the credit given for the basal score.

Intercorrelations of Scores at Different Age Levels. The small number of items is perhaps responsible, in part, for the low intercorrelations between the scores on the six visits. In addition, there were varying num-

TABLE 10-8
MEANS, STANDARD DEVIATIONS, AND NUMBER OF INFANTS TESTED WITH THE BAYLEY SCALES OF MENTAL AND MOTOR DEVELOPMENT AT SIX AGE LEVELS

Age when tested	x	S.D.	n
	Scale of Mental Development		
3 months	34.89	7.76	118
6 months	78.19	9.14	145
9 months	105.03	3.57	145
12 months	121.58	4.41	160
18 months	147.86	4.37	142
24 months	175.64	3.24	137
	Scale of Motor Development		
3 months	14.63	2.22	116
6 months	26.28	3.74	145
9 months	39.95	3.81	145
12 months	46.84	1.92	158
18 months	51.25	1.01	142
24 months	55.82	3.17	83*

* For a period of several months in the summer and fall of 1964 many infants were not examined with the Motor Scale as a result of changes in personnel.

bers of cases missing at each of the visits. The values of these intercorrelations are shown in Table 10-9. It may be seen that in general the intercorrelations are higher for the scores on the Mental Development scale at various ages, than for the Motor Development scale. Although it might be expected that the intercorrelations between the two scales would be especially low, this is not the case. Assessment of infant mental processes at this preverbal level must necessarily rely on observation of motor activities with test objects. In the Bayley Scales, for example, only one item does not depend on motor ability at the ten month level, and at the eleven month level all items are dependent on motor skills.

Importance of Hereditary Components in the Bayley Scales. To determine the importance of hereditary components in the variance observed in these test scores, F ratios were calculated between the fraternal (DZ) and the identical (MZ) within-pair variances at each of the six ages at which the Bayley Scales were administered. These results are shown in Table 10-10.

Because the intercorrelations are so low, it was decided to calculate for each infant an average of his scores on all Mental Development Tests he had taken. In order to do this some arbitrary decisions had to be made and the following steps were taken.

Table 10-9
Intercorrelations between Scores on the Bayley Scale of Mental Development (on the Top Left) and the Bayley Scale of Motor Development (on the Bottom Right), and between the Two Scales (on the Top Right) at Ages 3, 6, 9, 12, 18 and 24 Months. The Number of Cases for Each Correlation Is Shown below the Diagonal

	MENTAL SCALE						MOTOR SCALE					
	3 mos.	6 mos.	9 mos.	12 mos.	18 mos.	24 mos.	3 mos.	6 mos.	12 mos.	9 mos.	18 mos.	24 mos.
Mental												
3 mos.	—	.58	.25	.23	−.11	.18	.36	.51	.41	.05	.00	.49
6 mos.	(99)	—	.45	.33	.18	.36	.22	.69	.63	.41	.19	.35
9 mos.	(98)	(117)	—	.33	.24	.36	−.10	.40	.41	.23	.09	.28
12 mos.	(107)	(126)	(130)	—	.41	.34	.09	.42	.35	.25	.13	.34
18 mos.	(99)	(114)	(114)	(126)	—	.43	−.04	.18	.22	.16	.18	.20
24 mos.	(92)	(109)	(113)	(118)	(109)	—	−.02	.39	.29	.15	.05	.40
Motor												
3 mos.	(116)	(97)	(96)	(104)	(99)	(90)	—	.30	.20	.20	−.08	.26
6 mos.	(99)	(145)	(117)	(126)	(114)	(109)	(97)	—	.60	.47	.27	.35
9 mos.	(98)	(117)	(145)	(130)	(114)	(113)	(96)	(117)	—	.52	.14	.23
12 mos.	(105)	(126)	(130)	(158)	(124)	(118)	(104)	(126)	(130)	—	.31	−.02
18 mos.	(99)	(114)	(114)	(126)	(142)	(109)	(99)	(114)	(114)	(124)	—	.12
24 mos.	(55)	(75)	(71)	(68)	(73)	(83)	(53)	(75)	(71)	(68)	(73)	—

TABLE 10-10

F RATIOS BETWEEN FRATERNAL AND IDENTICAL WITHIN-PAIR VARIANCES ON THE BAYLEY SCALES OF MENTAL AND MOTOR DEVELOPMENT, FOR VISITS AT SIX DIFFERENT AGES, AND FOR AN AVERAGED MENTAL DEVELOPMENT SCORE

Scale and age of test		F ratio for girls	F ratio for boys	F ratio for all cases
Mental scale	3 months	1.04	17.32**	3.89**
	6 months	.69	2.06	1.35
	9 months	1.38	3.27*	1.92*
	12 months	2.36*	1.47	1.91*
	18 months	.93	2.82	1.13
	24 months	.41	9.72**	1.24
Average mental score †		.82	1.47	1.17
Motor scale	3 months	1.57	6.60**	2.29**
	6 months	1.81	.71	1.21
	9 months	3.67**	7.87**	4.43**
	12 months	4.82**	34.00**	9.97**
	18 months	.74	.00	.58
	24 months	1.15	1.45	1.26

* $p < .05$
** $p < .01$
† For explanation see text.

The raw score of each infant's examination at each age level was the number of items passed. These raw scores were converted to percentage scores by dividing the raw score by the maximum number of items presented to *any* infant in our sample at that age. Then these percentage scores were averaged for each infant. (The minimum number of test scores thus averaged was 3, and the maximum 6.)

The F ratios for these averaged percentage scores are included in Table 10-10, while F ratios for the percentage scores for *separate* visits are shown in Table 10-11. The F ratios for these latter scores closely resemble the values for the original scores at the six age levels as might be expected.

3. *Results on the Pacific Multifactor Test Battery.* The version of the Pacific Multifactor Test level two that we used consists of 17 subtests. Twelve of these subtests have been described in detail by Meyers, Dingman, *et al.* (1962, 1964), but in our study we have also employed three measures of memory and two of number concepts, that had been developed by the same authors.

These tests were administered to 63 pairs of twins at age three, and to 25 pairs of twins at age two. We have stopped administering this battery at age two, because the tests frequently seemed to be too difficult for our subjects, and it was necessary to avoid taxing the infant's attention since

TABLE 10-11
F RATIOS BETWEEN FRATERNAL AND IDENTICAL WITHIN-PAIR VARIANCES FOR PERCENTAGE SCORES ON THE BAYLEY SCALE OF MENTAL DEVELOPMENT AT SIX AGE LEVELS

Age when tested	F ratio for girls	boys	all cases
3 months	1.28	7.36**	3.89*
6 months	.64	.79	.77
9 months	1.35	4.00	2.03*
12 months	2.21*	1.63	1.92*
18 months	1.00	2.13	1.20
24 months	.46	10.20**	1.31

* $p < .05$
** $p < .01$

the Bayley scales are also administered at that age. The 17 subtests are said to measure six separate abilities. The means and standard deviations for these 17 subtests, as well as the number of cases for each of the tests, are shown in Table 10-12 separately for male and for female infants, but with the two and three year old children combined. Except for the last test, none of the differences between the means for the two sexes reaches the .05 level of significance. For this reason correlations have been calculated for both sexes combined, but separately for the two year and three year old children.

TABLE 10-12
MEANS, STANDARD DEVIATIONS AND NUMBER OF CASES FOR 17 SUBTESTS OF THE PACIFIC MULTIFACTOR TEST

Name of test	Females			Males		
	$\bar{x}$	S.D.	n	$\bar{x}$	S.D.	n
Cube stacking	10.27	5.28	104	9.94	4.25	65
Disc stacking	144.15	55.69	106	139.91	47.33	68
Bead stringing	22.48	5.99	105	22.84	5.37	68
Form, color & size matching	7.54	4.85	99	7.25	3.59	60
Form & color matching	6.48	4.02	87	6.27	4.22	59
Form matching	5.57	4.93	83	5.73	4.67	55
Expressive vocabulary	10.58	4.61	107	10.68	5.02	66
Receptive vocabulary	8.47	3.06	105	7.92	3.78	65
Identifying objects	7.18	2.79	101	6.80	3.31	65
Pattern completion	10.54	2.72	104	10.37	2.82	65
Form & picture completion	10.77	2.76	104	10.92	3.21	66
Design copying	2.90	2.48	99	3.03	2.06	62
Form & color memory	4.86	3.13	94	5.52	3.01	58
Picture memory	7.93	2.92	99	7.47	3.60	60
Knox cubes	1.68	2.04	87	1.90	1.77	58
Concept of 2	1.22	1.87	89	.95	1.49	59
Concept of 3 and 4	.47	.94	55	.06	.24	35

The intercorrelations between 16 of the 17 subtest scores are shown in Table 10-13, for the three year olds only. In this analysis the scores on subtest 17—the concept of three and four—have been excluded as being too infrequently passed or even attempted.

The F ratios for the six abilities measured in three year old twins are given in Table 10-14. For two of these abilities, Perceptual Speed and Number Ability, the within-twin variance is significantly greater in fraternal twins indicating an important influence by hereditary components.

It is not known whether this Perceptual Speed ability is the same factor, or is related to the Spatial Visualization factor measured in Thurstone's Primary Mental Ability tests or in the DAT Spatial Reasoning test. If it is, then these results could fit in with earlier reports by Vandenberg (1965, 1966a, 1966b) of an important hereditary component in tests of spatial visualization in studies at the high school level. The fact that the results on Number Ability are also significant, definitely fits in with those findings.

Are the Tests Measuring Independent Abilities? The intercorrelation matrix for the three year olds, the two year olds, and the combined sample were all factor analyzed. Because subtest 16—the concept of two—was expected to appear as a singlet it was dropped. Only the results for the three year olds are discussed here. In spite of the varying number of cases for each correlation, it was possible to obtain communality estimates by the squared multiple correlation method. The squared multiple correlation of a variable with all the others has been shown by Guttman (1957) to provide a lower bound for the communality. Although there were more than five positive eigen values, only five were retained for rotation.

Two rotational solutions were sought. First we rotated to the Varimax criterion (Kaiser, 1958) and later we obtained a Procrustes solution (Hurley and Cattell, 1962). The Varimax solution is shown for the three year old infants in Table 10-15. It can be seen from the Varimax solution that the tests come close to measuring five orthogonal factors.

The results of the attempt to force the rotation into a neater solution by the Procrustes program are shown in Table 10-16. The attempt was moderately successful, but led to factors which were intercorrelated, as shown at the bottom of Table 10-16.

In either case it is clear that Meyers, Dingman, and their associates have been remarkably successful in constructing nearly independent measures of five abilities. The sixth one—number ability—appears to develop somewhat later than the others, but may be expected to be independent, too.

TABLE 10-13
INTERCORRELATIONS BETWEEN 16 SUBJECTS OF THE PACIFIC MULTIFACTOR TEST FOR THREE YEAR OLD TWINS

	1	2	3	4	5	6	7	8	9	10	11	12	13	14	15	16
1. Cube stacking		37	26	25	18	07	06	16	23	29	13	17	19	35	09	17
2. Disc stacking			37	13	16	10	10	21	23	12	30	08	21	24	31	17
3. Bead stringing				13	21	25	18	26	23	30	39	23	20	39	16	15
4. Form, color, size matching					50	35	07	14	14	14	05	22	22	34	20	31
5. Form & color matching						81	20	36	18	03	24	13	42	48	33	03
6. Form matching							20	27	24	02	37	22	58	52	30	09
7. Expressive vocabulary								70	63	05	25	12	20	36	19	22
8. Receptive vocabulary									48	07	24	08	21	38	40	34
9. Identifying objects										12	31	29	30	36	13	24
10. Pattern completion											12	05	07	00	−01	33
11. Form & picture completion												13	23	43	13	08
12. Design copying													38	13	03	22
13. Form & color memory														40	26	19
14. Picture memory															29	12
15. Knox cubes																45
16. Concept of 2																
$\bar{x}$	11.73	162.77	24.24	8.70	7.44	6.73	12.31	9.32	8.06	11.18	11.91	3.66	5.94	8.48	2.19	13.42
S.D.	4.29	29.82	5.41	3.92	3.70	4.67	3.64	2.82	2.28	1.84	1.78	2.16	2.79	2.71	2.01	19.21
N	125	128	127	120	115	110	127	124	124	127	127	123	116	124	110	111

TABLE 10-14
F RATIOS OF FRATERNAL AND IDENTICAL WITHIN-TWIN VARIANCES FOR SIX ABILITIES MEASURED IN THREE YEAR OLD TWINS WITH THE PACIFIC MULTIFACTOR TEST

Motor ability	1.67	34 DZ	28 MZ
Perceptual speed	2.33*	30 DZ	23 MZ
Language	.83	30 DZ	26 MZ
Reasoning	1.09	33 DZ	26 MZ
Memory	1.32	27 DZ	27 MZ
Number ability	1.99*	29 DZ	24 MZ

* $p < .05$

TABLE 10-15
VARIMAX ROTATIONS OF 5 FACTORS FOUND IN 15 SUBTESTS OF THE PACIFIC MULTIFACTOR TESTS FOR 128 THREE YEAR OLD TWINS

	Perc. speed	Lang.	Motor speed	Perc. Reason.	Memory
Motor speed					
Cube stacking	14	07	60	10	08
Disc stacking	12	11	45	−08	36
Bead stringing	11	13	39	10	47
Perceptual speed					
Form, color, size matching	51	04	30	14	−13
Form & color matching	86	13	04	08	12
Form matching	78	08	−13	30	34
Language					
Expressive vocabulary	07	83	−02	12	12
Receptive vocabulary	27	75	11	−10	14
Identifying objects	06	62	18	34	18
Perceptual reasoning					
Pattern completion	−03	01	41	09	06
Form and picture completion	14	18	11	13	56
Design copying	11	08	16	51	06
Memory					
Form & color memory	43	13	08	44	24
Picture memory	47	30	20	12	39
Knox cubes	39	24	12	−18	17

Decimals omitted.

4. *Interrelations between Measures*

Intercorrelations between the Bayley Scores and the Six "Pacific" Abilities. Correlations were obtained between (1) the scores on the Bayley Scale of Mental Development at each of the six ages, and (2) the scores on the six abilities measured by the Pacific Multifactor Test battery. For this analysis the scores on the three tests measuring the Motor, Perceptual

TABLE 10-16
PROCRUSTES ROTATION OF FIVE FACTORS FOUND IN 15 SUBTESTS OF THE PACIFIC MULTIFACTOR TEST

Subtests	Motor speed	Perc. speed	Lang.	Perc. Reason.	Memory
Cube stacking	45	03	−07	27	06
Disc stacking	37	−02	05	24	15
Bead stringing	38	−10	−05	28	30
Form, color, size matching	08	48	−03	04	13
Form & color matching	−07	61	−01	−13	51
Form matching	−10	46	−06	04	62
Expressive vocabulary	−13	−05	57	03	14
Receptive vocabulary	05	−02	46	−09	31
Identifying objects	−01	−05	42	23	13
Pattern completion	30	03	05	19	01
Form & picture completion	30	−07	01	32	34
Design copying	08	00	06	42	06
Form & color memory	−03	28	06	25	26
Picture memory	21	16	03	07	46
Knox cubes	06	15	15	−10	36
			Intercorrelations between factors		
		55	65	10	−08
	55		62	50	04
	65	62		46	19
	10	50	46		32
	−08	04	19	32	

Decimals omitted.

Speed, Language, Reasoning and Memory abilities and the two tests for the Number Ability were merely added.

At some future time it may be worth while to obtain standard scores on each test, so that each test would have equal weight within an ability score and to use these scores. We definitely plan to get a total Pacific score for future correlational studies.

The correlations with the Bayley Mental Development Scale are shown in Table 10-17 and are presented graphically in Figure 10-5a.

Several conclusions are apparent:

a. There are not significant correlations between the Bayley Mental score at three months and the six abilities measured by the Pacific subtests at three years, although there is a suggestion of a negative correlation with Memory and Number.

b. In general there is an increase in the size of all correlations with the age at which the Bayley Mental Scale was administered.

c. The correlations with Pacific Motor Ability scores are generally not significantly different from zero.

TABLE 10-17

CORRELATIONS BETWEEN SIX ABILITIES MEASURED AT AGE THREE WITH THE PACIFIC MULTIFACTOR TEST (PMT) AND THE SCORE ON THE BAYLEY SCALE OF MENTAL DEVELOPMENT ADMINISTERED AT SIX DIFFERENT AGES (AND THE NUMBER OF CASES ON WHICH EACH CORRELATION IS BASED, IN PARENTHESES)

PMT ABILITY	BAYLEY SCALE OF MENTAL DEVELOPMENT administered at					
at age 3 years	3 months	6 months	9 months	12 months	18 months	24 months
Motor	08 (86)	17 (103)	22 (105)	25** (113)	17 (100)	19 (95)
Perception	09 (64)	08 (76)	21 (80)	21 (84)	30** (75)	43** (72)
Language	−18 (79)	03 (96)	28** (101)	19 (104)	43** (93)	45** (89)
Reasoning	01 (82)	28** (100)	25 (102)	17 (108)	36** (98)	43** (94)
Memory	−24 (68)	21 (85)	36** (86)	20 (90)	58** (80)	37** (77)
Number	−33 (31)	−02 (36)	33 (39)	34 (39)	57** (35)	36 (37)

** $p < .01$ Decimals omitted.

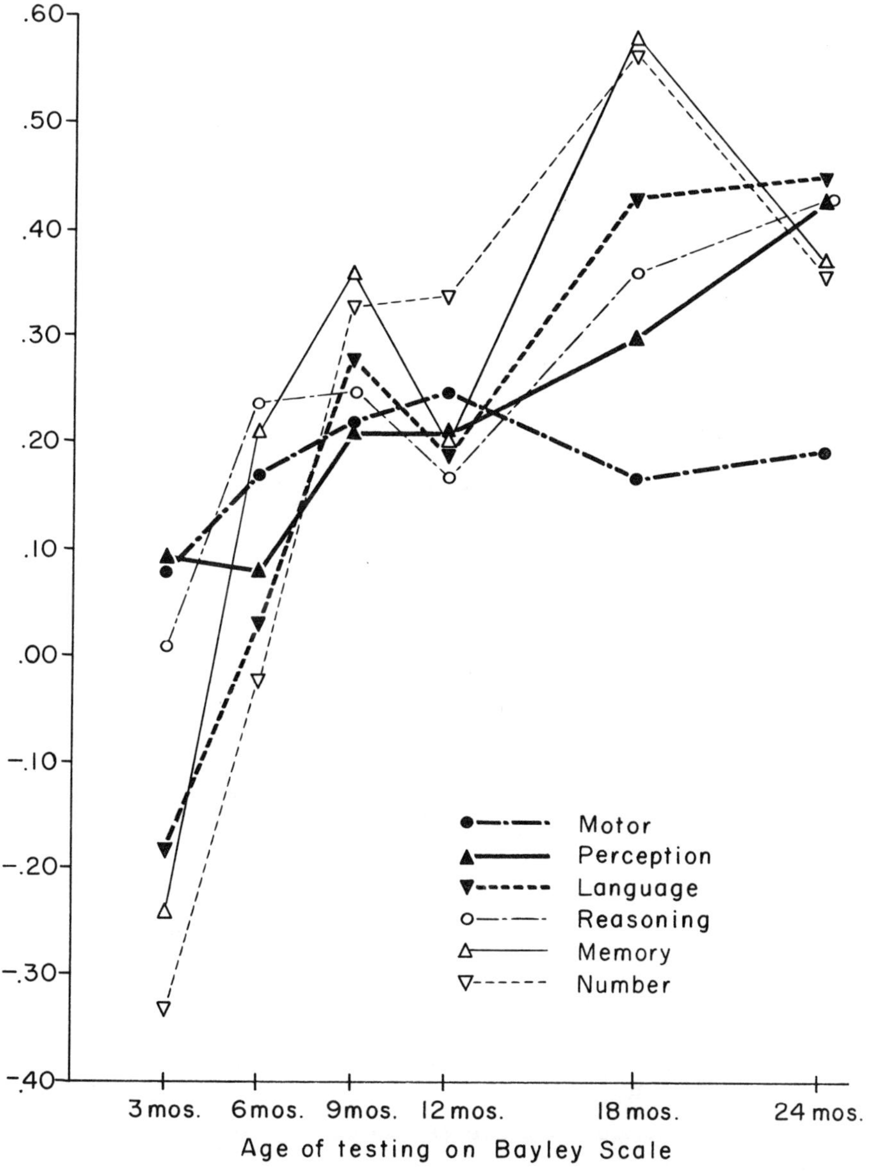

FIG. 10-5A. CORRELATIONS BETWEEN SIX ABILITIES MEASURED AT AGE THREE AND THE SCORE ON THE BAYLEY SCALE OF MENTAL DEVELOPMENT ADMINISTERED AT SIX DIFFERENT AGES.

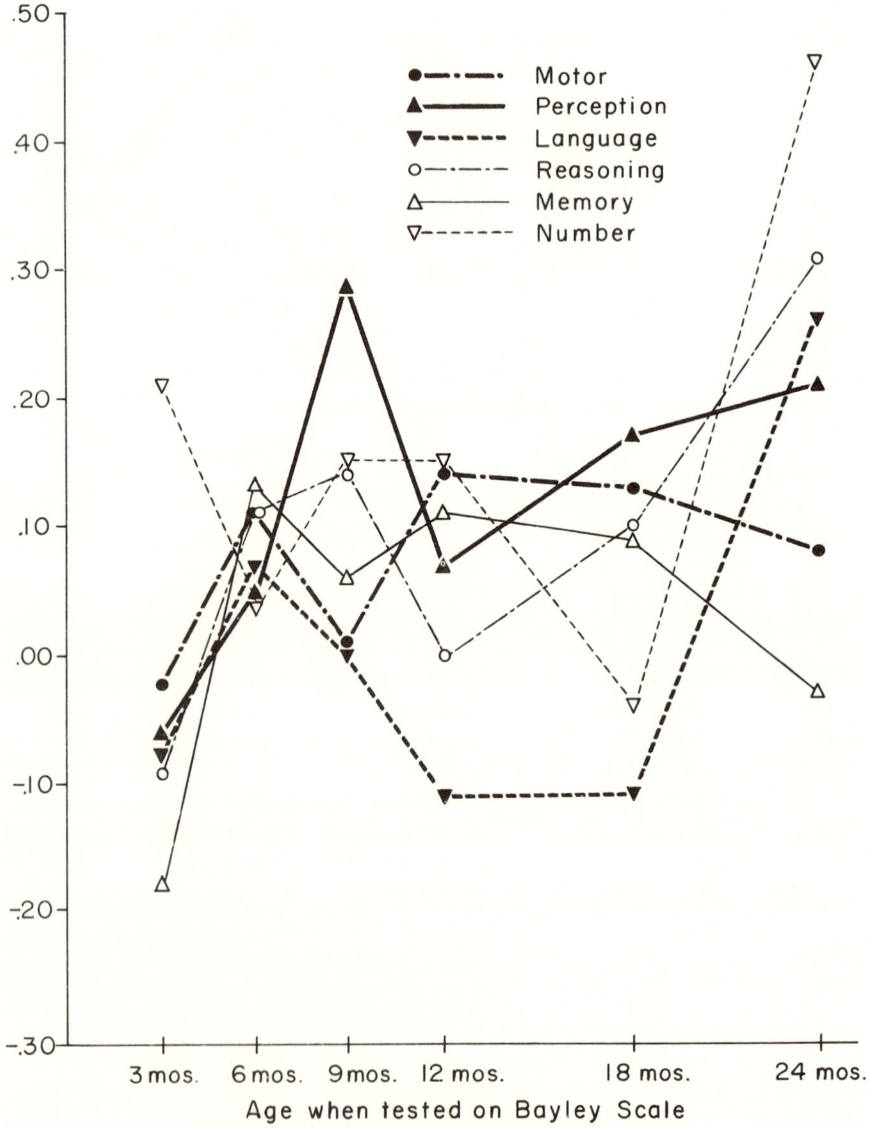

FIG. 10-5B. CORRELATIONS BETWEEN SIX ABILITIES MEASURED AT AGE THREE AND THE SCORE ON THE BAYLEY SCALE OF MOTOR DEVELOPMENT ADMINISTERED AT SIX DIFFERENT AGES.

It will be interesting to see whether these results will change in any way when our sample size is doubled after the remainder of the infants are blood typed.

The correlations between the Bayley Motor Scale at six ages, and the six Pacific abilities are shown in Figure 10-5b.

The results can be rather quickly summarized: in general there are no significant correlations with the Bayley Motor scale except when administered at 24 months. This is somewhat surprising because there are so few items in the motor scale at this level.

Relations between Physical Growth and the Bayley Scales. An earlier paper (Vandenberg and Falkner, 1965) described the fitting of growth curves to the anthropometric data obtained for as many as ten visits. It was found that for the first few years of life a parabolic curve segment fits very well. The formula for this curve is:

$$y = a + bx + cx^2$$

where y is height (or whatever measure is being fitted), x is the age, a is the estimated value at birth, b is the rate of growth, and c is the change in the rate of growth. The value of c is always negative, to reflect the fact that the initial high rate of growth is slowing down as a function of age. Because values of a, b, and c are available for all infants for whom growth curves were fitted it became possible to correlate these with the results on the Bayley Scales.

The observed values for height of a pair of identical girls and a pair of fraternal boys are shown in Figure 10-6, as well as the values for the parameter a, b, and c.

The correlations between the scores of the Bayley Scale of Mental Development at ages 3, 6, 9, 12, 18 and 24 months with the three parameters for the growth curve for height are shown in Table 10-18. It can be seen that no correlation reaches a high enough value to be useful for prediction, although two correlations are statistically significant beyond the one per cent level of probability. Table 10-19 shows the correlations with the scale of Motor Development. The same lack of high values is apparent.

When interpreting these findings it must be kept in mind that the range of variation in growth in height was perhaps somewhat limited in our sample. One indication of this is furnished by looking at the means and standard deviations of the three parameters which are as follows:

	$\bar{x}$	S.D.	Coefficient of variation
Estimated birth length	52.45	3.35	15.64
Rate of growth in height	.6835	.1959	3.49
Rate of change in growth rate	.3822 (10^{-4})	.2453 (10^{-4})	1.56

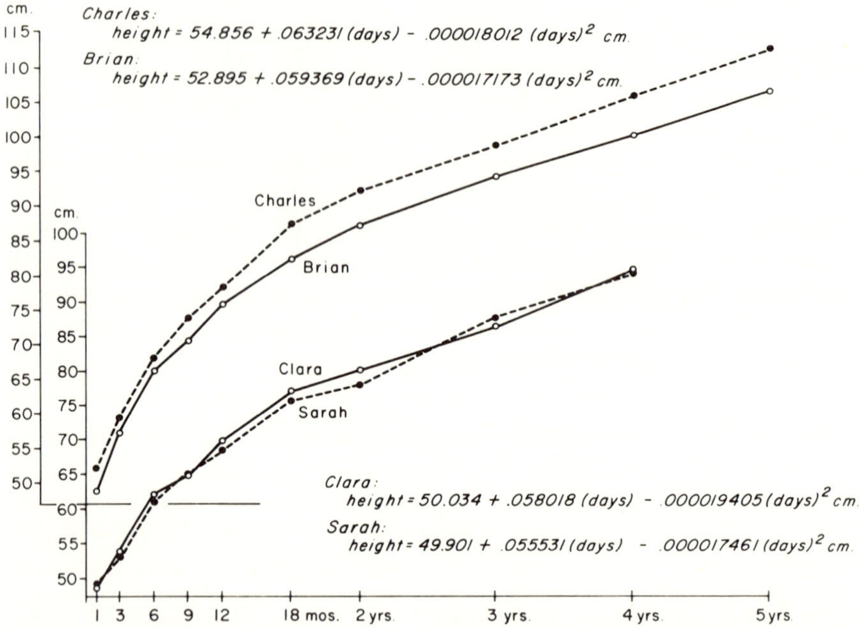

FIG. 10-6. OBSERVED VALUES (IN CM) FOR HEIGHT OF FRATERNAL TWINS, CHARLES AND BRIAN; AND IDENTICAL TWINS, CLARA AND SARAH; AND THE VALUES OF A BEST FITTING PARABOLIC SEGMENT FOR EACH CHILD. (THE CURVES FOR CHARLES AND BRIAN WERE MOVED UP FOR CLEARER PRESENTATION.)

The ratio of the mean to the standard deviation, which is called the coefficient of variation, gives some indication of the amount of variability of a measure around its mean. Clearly this variability is quite narrow for the estimated length at birth, about average for biological phenomena for the rate of growth, and somewhat more than average for the change in the rate of growth. These judgments are based on the report by Wechsler that the coefficient of variation tends to be around 3.0 for many biological distributions (Wechsler, 1952).

Relations between Physical Growth and the Pacific Multifactor Test. The correlations between the three parameters of the individual growth curves for height and the scores on six parts of the Pacific Multifactor Test administered when the children were three years old are shown in Table 10-20.

To obtain these six scores we added the three subtests for Motor

TABLE 10-18
CORRELATIONS BETWEEN THREE PARAMETERS OF THE GROWTH CURVE FOR HEIGHT AND THE SCORES ON THE BAYLEY SCALE OF MENTAL DEVELOPMENT AT SIX DIFFERENT AGES

Growth curve parameters	Score on Bayley Mental Scale administered at age						Averaged percentage score
	3 mos.	6 mos.	9 mos.	12 mos.	18 mos.	24 mos.	
a. Estimated value at birth	.003	.302**	.017	.033	−.034	.227	.129
b. The rate of growth	.254	.039	.045	−.066	−.067	.074	.060
c. The rate of change in the growth rate	.385**	.056	.168	.138	.104	.048	.240**
Number of infants	116	143	143	152	136	131	173

** $p < .01$

TABLE 10-19
CORRELATIONS BETWEEN THREE PARAMETERS OF THE GROWTH CURVE FOR HEIGHT AND THE SCORES ON THE BAYLEY SCALE MOTOR DEVELOPMENT AT SIX DIFFERENT AGES

Growth curve parameters	Score on Bayley Motor Scale administered at age					
	3 mos.	6 mos.	9 mos.	12 mos.	18 mos.	24 mos.
a. Estimated value at birth	.283**	.221	.068	.151	−.108	−.083
b. The rate of growth	.059	−.006	.122	−.106	−.017	.213
c. The rate of change in the growth rate	−.057	.092	.142	−.075	.018	.322**
Number of infants	114	143	143	150	136	79

**p < .01

TABLE 10-20
CORRELATIONS BETWEEN VALUES OF THREE PARAMETERS OF CURVE FOR GROWTH IN HEIGHT BASED ON UP TO NINE VISITS, AND SIX ABILITIES MEASURED AT AGE THREE WITH THE PACIFIC MULTIFACTOR TEST

Growth curve parameters	Ability measured:					
	Motor	Perc.	Lang.	Reas.	Memory	Number
a. Estimated length at birth	.252	−.225	−.137	.078	.060	−.135
b. Rate of growth in height	−.188	.156	.052	−.103	−.152	−.204
c. Rate of change in growth rate	.208	−.177	−.195	−.047	−.029	−.068
Number of cases	117	90	110	114	96	45

Ability, Perceptual Speed, Language, Reasoning and Memory. The score for Number Ability is based on two subtests.

The correlations are uniformly low and generally not even significantly different from zero, confirming the findings with the Bayley Scales that in our sample the growth in height during the first three or four years is not closely related to mental ability.

The intercorrelations between the six scores of the Pacific Multifactor Test are shown in Table 10-21. The memory score correlates highest with each of the other five scores; four of these correlations are significant beyond the one per cent level of significance. Factor analysis of the 17 individual subtests does not show such a crucial position for the memory tasks. Lack of time has prevented us from tracking down the reason for this discrepancy.

TABLE 10-21
INTERCORRELATIONS BETWEEN THE SIX ABILITIES MEASURED BY THE PACIFIC MULTIFACTOR TEST OBTAINED BY SIMPLE SUMMATION OF SUBTESTS SCORES. CORRELATIONS ABOVE DIAGONAL, NUMBER OF CASES BELOW DIAGONAL

	Motor	Perc.	Lang.	Reas.	Memory	Number
Motor		.171	.046	.182	.234	.252
Perceptual	(95)		.188	.276	.525**	.179
Language	(115)	(92)		.240	.436**	.401**
Reasoning	(119)	(32)	(113)		.334**	.262
Memory	(102)	(87)	(99)	(99)		.510**
Number	(46)	(44)	(46)	(44)	(42)	

** $p < .01$

THE HIGH SCHOOL STUDY

Reason for This Work

Beginning in the spring of 1961 group tests have been administered each year to twins in the public and private schools in Louisville and Jefferson County, Kentucky; in Brandenburg, Kentucky; and in Southern Indiana. This program of studies was added for several reasons; longitudinal studies, in general, do not permit rapid analysis and frequent reporting of results. In our study this handicap is aggravated by the necessity to wait until the twins are old enough to be blood typed. Cross-sectional twin studies at the high school level offer an opportunity to report some findings each year at professional meetings and to prepare papers for publication without this need to wait.

In addition, these studies made it possible to start sampling wider areas of cognition and personality than can readily be studied in infants, to determine where hereditary components contribute a significant portion of the variance. By adding each year to a more and more comprehensive survey, we hoped to find guidelines for future choice of variables to be included in the longitudinal study.

Recruitment of Twins

Twins are recruited for the high school study as follows. After initial approval of the project by the several superintendents, lists of twins and twin parents' addresses and phone numbers are obtained from the Board of Education (or in some instances, from individual schools). Next, a letter in which the study is briefly described and permission is asked from the parents for participation by the twins in the current year's twin study is sent to the parents of each twin pair. Return postcards are included for use by the parents. As these postcards come back, we check the names. Finally, we contact by phone those parents from whom we have not heard after a reasonable length of time. Upon completion of these phone calls, we send a list of twins whose parents have given permission to the principals of each school to indicate which twins will be participating.

Typically, cooperation in each school varies between 85 to 100 per cent. While we thus do not have complete participation by all twins, it is our impression that no specific selection factor seems to be operating, other than age. Seniors, that is students who are graduating that year, are less likely to participate because they are concerned about the loss of one whole school day, and because of conflicts with course quizzes or special tests. Occasionally, a twin is married and feels embarrassed by joining a group of younger twins for the group testing. A few of these have come to the Child Development Unit separately during the summer.

Summary of Twins Studied in 1961-1966

The number of twins seen each year are shown in Table 10-22.

We decided to include boy-girl pairs because:

a. Occasionally exclusion of unlike sexed twins was misinterpreted or misunderstood by some persons in a school and taken to mean that fraternal twins (unlike twins?) were not wanted, as shown by the low numbers of like sexed fraternals reported. After some checking we found several unreported pairs of like sexed fraternals in these schools and the reason why they were not reported.

b. Excluding some twin pairs leads to feelings on the part of some twins which might be phrased: "Why do we have to do it, if they don't?" Fortunately such negative feelings toward the testing session are rare, but it is better not to create an opportunity for such feelings. For a similar

Table 10-22
Number of Identical Like-sexed Fraternal Twins and the Number of Boy-Girl Pairs Tested in 1961-66

	Total	DZ	MZ	Boy-Girl
1961	139	27	67	47
1962	153	30	76	47
1963	212	43	89	80
1964	327	87	135	105
1965	307	90	111	106
1966	300+	?	?	?

reason twins whose partners are ill or otherwise absent are encouraged to participate by themselves. In some cases the missing twin later showed up on his own, or could be persuaded to come to the twin testing in another school. The absence of negative feelings is not an accident, but is due to our special efforts to maintain interest, which are discussed below.

c. The data on boy-girl pairs allows us to study sex differences when age, socioeconomic status and many other variables are perfectly matched and to compare these with other sex differences reported for the variables under study. Occasionally the boy-girl pairs can be used in correlation studies.

Longitudinal Analysis. Because many of the twins participate year after year, it will be possible to analyze certain aspects of the data longitudinally. This will require a common format for all IBM cards punched. We are hoping to accomplish this in the near future. This is of special importance because an interlocking program of replication of twin studies and the use of different tests with the same twin samples will be necessary before a clear delineation can be made of the areas in cognition and personality where heredity plays the clearest role. Some recent results are reported below, but should be replicated before too much weight is given to the conclusions drawn.

Selected Results

1. *Hereditary Factors in Ability Measures.* Replicated findings with the PMA and DAT have been presented elsewhere, including multivariate analyses of twin differences (Vandenberg, 1965a, 1965b; 1966a, 1966b). Since that time two studies have been conducted in which the search was broadened. In one of these studies, 20 tests selected from Thurstone's first psychometric monograph were used. The F ratios for these 20 tests are shown in Table 10-23. A warning must be issued here—we discovered some scoring errors and all this is being carefully rechecked before the precise values can be definitely accepted. In the other study the emphasis was on spatial ability and number ability.

Some months later, Travis Osborne of the University of Georgia administered the same test battery to twins in Georgia, and we are cooperating on the analysis of these tests. To avoid errors some of the tests are also being rescored. This study will allow answers to the following questions: 1. Are various spatial ability tests measuring the same thing? 2. What is the influence of differences in format on the hereditary component of a test? 3. Is the hereditary component in these different tests the same?

TABLE 10-23
F RATIOS BETWEEN FRATERNAL AND IDENTICAL WITHIN-PAIR VARIANCES ON 20 TESTS FROM THURSTONE'S FIRST PSYCHOMETRIC MONOGRAPH

Number and name of test		F	Number of pairs DZ	MZ
PMA 5	Reading II	1.854**	61	80
PMA 11	Verbal Completion	2.171**	63	91
PMA 58	Vocabulary	1.878**	70	96
PMA 41	Verbal Analogies	1.292	68	94
PMA 6	Verbal Classification	1.334	70	95
PMA 55	Sound Grouping	1.827**	65	88
PMA 40	Reasoning	1.276	69	95
PMA 43	Code Words	1.480*	67	87
PMA 24	Punched Holes	1.093	69	94
PMA 20	Flags	1.136	70	96
PMA 21	Form Board	1.833**	70	96
PMA 18	Cubes	1.627*	70	91
PMA 39	Arithmetic Reasoning	1.829**	60	90
PMA 31	Addition	1.790**	70	94
PMA 33	Multiplication	1.505*	70	96
PMA 30	Number Code	1.766**	66	89
PMA 46	Word Number Memory	.835	66	87
PMA 50	Figure Recognition	.867	69	97
PMA 26	Identical Forms	1.329	70	94

** $p < .01$ * $p < .05$

The following tests were used:

1) Object Aperture A, B (Dubois and Gleser, 1948); 2) Newcastle Spatial Test, six parts (Macfarlane Smith, 1954, 1960); 3) Cube Comparison I & II; 4) Surface Development I & II; 5) Form Board I & II; 6) Paper Folding I & II; 7) Card Rotation I & II (all from ETS kit, French *et al.,* 1963).

Other tests administered at that time were:

8) Calendar Test (Remolino, 1962); 9) How well do you know yourself? (Jenkins, 1959, 1961, 1962); 10) Mazes (from the laboratory manual by MacKinnon and Henle, 1948); 11) Self-judging Vocabulary Test (Heim, Povey and Watts, 1965); 12) Identical Pictures (ETS kit); 13) Bourdon-Wiersma Cancellation Test (Kamphuis, 1963; Vander Ven, 1964, 1965); 14) Spelling (Metropolitan Achievement Test); 15) Social Perception Test (Whiteman, 1954); 16) Arithmetic Test (seven parts of decreasing complexity) (Mukherjee, 1965); 17) Faces Closure Test (Mooney, 1957); 18) Draw-A-Man; 19) Ship Destination (Christensen and Guilford, 1955); and 20) Logical reasoning (Hertzka and Guilford, 1955).

Osborne has reported on the results from his small sample (Osborne and Gregor, 1966, Osborne, Gregor and Miele, 1967).

2. *Divergent Thinking.* A small beginning has been made by our group in the investigation of hereditary factors in the area of divergent thinking, as defined by some of Guilford's tests. The results shown in Table 10-24 are based on a small number of twins and should be taken as preliminary only. Unfortunately the scoring of this type of test is time consuming; replication on a larger sample will be a laborious task. Nevertheless, the recent interest in creativity and originality makes such an undertaking challenging. These preliminary results suggest that environmental factors rather than hereditary components contribute the major variances. A replication which would permit simultaneous study of the within-sibship, parental, school, and other environmental factors associated with a high performance on such tasks would make such a study a less risky investment, but would also require a larger one.

3. *Cognitive Styles.* A number of test procedures designed by Dr. Gardner of the Menninger Foundation have been administered individually during the last few summers. These procedures take considerable time to score and in addition require some special training. The assistant in charge of these studies, Mr. Maurice LeCroy, left us for the Army, which has delayed the completion of this analysis. The following tests are individually administered:

1) Color Form Movie (Thurstone, 1952); 2) WAIS;

Cognitive Style Variables—3) Object Sorting; 4) Photo Sorting; 5) Behavior Sorting; 6) Schematizing; 7) Rod and Frame;

Motor Skills Tests—8) Beam Balancing; 9) Card Sorting; 10) Rotary Pursuit;

TABLE 10-24

F RATIOS BETWEEN FRATERNAL AND IDENTICAL WITHIN-PAIR VARIANCES FOR NINE OF GUILFORD'S TESTS OF DIVERGENT THINKING FOR 24 LIKE-SEXED DZ AND 67 MZ TWIN PAIRS

Name of test	F
1. Pertinent Questions	1.85*
2. Different Uses	1.53
3. Social Institutions	1.39
4. Seeing Deficiencies	1.35
5. Making a Plan	1.11
6. Similar Words	1.10
7. Associations	1.08
8. Figure Production	1.03
9. Picture Arrangements	.94

* $p < .05$

Test of Lateral Dominance—11) Pulfrich Pendulum.

4. *Social Intelligence.* In view of the implication of heredity in schizophrenia, two measures of social intelligence from the behavioral content section of Guilford's cube model of factors in intelligence, have been administered to twins as well as two forms of another test of social sensitivity developed by Vandenberg and Mattson (1961). The latter test was administered in two successive years. The results are shown in Table 10-25. They indicate no significant hereditary component in these measures. It is possible that this result, which we did not expect, is due to unreliability of the measures, or perhaps the measures were inappropriate for this age range. We hope to analyze these results further. Intercorrelations between these measures are not yet calculated, nor has the reason for the discrepancy between the results for form F and form M of the Faces test been investigated.

TABLE 10-25
F RATIOS BETWEEN FRATERNAL AND IDENTICAL WITHIN-PAIR VARIANCES FOR FOUR MEASURES OF SOCIAL INTELLIGENCE

Name of test	F	Number of pairs DZ	Number of pairs MZ
Faces (Guilford)	.96	63	78
Expressions (Guilford)	1.18	61	71
Faces F (Vandenberg-Mattson)	1.60*	43	71 (in 1961)
	1.53	29	60 (in 1962)
Faces M (Vandenberg-Mattson)	.61	44	70 (in 1961)
	.90	29	60 (in 1962)

*$p < .05$

5. *Learning.* In the spring of 1966 we administered several learning tasks. Some of these were chosen from the tasks described by Duncanson (1964) to whom we are indebted for a loan of 100 teaching machines and samples of the other materials. The tasks chosen are:

1) Word-Number, a paired associates task of 10 trials of 8 pairs; 2) Word-Nonsense Word, a paired associates task of 10 trials of 8 pairs (these two tasks are presented by teaching machines); 3) Memorizing Words, a serial learning task, 10 trials of 10 words presented each twin in the same order by tape recorder; 4) Figure Concepts, a concept formation task with 10 sets of 10 stimuli.

To these we added: 5) a motoric serial learning task constructed by making up a booklet with 10 sections in different colors, each containing 5 pages with the same trail finding test consisting of numbers from 1 to 25 scattered over the page. Subjects were given 50 seconds for each trial in which they could go from page to page drawing lines between the numbers until told to stop and to go to the first page of a different color. In this way we obtained a score on each of 10 trials.

Other measures used are: 6) a test of the understanding of three passages from the STEP test 3A recorded at the American Printing House for the Blind by professional readers at 375 words per minute under the supervision of Emerson Foulke of the University of Louisville for use in his studies of compressed speech; 7) and 8) the Elithorn maze test, with and without an indication of the maximum possible score for each maze (Elithorn, 1964) ; 9) a block design test; 10) and 11) two arithmetic tests; 12—14) three parts of Carroll's language aptitude test; 15) and 16) two perceptual speed tests; and 17) and 18) new forms of the Faces tests in which every wrong alternative is matched once with every right one. In addition we administered 19) a questionnaire concerning motion sickness, and 20) and 21) a modification of a questionnaire by Schaefer (1965) aimed at getting a child's perception of father and mother. It will take many months to score these tests, so we are only raising expectations at this time.

Curves are to be fitted to the learning performances and the parameters of these curves will be treated as the variables to be studied for concordance. In addition, we hope to study the relation between learning and the ability measures obtained in earlier years which are available for many of the twins who participated this year. We will welcome, as usual, any offers of help in analyzing these results.

6. *Personality, Attitudes, and Interests.* The results of a twin study of the Myers-Briggs Type Indicator have been mentioned before, but we have not described them in detail because the number of cases does not seem adequate for firm conclusions. Nevertheless, the results are highly suggestive. Table 10–26 summarizes our findings. It appears from these preliminary results that only the introversion-extroversion dimension has a significant hereditary component. The Myers-Briggs Type Indicator has

TABLE 10-26
F RATIOS FOR FRATERNAL AND IDENTICAL WITHIN-PAIR VARIANCES OF THE FOUR SCALES OF THE MYERS-BRIGGS TYPE INDICATOR FOR 27 LIKE-SEXED DZ AND 40 MZ TWIN PAIRS

Name of scale	F
Extroversion-Introversion	1.84*
Sensing-Intuition	.70
Thinking-Feeling	.80
Judgment-Perception	.76

* $p < .05$

been described by Myers (1960), Saunders (1960), and Stricker and Ross (1963, 1964 a, 1964 b). The suggestion by Saunders (1960) of a rational correspondence between the typologies of Spranger and of Jung should be further investigated by using the Allport-Vernon-Lindsey study of Values Scale and the Myers-Briggs Type Indicator in the same twin study.

Scales developed by Comrey (1966) to measure personality and attitude variables were administered in 1965. The results of the comparison of fraternal and identical within pair variances are shown in Table 10-27.

These results provide further evidence for an hereditary component in shyness, or its polar opposite sociability, which may be related to schizophrenia. The high F value for achievement need is a new finding which may reflect differences in ability which would explain the present findings as a result of the well established fact of hereditary factors in ability.

In 1964 we administered the Minnesota Vocational Interest Inventory (MVII) which was developed by Clark (1958, 1961) to measure interests in non-professional jobs. The F ratios between the fraternal and identical within-pair variances for the scores on this test are shown in Table 10-28.

We administered this test to determine whether interests in jobs which do not require extensive science training would show as much evidence of hereditary factors as do some of the scales of the Strong Vocational In-

TABLE 10-27

F RATIOS BETWEEN FRATERNAL (DZ) AND IDENTICAL (MZ) WITHIN-PAIR VARIANCES OF 12 SCORES ON THE COMREY PERSONALITY AND ATTITUDE FACTOR SCALES

Name of scale		Boys	Girls	All cases
Empathy		1.46	1.18	1.28
Neuroticism		.77	1.32	1.23
Welfare State Attitude		1.71	1.04	1.27
Achievement Need		1.91*	2.10**	2.20**
Dependence		1.25	1.15	1.15
Compulsion		1.49	1.50*	1.50*
Self-control		1.24	1.24	1.28
Religious Attitudes		.84	1.86**	1.49*
Hostility		.93	.72	.82
Punitive Attitudes		.89	1.61*	1.27
Shyness		2.83**	1.51*	1.94**
Ascendance		.50	.89	.80
Number of pairs	DZ	27	63	90
	MZ	52	59	111

* $p < .05$
** $p < .01$

Table 10-28
Comparison by F Test of Fraternal and Identical Within-pair Variances on 30 Scores on the Minnesota Vocational Interest Inventory

Score	Girls 34 DZ 42 MZ	Boys 19 DZ 29 MZ	All Cases 53 DZ 71 MZ
Baker	1.17	2.68**	1.79**
Carpenter	1.38	1.95*	1.58*
Electrician	1.41	1.39	1.33
Food Service Man	1.13	1.52	1.27
Hospital Attendant	2.34**	2.46*	2.44**
Industrial Education Teacher	.94	1.03	.98
IBM Operator	1.98*	2.69**	2.25**
Milk Wagon Driver	2.20**	1.35	1.62*
Machinist	1.25	1.45	1.31
Painter	1.28	.67	.97
Plasterer	1.13	1.62	1.31
Plumber	.94	1.00	.93
Pressman	1.43	1.36	1.42
Printer	1.17	2.16*	1.51
Retail Sales Clerk	3.79**	1.99*	2.29**
Radio & TV Repairman	1.05	1.85	1.37
Shipping & Stock Clerk	.85	1.42	1.07
Sheet Metal Worker	1.40	1.11	1.24
Truck Driver	1.31	2.74**	1.82**
Truck Mechanic	1.65	1.58	1.53*
Warehouseman	1.24	3.31**	1.94**
Factor Score			
H1 Machine Repairs	3.83**	2.50*	2.50**
H2 Medical Hospital Service	1.57	.99	1.35
H3 Office Work, Accounting	1.71*	.79	1.40
H4 Radio etc.	1.61	1.23	1.29
H5 Food Preparation & Menu Planning	1.11	1.19	1.14
H6 Carpentry & Furniture Making	1.12	.92	1.00
H7 Verbal Activity, Aesthetic	1.01	.66	.81
H8 Clean Hands	1.41	.90	1.11
H9 Athletics, Outdoor, Masculine	.77	1.09	.90

* $p < .05$
** $P < .01$

terest Blank. A summary of findings by Carter (1932) and Vandenberg (1964) with the SVIB are shown in Table 10-29. It will be of interest to remove variance due to ability from the vocational interests to see if hereditary components are still significant. In the absence of such definitive information it does appear that ability factors are not the sole explanation of hereditary components in vocational interests, because the results with the MVII show as high a proportion of significant F ratios as do the results on the SVIB.

TABLE 10-29
"AVERAGED" HOLZINGER'S H^2 FOR FOUR TYPES OF SCALES OF THE STRONG VOCATIONAL BLANK BASED ON AVERAGED INTRACLASS CORRELATIONS (AFTER Z TRANSPORTATION)

	Carter (1932)	Vandenberg (1964)
8 Science scales	.42	.39
4 Language scales	.20	.10
4 Dealing with people scales	.41	.20
5 Business scales	.24	.26
	43 DZ	34 DZ
	43 MZ	43 MZ

Only one of the factor scores, H1, interest in mechanical things and machine repairs, showed a significant hereditary component. It may be that a score which is a derived one, rather than a direct one, misses some of the precision in measuring an individual's interests which may, in fact, be idiosyncratic. Some of this may be highly specific variance as opposed to the more general variance measured by the factor score. But it could be precisely this specific variance which is in part of genetic origin.

MEASURES TO MAINTAIN INTEREST IN THE HIGH SCHOOL AGE TWIN STUDIES

As soon as the scoring of the tests is completed, usually in the fall after the spring testing, reports are sent to each school listing the scores on all tests for the students in that school. When necessary a brief description of the tests is included.

Individual reports are also prepared for the parents of the twins. These consist of forms on which no numerical information is used, such as a percentile score or raw score. Instead the score is reported as below average, average, above average, or superior. In assigning these labels, the student is generally given the benefit of the doubt.

Frequently parents call us after receiving these reports. This offers an opportunity to discuss the results in more detail. In some instances parents have called to obtain advice on specific problems. These are generally referred elsewhere after sufficient discussion to prevent a feeling on the part of the parents that we are not interested in helping them.

One of the most effective ways of maintaining interest in the study has proved to be provision for enough variation in the measures used each year to make the twins look forward to participating another year. This may at times conflict with the efficient design of the battery. We have found that up to two-thirds of the measures can be rather difficult tests, as long as they are interspersed from time to time with interesting, enjoyable or easier tasks.

POSSIBLE FUTURE RESEARCH

Before Vandenberg and Stafford decided to leave the Louisville Twin Study several possibilities were being considered for future research. Some of these may still be carried out in Louisville under Wilson's direction, while others may be undertaken by Vandenberg in Colorado. A few of these ideas do not require the collection of new information, but only a more extensive analysis of data on hand.

Analysis of Sex Differences. We hope to compare for each test administered the sex differences of the boy-girl twin pairs with the sex differences found for the other twins and with sex differences reported elsewhere. This will help to throw some light on the effect of the twin situation on the abilities and personalities of twins. In view of the special nature that is often attributed to this relationship it will be useful to see which traits are more influenced by having a brother or sister of the same age.

Longitudinal Analysis of the High School Data. Because a number of twins participated in more than one testing program it will be possible to obtain correlations for a number of them between scores on tests administered 1, 2, 3, or even 4 years apart.

Follow-up of High School Graduates. In the future a follow-up program may be attempted to locate twins who participated in one or more of the high school testing programs, to see what their occupational, marital, and mental health status is. Because the twins form a relatively unselected group with respect to these variables, valuable information may be obtained by comparing the test scores of contrasting groups such as married and single individuals, those with steady employment records with those unemployed, etc.

Use of School Records. In a similar manner, it may be worthwhile to obtain information about high school grades and to correlate this with the test scores. However, this may require more effort than the outcome would warrant, because of differences in standards of grading between schools and between teachers in the same school.

Experimental Nursery Program. In the summer of 1966 we organized an experimental nursery for six pairs of identical twins. One twin from each pair was given some training with pre-reading games, while the other received experience with games oriented around number concepts. Tests were administered at the beginning and after eight weeks of the program. Unfortunately, we had to discontinue this program at the end of the summer. At some future time it will be worthwhile to repeat this attempt to produce differences in ability patterns, in order to see how lasting the effect of differential training will be.

Attempts to Improve Ability Test Scores at the High School Age Level. Some thought was given to the possibility of giving intensive training to selected high school age students, not necessarily twins, to see whether it is possible to get lasting improvement in ability test scores, especially in spatial ability or reasoning. For this purpose a start was made toward the collection of moveable models or film strips for use in spatial visualization training and in reasoning exercises for use in efforts to improve scores on abstract, verbal, numerical or mechanical reasoning tests.

Observation of Free Play Behavior. Finally, modest pilot efforts were made to collect data on the development of sex differences by observing boy-girl twin pairs during visits at ages 3, 6, 9, 12 and 18 months. This was discontinued because no suitable space in which to observe free play activity was available.

The Need for Cooperation. The expenses for a carefully considered and executed twin study are such that individual investigators can only make small contributions. Often such studies will be isolated and perhaps ignored unless they fit into some over-all program consisting of an interlocking system of replications as well as new studies. The demands on the subjects' time, the cost of blood typing, the need for knowledge about the best measures available, and the time necessary for test scoring, coding, statistical analyses, and the availability of suitable computer facilities are all limiting factors. Much can be gained by sharing any or all of these, in spite of the obvious difficulties involved. We have been exceptionally fortunate by having friends and associates who have been generous in providing some of these kinds of help. Data collected in any twin study may warrant storage in an archive for further analysis in the future. Much of the blood typing information on the fraternal twins, for instance, may be of considerable interest for future linkage studies. Replication of findings in this country by investigators in other parts of the world and vice versa could be especially valuable because such findings could simultaneously contribute to knowledge of population genetics and of the cross cultural congruence of frequencies, correlations and patterns of relationships. Because genes know no racial or geographical boundaries, such cross cultural studies would lead to clearer understanding of the way in which cultural, i.e., historical and personal, influences modify the expression of genetic influences. Even within the United States considerable variation may be expected in the importance of genetic components in various psychological variables as determined by concordance studies of twins from different ethnic or cultural groups.

ACKNOWLEDGMENTS

We are deeply grateful for the continued help and cooperation of the twins and their parents, which has made our work possible. Financial support for the Louisville Twin Study has been provided by grants from the Institutes of Health of the U. S. Public Health Service: RG 5527, M 6203, MH 07033, K3-MH-18, 382, HD 00843, and by NSF grant GB 466. We have received help with our calculations from the following computer centers and individuals: University of California, Los Angeles (Andrew L. Comrey), University of Chicago (R. Darrell Bock), University of Georgia (R. Travis Osborne), University of Kentucky (John W. Donahue and Silvio Navarro), University of Lousiville (Alfred T. Chen), University of Michigan (James C. Lingoes), University of North Carolina (R. Darrell Bock), University of Rochester, N.Y. (Kenneth E. Clark), University of Texas (John C. Loehlin), and the Western Data Processing Center (Harvey F. Dingman and Curtis Miller).

For their help, especially in the early planning of the high school study, we are grateful to Mr. Richard VanHoose, Superintendent, Mr. James E. Farmer, Assistant Superintendent, and Mr. O. L. Shields, Director of Psychological Testing, of the Jefferson County Public Schools; to Mr. Samuel V. Noe, Superintendent, Mr. Eddie Beecher, Assistant Superintendent, and Mr. Ben X. Freeman, Director of Pupil Personnel, of the Louisville Public Schools; to the Very Rev. Msgr. Alfred W. Steinhauser, Director of the Parochial Schools, and Brother Edward Daniel, C. F. X., Principal of St. Xavier High School. In addition, we are indebted to the superintendents of schools and principals in Hanover, New Albany, Clarksville, and Jeffersonville, Indiana, and several other counties in Kentucky and southern Indiana, as well as of the parochial schools.

REFERENCES

Christensen, P. R. and Guilford, J. P. 1955. *Ship Destination Test.* Los Angeles: Sheridan Supply Co.

Clark, K. E. 1958. *Minnesota Vocational Interest Inventory.* New York: Psychological Corporation.

——. 1961. *The vocational interests of non-professional men.* Minneapolis: University of Minnesota Press.

Comrey, A. L. 1966. Comparison of personality and attitude variables. *Educ. Psychol. Meas.* **26:** 853-60.

Dubois, P. H. and Gleser, Goldine. 1948. The object-aperture test; a measure involving visualization in three dimensions. *Amer. Psychol.* **3:** 363.

Duncanson, J. P. 1964. *Intelligence and the ability to learn. Research Bulletin* 64-29. Princeton, N.J.: Educational Testing Service.

Elithorn, A. 1964. Subjective difficulty as a function of complexity; a logical approach to the design of tests of intellectual functions. Copenhagen: Proc. Internat. Copenhagen Congress on the Scientific Study of Mental Retardation.

French, J. W.; Ekstrom, R. B.; and Price, L. A. 1963. *Manual for a kit of reference tests for cognitive factors.* Princeton, N.J.: Educational Testing Service.

Gottesman, I. I. 1963. Heritability of personality: A demonstration. *Psychol. Monogr.* **77:** No. 9 (Whole No. 572).

Guttman, L. 1957. Simple proofs of relations between the communality problem and multiple correlation. *Psychometrika* **22:** 147-57.

Heim, Alice W.; Povey, R. N.; and Watts, K. P. 1965. An attempt to measure some aspects of temperament by means of the word-in-context and self-judging vocabulary tests. *J. Gen. Psychol.* **72:** 285-94.

Hertzka, A. F. and Guilford, J. P. 1955. *Logical Reasoning.* Los Angeles: Sheridan Supply Co.

Hurley, J. R. and Cattell, R. B. 1962. The Procrustes program: producing direct rotation to test a hypothesized factor structure. *Behav. Sci.* **7:** 258-82.

Jenkins, T. N.; Coleman, J. H.; and Fagin, H. T. 1959. *How well do you know yourself?* Manual of Instructions. New York: Executive Analysis Corporation.

Jenkins, T. N. 1961. *How well do you know yourself?* Secondary school edition. New York: Executive Analysis Corporation.

———. 1961. The second order components of human personality. *J. Psychol. Studies* **12:** 237-60.

———. 1962. Efficiency of the Jenkins global personality inventory. *J. Psychol. Studies* **13:** 11-20.

Kaiser, H. F. 1958. The Varimax criterion for analytic rotation in factor analysis. *Psychometrika* **23:** 187-200.

Kamphuis, G. H. 1963. Een onderzoek naar de reliability van de Bourdon-Wiersma (A study of the reliability of the Bourdon-Wiersma test). *Nederl. Tijdschr. Psychol.* **17:** 269-75.

MacKinnon, D. W. and Henle, M. 1948. *Experimental Studies in Psychodynamics.* Cambridge: Harvard University Press.

Meyers, C. E.; Dingman, H. F.; Orpet, R. E.; and Attwell, A. A. 1962. Primary Abilities at Mental Age Six. *Monogr. Soc Res. Child Develpm.* **27,** 1, serial number 82.

———; ———; and ———. 1964. Four ability-factor hypotheses at three preliterate levels in normal and retarded children. *Monogr. Soc. Res. Child Develpm.* **29,** 4, serial number 96.

Mooney, C. M. 1957. Age in the development of closure ability in children. *Canad. J. Psychol.* **11:** 219-26.

Mukherjee, B. N. 1965. Derivation of likelihood ratio tests for Guttman quasi-simplex covariance structure. University of North Carolina doctoral dissertation.

Myers, Isabelle Briggs, 1962. *The Myers-Briggs Type Indicator.* Princeton, N.J.: Educational Testing Service.

Newman, H. H.; Freeman, F. N.; and Holzinger, K. T. 1937. *Twins; a study of heredity and environment.* Chicago: University of Chicago Press.

Osborne, R. T. and Gregor, A. J. 1966. The heritability of visualization, perceptual speed and spatial orientation. *Percept. Mot. Skills.* **23:** 379-90.

——; ——; and Miele, F. 1967. The heritability of numerical facility. *Percept. Mot. Skills.* **24:** 659-66.

Reiss, A. J., Jr., *et al.* 1961. *Occupations and Social Status.* Glencoe, Ill.: The Free Press.

Remondino, C. 1962. Recherche sur la signification du facteur numérique (An investigation into the psychological nature of the number factor). *Rev. Psychol. Appliquée.* **12:** 62-81.

Saunders, D. R. 1960. Empirical evidence for a rational correspondence between the personality typologies of Spranger and of Jung. *Amer. Psychol.* **15:** 459.

Schaefer, E. S. 1965. Children's reports of parental behavior; an inventory. *Child Develpm.* **36:** 413-24.

Shotwell, Anna M.; Dingman, H. F.; and Tarjan, G. 1956. A number test for mental defectives. *Amer. J. Ment. Def.* **60**: 589-94.

Smith, I. Macfarlane and Lawes, J. S. 1960. *Spatial test 3 (Newcastle Spatial Test).* London: National Foundation for Educational Research in England and Wales.

Smith, I. Macfarlane. 1964. *Spatial ability: its educational and social significance.* San Diego: Robert R. Knapp.

Stricker, L. J. and Ross, J. 1963. Intercorrelations and reliability of the Myers-Briggs Type Indicator scales. *Psychol. Reports.* **12:** 287-93.

——. 1964a. An assessment of some structural properties of the Jungian personality typology. *J. Abnorm. Soc. Psychol.* **68:** 62-71.

——. 1964b. Some correlates of a Jungian personality inventory. *Psychol. Reports.* **14:** 623-43.

Thurstone, L. L. 1952. *Progress Report on a Color-form Test.* Chicago: Report No. 80 from the Psychometric Laboratory, University of Chicago.

Thurstone, Thelma G.; Thurstone, L. L.; and Strandskov, H. H. 1955. *A Psychological Study of Twins.* Chapel Hill, N.C.: Report No. 4 from the Psychological Laboratory, University of North Carolina.

Vandenberg, S. G. and Mattson, E. 1961. The interpretation of facial expressions by schizophrenics, other mental patients, normal adults and children. *Acta Psychol.* **19:** 495-96.

Vandenberg, S. G. 1962. The hereditary abilities study: hereditary components in a psychological test battery. *Amer. J. Hum. Genet.* **14:** 220-37.

Vandenberg, S. G. and Falkner, F. T. 1965. Hereditary factors in physical growth. *Hum. Biol.* **37:** 357-65.

Vandenberg, S. G. 1965a. Innate abilities, one or many? A new method and some results. *Acta Genet. Med. Gemell.* **14:** 41-7.

———. 1965b. Multivariate analysis in human genetics. In *Methods and Goals in Human Behavior Genetics,* ed. S. G. Vandenberg, pp. 29-40. New York: Academic Press.

———. 1966. Contributions of twin research to psychology. *Psychol. Bull.* **66:** 327-52.

———. 1967. Hereditary factors in psychological variables in man with special emphasis on cognition. In *Genetic Diversity and Human Behavior,* ed. J. N. Spuhler, pp. 99-133. Chicago: Aldine.

Vander Ven, A. H. G. S. 1964. Een curve analysis van de Bourdon-Wiersma test (A curve analysis of the Bourdon-Wiersma test). *Nederl. Tijdschr. Psychol.* **19:** 300-08.

———. 1965. De qualitatieve gegevens van de Bourdon-Wiersma test (Qualitative data in the Bourdon-Wiersma test). *Nederl. Tijdschr. Psychol.* **20:** 219-27.

Whiteman, M. 1954. The performance of schizophrenics in social concepts. *J. Abnorm. Soc. Psychol.* **49:** 266-71.

Wechsler, D. 1952. *The Range of Human Capacities,* second ed. Baltimore: Williams & Wilkins.

11

SANDRA SCARR
Graduate School of Education
University of Pennsylvania

ENVIRONMENTAL BIAS IN TWIN STUDIES

The twin study method is useful in attempting to estimate genetic contributions to variation in human behavior by comparing intrapair differences of monozygotic (MZ) and same-sex dizygotic (DZ) pairs. The comparisons of intraclass correlations and resulting estimates of heritability are based upon the perhaps questionable assumption that the environments of DZ co-twins are no more dissimilar than the environments of MZ co-twins. Since excess MZ correlations are interpreted as genetic in origin, additional DZ differences created by greater environmental variance would lower DZ intraclass correlations and hence would bias the results in favor of high genetic estimates.

Several investigators have already shown that MZ co-twins are indeed more similar in the treatment they receive from their parents, in the development of mutually interdependent roles (Jones, 1955), choice of friends, sports attendance, and some food preferences (Smith, 1965). Data to be presented in this paper support the contention that the home environments of MZ co-twins are in general more similar than those of DZ co-twins, but raise questions about the assumption that this is necessarily *prima facie* evidence for environmental bias in twin studies.

BEHAVIORAL SIMILARITIES OF CO-TWINS

Smith (1965) interviewed 164 pairs of adolescent twins about work, school, sports, leisure, sleep, dress and study habits, and food and bever-

---

The author gratefully acknowledges the guidance of Irving I. Gottesman, who directed the dissertation from which this paper is derived. The research was supported in part by a U. S. Public Health Service Fellowship (MH-16841) and by a generous grant from the Laboratory of Social Relations, Harvard University.

The sample of twins was selected from those being studied by Dr. Coenraad F. A. Moorrees and his associates. The longitudinal studies of twins and their age-matched siblings conducted at the Forsyth Dental Center, Boston, Massachusetts, under the direction of Dr. Moorrees, are supported by USPHS Research Grant DE-01309 from the National Institute of Dental Research.

Mrs. Elizabeth Nadeau was the invaluable assistant without whose efforts the research would not have been completed.

age preferences. He found that, in general, MZ co-twins were more similar than DZ co-twins, with the results more positive for females than for males. For questions about the aforementioned habits, two of the eight MZ correlations were significantly greater than DZ correlations. MZ co-twins were more likely to have the same friends and the same patterns of attendance at sports events. Habits of work, school, sports participation, sleep, and study did not significantly differentiate between MZ and DZ pairs. For household chores, food and beverage preferences, MZ females tended to be more similar than DZ females, but the effect was not as strong for males. Adolescent MZ co-twins were more likely to perceive no differences between themselves in habits and activities, and the MZ females were more likely to dress alike. For males, there were no differences in dressing alike for MZ and DZ pairs. Smith concluded that these results cast doubt upon the validity of the assumption of equal environmental variances for MZ and DZ twins.

In a study of 61 pairs of grade school age twin girls in the Boston area, MZ co-twins were also found to be somewhat more similar than DZ pairs in their behaviors and in the parental treatment they received (Scarr, 1964; 1966a; 1966b). Unlike Smith's (1965) sample, there were no differences in socioeconomic status or educational level of the parents of MZ and DZ twins. The modal parents of both kinds of twins were high school graduates in lower white collar occupations, such as policemen, firemen, salesmen, office workers, and so forth. The twin groups showed no differences in IQ (total group mean = 100.4) and age (total group mean = 95 months).

Both the MZ and DZ co-twins were likely to prefer similar kinds of activities, but the MZ co-twins were more similar in the number of activities they engaged in. This finding was interpreted as variation in activity motivation, for which genetic contributions were found, not only in ratings but behavioral measures as well (Scarr, 1966a). The behavioral similarities and differences of MZ and DZ co-twins doubtless result from both genetic and environmental factors, but are MZ co-twins more similar mainly because of environmental pressures for similarity, as Smith (1965) implies? Or are the greater similarities of MZ co-twins principally a reflection of greater genetic similarity, resulting in phenotypic similarity, behavioral and otherwise? From these data it is impossible to separate the sources of behavioral variations.

PARENTAL BEHAVIOR

Many of the arguments presented by critics of the twin study method have focused on the differences in parental treatment of identical and fraternal twins. Since MZ twins are supposed to be alike, parents may emphasize their similarities; and since DZ twins are not supposed to be alike, parents may concentrate on differentiating them. The measurement of parental behavior toward their children is a hazardous but worthy goal (Shaefer and Bell, 1958). The differences of treatment received within the family by MZ and DZ co-twins are difficult to measure from an "objective" point of view; and in the brief time we spent in the homes of our sample, objective measures were impossible to obtain. It was possible, however, to obtain a subjective evaluation of parental behavior by interviewing the mothers and by having them rate the attitudes and expectations they held for their twins.

The mothers were interviewed about their twins' present and past behaviors. They were asked, "How similar or different do you feel (Twin A) and (Twin B) are?" Their answers were coded on a scale from 1 (very different) to 5 (very similar). The Vineland Social Maturity Scale (Doll, 1947) was completed during the interviews, to measure the amount of responsibility and independence the mothers believed each twin could accept. The mothers were also asked to recall the twins' early development and behavior "problems," from which scales of co-twin similarity were constructed. The anamnestic data on the twins' early years are not necessarily accurate but rather reflect the mothers' selective recall of similarities and differences in their twins. The results of the mothers' interviews are given in Table 11-1.

The data are shown as percentages of pairs reported to be similar for each measure in order to make the scores comparable. The original calculations were either comparisons of intraclass correlations or Chi Square tests of the MZ and DZ distributions, all of which showed significant differences at or beyond the .05 level of probability, one-tailed, except early development.

It is abundantly clear that the mothers of MZ twins believe them to be

TABLE 11-1
PERCENTAGES OF MZ AND DZ PAIRS RATED AS SIMILAR FOR SEVERAL CHARACTERISTICS

	Percentage of Pairs Similar	
Mother's Ratings	MZ (N = 23)	DZ (N = 29)
Mothers Say Similar Now	78%	17%
Mothers Expect Similar Social Maturity	91%	62%
Dressed Alike	74%	48%
Similar Early Behavior Problems	83%	59%
Similar Early Development	78%	59%

more similar than the mothers of DZ pairs believe theirs to be, both at the present time and in the past. Of the 52 pairs for whom complete data and blood-grouping were available, the MZ pairs were more likely to be considered generally similar at the time of the study, and their mothers were more likely to expect the same levels of social responsibility and independence from them. MZ co-twins were also more likely to be dressed alike (or to choose to dress alike).

The mothers of DZ twins recalled more differences between their children at an early age. If one DZ twin had a behavior "problem" whether eating, sleeping, thumb sucking, toilet training, or other, the co-twin might or might not have had a similar "problem." But if one MZ twin was said to have had a "problem," her sister was very likely said to have shown the same behavior. The early development of most MZ co-twins was recalled by their mothers as somewhat more similar than the early development of DZ co-twins. In general, the mothers of MZ twins believed that their children had been similar and continued to be similar, while the mothers of DZ twins noticed more differences between their children.

The mothers were also asked to complete the Adjective Check List (Gough, 1960) separately for each twin. This instrument contains 300 adjectives which comprise 26 personality scales. Of the 20 scales which describe the twins, the MZ co-twins were rated as significantly similar on 11 scales, and the DZ pairs on five, as shown in Table 11-2. Three of the scales have significantly higher MZ than DZ intraclass correlations: *n* affiliation, *n* change, and counseling readiness, a measure that Gough calls "available anxiety."

The mothers of MZ twins again perceived their twins as relatively more similar than the mothers of DZ twins perceived their children, significantly so on measures of sociability (*n* affiliation), flexibility (*n* change), and anxiety. The results from the mothers' interviews, the Vineland Social Maturity Scale, and the Adjective Check List indicated that DZ co-twins were perceived and treated more differently by their mothers than MZ twins.

THE ASSUMPTION OF EQUAL ENVIRONMENTAL VARIANCES

Even if most investigators now agree that MZ co-twins experience generally more similar environments than DZ co-twins, does this imply that we have to abandon the twin study method? Not yet. For, the real problem with the assumption of equal environmental variances for MZ and DZ co-twins is that when parents are correct about their twins' zygosity, two important factors are confounded: (1) the greater genetic differences

TABLE 11-2
INTRACLASS CORRELATIONS OF MOTHER'S RATINGS FOR MZ AND DZ PAIRS ON THE ADJECTIVE CHECK LIST

	Intraclass Coefficients			
ACL Scales	MZ (N = 23)	*p*	DZ (N = 29)	*p*
Self-confidence	.24		.12	
Self-control	.51	∠ .01	.31	
Lability	.53	∠ .01	.40	.05
Personal Adjustment	.57	∠ .01	.40	.05
n Achievement	.17		−.05	
n Dominance	−.04		−.16	
n Endurance	.10		.11	
n Order	.29		.05	
n Intraception	.42	∠ .05	.34	
n Nurturance	.55	∠ .01	.50	∠ .01
n Affiliation	.83	∠ .001	.56	∠ .01
n Heterosexuality	.57	∠ .01	.54	∠ .01
n Exhibition	.39	.05	.09	
n Autonomy	.40	.05	.11	
n Aggression	.35		−.08	
n Change	.70	∠ .001	−.12	
n Succorance	−.16		−.02	
n Abasement	.00		.00	
n Deference	.19		.02	
Counseling Readiness (available anxiety)	.56	∠ .01	.03	

of DZ co-twins, with accompanying physical, intellectual, and behavioral differences; and (2) the greater differences of parental treatment of DZ pairs, which might create additional intrapair dissimilarities.

If parents are simply reacting to the existing differences between their DZ twins' behavior, then no bias is introduced into twin studies. But, if they effectively train differences, then these environmentally determined differences would bias the comparisons of intraclass correlations in favor of genetic hypotheses, by reducing the possible similarities of DZ co-twins. By the same token, the parents of MZ twins who know their twins are identical may react to existing similarities or seek to train greater similarities than would otherwise exist. When parents are correct about their twins' zygosity, it is impossible to distinguish between parental behavior that is a reaction to the phenotypic behavior of their twins and parental treatment that seeks to train greater differences or similarities.

A METHOD FOR ESTIMATING ENVIRONMENTAL BIAS

Not all parents of twins are correct about their twins' zygosity, however, and these parents offer a critical test of environmental bias in twin studies. By examining the cases of parents who are *wrong* about their twins' zygosity, it is possible to separate parental reactions to similarities and differences based on *genetic relatedness* from parental behaviors which arise from their *belief* that their twins should or should not be similar.

When parents are asked, "Are your twins identicals or fraternals?" a surprising number either do not know or are wrong about their twins' zygosity. Smith (1965) reported the following percentages of misclassification by parents, using blood grouping as the criterion diagnosis:

MZ males:	15% misclassified	(6 pairs)
MZ females:	12% misclassified	(6 pairs)
DZ males:	20.6% misclassified	(7 pairs)
DZ females:	35% misclassified	(14 pairs)

Results of the Boston female twin sample are similar: 17.4 per cent of the MZ pairs were believed by their mothers to be DZ; and 31.2 per cent of the DZ pairs were believed to be MZ. The total number of incorrect diagnoses is small, because twin samples are generally small, but the twelve error cases provide a critical test for environmental variance as a bias in twin studies.

A comparison of the behaviors and ratings of parents who were wrong about their twins' zygosity with those who were correct will yield results in one of the two following directions:

1. MZ pairs, misclassified as DZ, will be treated like correctly diagnosed MZ pairs; and DZ pairs, thought to be MZ, will be raised like correctly identified DZ pairs. From these results, we would conclude that the degree of *genetic relatedness* of the twins is a more important determinant of similar or different parental treatment than the parents' *belief* that their children should or should not be similar.
2. MZ pairs, misclassified as DZ, will be treated like correctly diagnosed DZ pairs; and vice versa for DZ pairs misclassified as MZ. From these results we would conclude that the parents' *beliefs* about zygosity determine the similarities and differences in their behavior toward their twins. If parental beliefs are important determinants of environmental similarity for co-twins, then differences are probably accentuated between presumed and real DZ co-twins and minimized for presumed and real MZ co-twins, thereby introducing environmental bias into genetic estimates from twin studies.

An approximate method of estimating the amount of environmental bias is the *direction* of results as predicted by the alternate hypotheses. With a large twin sample, it would be possible to calculate the degree to which results for correctly identified pairs deviate from those of misclassified pairs, and to correct for environmental variance in genetic estimates.

Unfortunately, these data will not permit such a refinement which would be appropriate to several hundred twin pairs.

RESULTS AND DISCUSSION

To test these hypotheses, the same measures of similarity and differences reported for the whole sample were calculated separately for the misclassified and correctly classified pairs. The results for four misjudged MZ pairs and seven misclassified DZ pairs are given with the results for the correctly classified pairs in Table 11-3.

TABLE 11-3
PERCENTAGE OF CORRECTLY AND INCORRECTLY CLASSIFIED PAIRS RATED AS SIMILAR FOR SEVERAL CHARACTERISTICS

	Percentage of Pairs Similar			
	Correctly Classified		Misclassified	
	MZ (N = 19)	DZ (N = 22)	MZ (N = 4)	DZ (N = 7)
Mothers Say Similar Now	79	9	75	43
Mothers Expect Similar Social Maturity	95	67	75	43
Dressed Alike	74	45	75	57
Similar Early Behavior Problems	79	59	100	57
Similar Early Development	79	54	50	71

The data generally confirm the first hypothesis: that genetic relatedness of the twins determines the similarity of parental treatment. Although the numbers are too small to yield statistical significance, the trends are clear. The mothers of MZ twins, whom they wrongly believe to be DZ, treat them more like correctly identified MZ twins. And the mothers of DZ twins, whom they believe to be MZ, treat them more like correctly classified DZ pairs. Despite the mothers' erroneous beliefs, the twins are recognized as having similarities and differences appropriate to their degree of genetic relatedness.

In the interview the mothers of MZ twins, wrongly believed to be DZ, tended to say that they are similar at the present time. The misclassified DZ pairs were said to be less similar despite their mothers' beliefs that they were identical twins. The Social Maturity scores more clearly reflect this trend: a larger percentage of the MZ pairs believed to be DZ were treated similarly by the mothers than the DZ pairs believed to be MZ.

The mothers expected independence and allotted responsibility similarly or differently according to their twins' actual zygosity. Actual DZ co-twins probably receive more differentiated parental treatment because they *are* different, not because the mothers believe they should be.

Dressing alike was more frequent among MZ females regardless of the mothers' diagnosis of zygosity. The proportions of MZ and DZ co-twins dressed the same was approximately the same in correctly and incorrectly diagnosed groups. Dressing alike does not seem to be a function of simply looking alike since the DZ twins who were mistaken for MZ did not dress alike as frequently as the MZ's who were mistaken for DZ.

Results from the mothers' recall of early "problems" indicated again that MZ twins are more similar than DZ twins, even when the mothers' beliefs were to the contrary. However, the mothers' recall of their twins' early development reversed the direction of previously reported findings. The mothers of DZ pairs, believed to be MZ, reported greater developmental similarity for their children than the mothers of MZ twins believed to be DZ. Perhaps developmental similarities and differences were an important basis for the parental diagnosis of zygosity when the twins were very young. The recall of early development was the only measure which reversed the direction of the findings.

The Adjective Check List scales corroborated previous results. Differences between misclassified DZ pairs were larger than those between misclassified MZ co-twins for the three scales which showed significantly greater DZ differences for the whole sample. The Vineland Social Maturity scores continue this trend, with DZ differences larger than MZ differences, regardless of the correctness of parental diagnosis. These results lend support to hypothesis 1, but this is not to say that no bias exists.

TABLE 11-4

MEAN CO-TWIN DIFFERENCE ON SIGNIFICANT ADJECTIVE CHECK LIST AND VINELAND SCALES FOR CORRECTLY AND INCORRECTLY CLASSIFIED GROUPS

	Mean Differences of Co-Twins			
	Correctly Classified		Misclassified	
Adjective Check List Scales	MZ (N=19)	DZ (N=22)	MZ (N=4)	DZ (N=7)
n affiliation	4.1	8.5	5.7	6.6
n change	5.6	15.8	7.3	9.7
Counseling Readiness (anxiety)	4.7	12.0	4.0	5.9
Vineland Social Maturity	0.4	1.1	1.3	1.8

Venturing farther out on the slim branch of small numbers, we might also note that DZ twins misclassified as MZ are treated more similarly than correctly classified DZ twins. Data from the Vineland Social Maturity Scale and the Adjective Check List suggest that beliefs about zygosity also have an effect on MZ pairs, whose reported differences are greater when they are misclassified as DZ. There is evidence for some bias toward minimizing differences between MZ pairs (and those DZ pairs believed to be MZ) and emphasizing differences between DZ pairs (and those MZ pairs believed to be DZ).

The comparisons of parental behavior for correctly and incorrectly classified pairs suggests, however, that environmental determinants of similarities and differences between MZ and DZ co-twins are not as potent as the critics charge. Differences in the parental treatment that twins receive are much more a function of the degree of their genetic relatedness than of parental beliefs about "identicalness" and "fraternalness."

The small numbers of twins reported in this paper limit the confidence that should be placed in the results, but the method of estimating environmental bias in twin studies may be useful to investigators with larger samples. Hopefully, we can answer some criticisms about differential parental treatment of MZ and DZ twins with a larger collection of misdiagnosed twins.

References

Doll, E. A. 1947. *Vineland Social Maturity Scale Manual.* Minneapolis: Educational Test Bureau.

Gough, H. G. 1960. The Adjective Check List as a personality assessment research technique. *Psychol. Rep.* **6:** 107-22.

Jones, H. E. 1955. Perceived differences among twins. *Eugen. Quart.* **5:** 98-102.

Scarr, Sandra. 1964. Genetics and Human Motivation. Unpublished Ph.D. Dissertation, Harvard University.

———. 1966a. Genetic factors in activity motivation. *Child Develpm.* **37:** 663-74.

———. 1966b. The Adjective Check List as a personality assessment technique with children: validity of the scales. *J. Consult. Psychol.* **30:** 122-28.

Schaefer, E. S. and Bell. R. Q. 1958. Development of the Parental Attitude Research Instrument. *Child Develpm.* **29:** 339-61.

Smith, R. T. 1965. A comparison of socioenvironmental factors in monozygotic and dizygotic twins, testing an assumption. In *Methods and Goals in Human Behavior Genetics,* ed. S. G. Vandenberg, pp. 45-62. New York: Academic Press.

12

STEVEN G. VANDENBERG and RONALD C. JOHNSON
The University of Colorado

FURTHER EVIDENCE ON THE RELATION BETWEEN AGE OF SEPARATION AND SIMILARITY IN IQ AMONG PAIRS OF SEPARATED IDENTICAL TWINS

Johnson (1963) argued that "if early environmental stimulation or deprivation has a significant effect on the measured intellectual ability of humans, then individuals who are genetically identical and who are exposed to a common early environment and thus also have shared the amount of stimulation that this environment offered, should resemble one another more closely in tested IQ than individuals who are genetically identical but who have not shared a common environment for any appreciable period of time." Johnson then obtained, from the previously published literature, 11 pairs separated prior to six months of age (median age of separation = one month) and 12 pairs separated at one year or later (median age of separation = eighteen months), and found that members of twin pairs in the late separation resembled one another significantly *less* closely than members of the twin pairs in the early separation group.

The publication recently of a Danish study of twins reared apart (Juel-Nielsen, 1962) provides an opportunity to check on this finding. Juel-Nielsen studied 12 pairs of identical twins who had been separated early in life. These twins were found by exhaustive search of the twins born between 1870 and 1910 and registered at the Institute of Human Genetics in Copenhagen. Information from the twin registry was matched with data from the Danish census, the Folke-registry, to find all twins in this age range who had been reared apart. The zygocity of the twins was determined by a battery of blood group tests performed at the Institute of Human Genetics in Copenhagen. Along with many other observations, IQ's were obtained on these twins; a Danish version of the WAIS was used. Table 12-1 shows the ages of separation and the IQ differences (along with the original source of the data) for Juel-Nielsen's 12 pairs, for the 23 pairs in the Johnson study, and for two pairs (Gates and Brash, 1941; Stephens and Thompson, 1943) that Johnson missed,

TABLE 12-1

AGE AT SEPARATION, SOURCE OF DATA, AND DIFFERENCES IN I.Q. FOR 37 PAIRS OF MZ TWINS FROM VARIOUS STUDIES. FOR ABBREVIATIONS SEE KEY AT BOTTOM OF TABLE.

1 day	(S & T)	4			
1 day	(J-N)	6	1 yr.	(J-N)	9
9 days	(B)	1	1 yr.	(J-N)	14
½ mo.	(M)	4	1 yr.	(NFH)	19
3 wks.	(J-N)	1	1 yr.	(NFH)	5
3 wks.	(J-N)	1	1 yr.	(NFH)	1
1 mo.	(S)	4	14 mos.	(NFH)	4
1 mo.	(G & N)	3	18 mos.	(NFH)	12
1 mo.	(NFH)	1	18 mos.	(NFH)	12
1 mo.	(NFH)	6	18 mos.	(NFH)	24
1 mo.	(NFH)	1	18 mos.	(NFH)	7
6 wks.	(J-N)	11	2 yrs.	(NFH)	10
2 mos.	(NFH)	2	2½ yrs.	(NFH)	2
3 mos.	(NFH)	15	3 yrs.	(NFH)	8
3 mos.	(G & B)	19			
5 mos.	(NFH)	17	3½ yrs.	(J-N)	8
6 mos.	(J-N)	1	3½ yrs.	(J-N)	6
7 mos.	(J-N)	4	5¾ yrs.	(J-N)	13
9 mos.	(J-N)	6	6 yrs.	(NFH)	9
10 mos.	(J-N)	3			

Key: S & T = Stephens & Thompson, 1943
J-N = Juel-Nielsen, 1964
B = Burks, 1942
M = Muller, 1925
S = Saudek, 1934
G & N = Gardner & Newman, 1940
NFH = Newman, Freeman & Holzinger, 1937
G & B = Gates & Brash, 1941

with pairs divided into those separated before as opposed to those separated at or after one year of age.

As was the case with those pairs discussed by Johnson (1963), the Juel-Nielsen pairs separated after one year differ from one another significantly more than do pairs in the early separation group. A sum of ranks test (Walker and Lev, 1953) yields a z of 2.19, $p = < .04$. For all pairs shown in Table 12-1, a sum of ranks test yields a z of 3.98 showing pairs in the late group to differ from one another significantly more than do pairs in the early separation group. The early separation group differs by an average of 5.50 IQ points; the late separation group by 9.59 points. The mean within pair difference for the entire 37 pairs is 7.64 points.

We also want to call attention to the results from the study by Shields (1962) of identical twins raised apart, and diagnosed by extensive blood typing. Shields did not determine an IQ. He used Raven's Dominoes Intelligence Test and the Synonyms part of the Mill Hill Vocabulary Scale, and reported points of difference in the scores for 48 pairs, so that only differences between raw scores can be compared.

Table 12-2 shows these differences for three groups of MZ twins: 1. those separated at nine months of age and before; 2. those separated at one year of age or later; and 3. those separated at birth, but reunited at ages varying between five and twelve years, plus one pair separated at 9 months and reunited at twelve years. The means and S.D. of test differences for these three groups are 8.81 and 6.89 (N = 19), 12.10 and 9.76 (N = 12), and 7.29 and 9.40 (N = 7). These means are not significantly different from one another. (A split between twins separated before or

TABLE 12-2
AGE AT SEPARATION AND DIFFERENCES IN SCORES ON THE DOMINOES TEST FOR IDENTICAL TWINS RAISED APART
(From Shields, 1962)

Age at separation	Differences in test scores	Age at separation	Differences in test scores
Birth	3	12 mos.	2
"	3	12 mos.	12
"	3	16 mos.	5
"	6	20 mos.	23
"	7	22 mos.	30
"	8	24 mos.	8
"	8	30 mos.	14
"	10	48 mos.	10
"	16	48 mos.	24
"	17	84 mos.	1
"	20	96 mos.	5
1½ mos.	22	108 mos.	1
3 mos.	0		
3 mos.	10		
3 mos.	23		
6 mos.	2		
6 mos.	5		
6 mos.	7		
9 mos.	1		

Age at separation	Age when reunited	Differences in test scores
Birth	5 yrs.	1
"	5 yrs.	4
"	9 yrs.	25
"	11 yrs.	6
"	12 yrs.	4
"	12 yrs.	6
9 mos.	12 yrs.	5

after one year did not give significance nor did a split before and after six months.)

Like Johnson's data, these data do not suggest that a longer period of common early environment produces a greater similarity in IQ. Rather, as long as actual IQ scores are used, the reverse appears to be true. It should be noted that the chief finding is that a longer common environment does not result in greater similarity between members of twin pairs, even though the negative obtained relation also appears to merit consideration and, if possible, explanation. One possible explanation has been advanced by Daniel G. Freedman (1966). He noted that the major difference in Table 12-1 really is between those separated at or before one month as opposed to after one month. He argued that twins placed in separate families prior to one month of age must have been more strong and vigorous than is common among twin pairs and, further, that they must both have been quite similar in this respect. Since it is common, in the case studies on separated identicals, to find very substantial differences in weight and in vigor between members of twin pairs, it may be that those twins separated at or prior to one month of age shared a much more similar intrauterine environment than twins in general—which, in turn, resulted in their being more similar in IQ than other twin pairs.

In short, two recent studies of separated identical twins allow a check of an earlier conclusion that members of early separated pairs of identical twins resemble one another more closely than do members of late separated pairs. One study, from which actual IQ scores are available (Juel-Nielsen), supports this earlier conclusion; the other study (Shields), from which raw score differences on two tests are available, shows no significant relation between age of separation and similarity in test scores.

References

Burks, B. S. 1962. A study of identical twins reared apart under differing types of family relationships. In *Studies in Personality,* ed. Q. McNemar and M. A. Merrill. New York: McGraw-Hill.

Freedman, D. G. 1966. Personal communication.

Gardner, I. C. and Newman, H. H. 1940. Mental and physical traits of identical twins reared apart, case XX. *J. Hered.* **31:** 119-26.

Gates, N. and Brash, H. 1941. An investigation of the physical and mental characteristics of a pair of like twins reared apart from infancy. *Ann. Eug.* **11:** 89-101.

Johnson, R. C. 1963. Similarity in I.Q. of separated identical twins as related to length of time spent in same environment. *Child Develpm.* **34:** 745-49.

Juel-Nielsen, N. 1964. Individual and environment. A psychiatric-psychological investigation of monozygotic twins reared apart. *Acta Psychiat. Scand.* 40 Suppl. **183:** 158-292.

Muller, H. J. 1925. Mental traits and heredity. *J. Hered.* **16:** 433-48.

Newman, H. H.; Freeman, F. N.; and Holzinger, K J. 1937 *Twins: a study of heredity and environment.* Chicago: University of Chicago Press.

Saudek, R. 1934. A British pair of identical twins reared apart. *Char. and Pers.* **3:** 17-39.

Shields, James. 1962. *Monozygotic twins brought up apart and brought up together.* London: Oxford University Press.

Stephens, F. E. and Thompson, R. B. 1943. The case of Millan and George. *J. Hered.* **34:** 108-14.

Walker, Helen M. and Lev, J. 1953. *Statistical inference.* Pp. 434-435. New York: Holt.

13

J. BRADFORD BLOCK
School of Medicine
University of Cincinnati

HEREDITARY COMPONENTS IN THE PERFORMANCE OF TWINS ON THE *WAIS*

Despite the interest in the relationship between heredity and environment, which has long been a controversial subject, there is a surprisingly small quantity of scientific literature on the subject. Although scientists have a large store of knowledge about hereditary factors in various abnormal conditions such as color blindness and hemophilia, almost nothing is known definitively about the inheritance of normal mental traits.

One reason for the long neglect of this area has been the lack of a suitable experimental model. The inheritance of intelligence in man has been studied largely by means of correlations between relatives. The correlation technique gives a quantitative value for the degree of resemblance between relatives, but it is inconclusive with respect to heredity-environment interpretation because the influence of assimilation cannot be removed. Thus, investigators have turned to studies of twins in an effort to eliminate the heredity-environment correlation which has been the source of much confusion in family studies.

By comparing the average within-pair differences of monozygous twins (MZ) with the average within-pair differences of dizygous twins (DZ), it is possible to obtain an estimate of the importance of heredity in the determination of intelligence.

Differences between MZ twins can only be due to nongenetic (i.e., environmental) factors, because MZ twins have identical genetic makeups. On the other hand, the within-pair differences of DZ twins are due to hereditary differences as well as environmental ones. Therefore, DZ twins should be more dissimilar than MZ twins on traits which are under genetic control. Thus, the discrepancy between the two sets of differences for any specific trait will constitute an indication of the degree of genetic control over that trait.

The use of twins in the study of intelligence dates back to Galton (1883). Before techniques of establishing zygosity became well standardized, most studies dealt with comparisons between like-sexed and unlike-sexed twins. Typical of these are the studies of Merriman (1924) and

TABLE 13-1
EXPERIMENTAL DESIGN SHOWING SAMPLE DISTRIBUTION

	Monozygous		Dizygous	
Age	Male	Female	Male	Female
13	5	5	5	5
14	5	5	5	5
15	5	5	5	5
16	5	5	5	5
17	5	5	5	5
18	5	5	5	5
Total	30	30	30	30

Lauterbach (1925). Both concluded that environment has little effect on the size of the correlations on standard intelligence tests.

One of the first studies to use MZ and DZ like-sexed twins was that of Tallman (1928). His early work was followed by that of Kramer and Lauterbach (1928), Wingfield and Sandiford (1928), Holzinger (1929), Stocks and Karn (1933), Hermann and Hogben (1933), Newman, Freeman, and Holzinger (1937), Thurstone, Thurstone, and Strandskov (1953), Cattell, Blewett, and Beloff (1955), Vandenberg (1962), and Vandenberg and McGinty (1964).

In spite of its general acceptance and common usage as a highly refined measure of intelligence, the Wechsler Adult Intelligence Scale (WAIS) has not been used in such studies. Perhaps that is because of its long administration time. It is the purpose of this study to investigate the hereditary influence, if any, on the eleven subtests, the Verbal Score, the Performance Score and the Total Score of the WAIS.

METHOD

A sample of 120 pairs of like-sexed twins was administered the WAIS. The sample consisted of 60 pairs of MZ twins and 60 pairs of DZ twins with an equal number of male and female pairs in each of six age groups from thirteen to eighteen. The design is shown in Table 13-1.

The twins were obtained from junior and senior, public and private high schools in the Louisville Metropolitan area.* They were contacted through their schools.

The zygosity of the twins was established by serological tests performed by Jane Swanson of the Minneapolis War Memorial Blood Bank. The following factors were tested for: A, B, O, M, N, S, s, P_1, P_2, Rho, rh′, rh″, Miltenberger, Verweyst, Lewis, Lutheran, Duffy, Kidd, Sutter, Martin, Kell, Cellano, and occasionally some others. Twins that differed on

* This study was supported by USPH grants HD 00843 and K3-MH-18,382 to S. G. Vandenberg and by a summer research scholarship from the University of Louisville Medical School to J. B. Block.

one or more of these serological tests were considered dizygous. It should be noted that any errors in diagnosis will lead to an underestimate of genetic control.

Each pair was tested on the same day, in the same room, and by the same examiner.

The raw scores of each individual subtest were converted to scaled scores which were used in determining within-pair differences. Thus the range for each subtest became zero to nineteen. It should be noted that the use of scale scores has the effect of minimizing raw score differences.

The method used to evaluate the significance of the hereditary component is based on one first used by Dahlberg (1926). The within-pair variance of the DZ twins σ^2wDZ is compared with the within-pair variance of the MZ twins σ^2wMZ and the ratio evaluated for statistical significance by the Fisher's F test; with N_{DZ} and N_{MZ} degrees of freedom,

$$F - \frac{\sigma^2_{wDZ}}{\sigma^2_{wMZ}}$$

where N is the number of pairs. To allow comparison with previous studies, Holzinger's h^2 measure of heritability was also calculated.

RESULTS

The within-pair variance of the DZ twins was significantly greater than the within-pair variance of the MZ twins on nine of the eleven subtests, as well as on the Verbal Scores, Performance Score and the Total Score. The results are shown in Table 13-2.

Three subtests, Information, Arithmetic, and Vocabulary, are significant at $p = .001$ level. Information and Vocabulary are two of the most reliable subtests with reliability coefficients of 0.91 and 0.94 respectively (Wechsler, 1955). They also have the highest intercorrelation of all subtests with an r of 0.83 between subtest and total score.

On the other hand, Object Assembly, one of the two subtests not reaching statistical significance, is the least reliable of the eleven subtests and also has the lowest intercorrelations with the other subtests. Picture Completion, the other non-significant subtest, missed significance at the $p = .05$ level by only 0.03 points.

This raises a question of considerable theoretical interest, whether the underlying hereditary contribution to each subtest may be regarded to be the same, or whether each subtest is under the control, at least in part, of separate hereditary mechanisms. Vandenberg (1964) has proposed a test for this, which might be considered a multivariate extension of the F ratio. It involves solution of the characteristic equation where the Cov_w are

$$| Cov_{wDZ} - \lambda \, Cov_{wDZ} | = 0$$

TABLE 13-2
F RATIOS FOR DZ AND MZ WITHIN-PAIR VARIANCES, LEVELS OF SIGNIFICANCE AND HOLZINGER'S H^2 VALUES FOR THE SCALE SCORES OF 60 PAIRS OF DZ AND 60 PAIRS OF MZ TWINS ON THE WAIS

Subtest	F	N_{DZ}	N_{MZ}	p	h^2
1. Information	3.88	60	60	.001	.74
2. Comprehension	2.25	60	60	.01	.55
3. Arithmetic	2.78	60	60	.001	.64
4. Similarities	1.81	60	60	.05	.45
5. Digit Span	1.53	60	60	.05	.35
6. Vocabulary	3.14	60	60	.001	.68
7. Digit Symbol	2.06	60	60	.01	.51
8. Picture Completion	1.50	60	60	NS	.33
9. Block Design	2.35	60	60	.01	.57
10. Picture Arrangement	1.74	60	60	.05	.43
11. Object Assembly	1.36	60	60	NS	.26
Verbal Score	3.38	60	60	.001	.70
Performance Score	3.41	60	60	.001	.71
Total Score	3.47	60	60	.001	.71

the within-pair covariances $\Sigma\Delta_i\Delta_j/N$ on tests i and j and the Δ's are the pair differences. The number of significant roots of this equation is interpreted as the number of independent hereditary components. Plans are being made to apply this to our data and the results will be reported in a future paper.

An F ratio was also computed separately for MZ vs. DZ male twins and for MZ and DZ female twins. The results are shown in Table 13-3.

The conclusion that the differences between MZ and DZ twins is determined by heredity is based on the assumption that differences in the environmental influences on MZ and DZ twins on an average are of equal importance. This assumption, however, is considered by some as being open to doubt. It has been pointed out that the social psychological structure assumes a different form for MZ than for DZ twins, Schulte (1928) and Poll (1930).

Stocks (1930) suggests that because DZ twins are often different in "general body build, healthiness, tastes, and temperament" they naturally tend to subject themselves, or to be subjected, to differences in environment to a greater degree than MZ twins. Hence the mean difference in any factor due to environment alone may be greater in DZ twins.

Kohn (1931), on the other hand, asserts that according to his experience differences determined by environment are greater in MZ than in DZ twins. In his opinion the urge to self-assertion and self-expression makes itself more clearly felt in MZ twins.

Bleuler (1932) and Wilson (1934) stress the fact that MZ twins, being

TABLE 13-3
F RATIOS FOR THE WITHIN-PAIR VARIANCES BASED ON THE SCALE SCORES OF 30 PAIRS OF DZ AND 30 PAIRS OF MZ MALE TWINS AND OF 30 PAIRS OF DZ AND 30 PAIRS OF MZ FEMALE TWINS AND LEVELS OF SIGNIFICANCE ON THE WAIS

Subtest	Males		Females	
	F	p	F	p
1. Information	4.38	.001	3.38	.01
2. Comprehension	1.57	NS	3.53	.001
3. Arithmetic	2.86	.01	2.44	.01
4. Similarities	1.71	NS	1.88	.05
5. Digit Span	1.59	NS	1.44	NS
6. Vocabulary	4.80	.001	2.50	.01
7. Digit Symbol	1.47	NS	2.88	.01
8. Picture Completion	1.31	NS	1.67	NS
9. Block Design	1.19	NS	4.20	.001
10. Picture Arrangement	1.05	NS	3.24	.01
11. Object Assembly	1.15	NS	2.06	.05
Verbal Score	3.29	.01	3.47	.001
Performance Score	2.26	.05	4.71	.001
Total Score	2.90	.01	4.06	.001

of identical heredity, are more likely to have similar social interests than DZ twins, and as a result seek and create for themselves an environment which is more uniform and similar than that acting on DZ twins.

Thus we find that the comparability of MZ and DZ twins may be disturbed by a number of sources of error, either diminishing or increasing the differences between MZ twins as compared with DZ twins. Each of these sources of error probably does not amount to much, and their interaction appears to minimize their effect. The almost universal finding that MZ twins' intrapair differences are significantly less than DZ differences on traits ranging from the number of dermal ridges to MMPI responses suggests that a similar mechanism (hereditary control) is at work. Actually when differences in genetic endowment lead to different choices of environment, it may well be that this interaction may properly be grouped with the hereditary portion of the variance.

This, however, does not mean to imply that such sources of error cannot act collectively in the same direction, resulting in an exaggeration of the likeness or difference within either MZ or DZ pairs. Perhaps MZ girl twins sometimes consciously or unconsciously strive at being alike to a greater extent than MZ boy twins, and perhaps DZ girl twins not infrequently consciously or unconsciously strive at being different. These two factors acting together, in the absence of balancing factors, might explain why for the girls, nine of the eleven subtests showed statistically significant hereditary components, while for the boys only three were statistically significant.

When these sex differences were tested for separately their effect was not seen (see Table 13-4). Only Block Design showed a significant difference (p = .05) for the DZ twins and only Comprehension (p = .05), and Picture Arrangement (p = .05) reached significance for the MZ twins. Thus, when tested separately the effect is not readily apparent, but when considered together their influence is quite noticeable.

Although part of the sample falls within the age range of the WAIS and part within the age range of the Weschler Intelligence Scale for Children (WISC), it was decided that the WAIS could justifiably be used for both populations. The fact that an equal number of both types of subjects were included in each sample, as well as the fact that within-pair differences on scale scores, rather than IQ's, were the units of measurement would tend to minimize the effect of using the WAIS only. The use of the WAIS only had the added advantage of eliminating the problem of equating two tests.

As a check, however, a test was run to determine what effect, if any, the use of the WAIS for the younger age range might have had. The sample was divided into two groups, ages thirteen through fifteen and ages sixteen through eighteen and four sets of within-pair variances computed (i.e. separately for MZ and DZ pairs within the two age ranges). Fisher's F test was then applied to test the homogeneity of the within-pair variances between the two age groups. The results are shown in Table 13-5.

The results did not reach statistical significance on any of the subtests for either group. The hypothesis that perhaps the use of the WAIS for the younger age range might result in exaggerated within-pair differences for that group is not supported by these findings.

TABLE 13-4

F RATIOS FOR THE WITHIN-PAIR VARIANCES BASED ON THE SCALE SCORES OF 30 PAIRS OF MALE AND 30 PAIRS OF FEMALE DZ TWINS, AND FOR 30 PAIRS OF MALE AND 30 PAIRS OF FEMALE MZ TWINS, AND LEVELS OF SIGNIFICANCE, ON THE WAIS

	Dizygous Twins		Monozygous Twins	
Subtest	F	p	F	p
1. Information	1.29	NS	1.00	NS
2. Comprehension	1.05	NS	2.15	.05
3. Arithmetic	1.53	NS	1.31	NS
4. Similarities	1.62	NS	1.47	NS
5. Digit Span	1.51	NS	1.37	NS
6. Vocabulary	1.20	NS	1.60	NS
7. Digit Symbol	1.64	NS	1.19	NS
8. Picture Completion	1.47	NS	1.15	NS
9. Block Design	2.03	.05	1.73	NS
10. Picture Arrangement	1.41	NS	2.18	.05
11. Object Assembly	1.64	NS	1.44	NS
Verbal Score	1.03	NS	1.09	NS
Performance Score	1.80	NS	1.15	NS
Total Score	1.37	NS	1.02	NS

TABLE 13-5
F RATIOS AND SIGNIFICANCE LEVELS FOR AGES 13-15 *vs.* 16-18 COMPUTED SEPARATELY FOR MZ AND DZ TWINS

Subtest	Monozygous Twins F	p	Dizygous Twins F	p
1. Information	1.24	NS	1.06	NS
2. Comprehension	1.56	NS	1.30	NS
3. Arithmetic	1.05	NS	1.67	NS
4. Similarities	1.04	NS	1.10	NS
5. Digit Span	1.20	NS	1.17	NS
6. Vocabulary	1.59	NS	1.50	NS
7. Digit Symbol	1.05	NS	1.24	NS
8. Picture Completion	1.00	NS	1.10	NS
9. Block Design	1.28	NS	1.29	NS
10. Picture Arrangement	1.34	NS	1.32	NS
11. Object Assembly	1.04	NS	1.48	NS
Verbal Score	1.40	NS	1.24	NS
Performance Score	1.20	NS	1.06	NS
Total Score	1.08	NS	1.12	NS

SUMMARY

A sample of 120 pairs of like-sexed twins was administered the WAIS. The sample consisted of 60 MZ and 60 DZ pairs with an equal number of male and female pairs in each of six age groups from thirteen to eighteen. Zygocity was established by serological tests.

Scale scores were used to determine within-pair differences. To evaluate the significance of the hereditary component or components of the individual subtests, the within-pair variance of the DZ twins was compared with the within-pair variance of the MZ twins and the ratio evaluated by Fisher's F and Holzinger's h^2 were also computed.

The within-pair variance of the DZ twins was significantly greater than the within-pair variance of the MZ twins on nine of the eleven subtests, as well as on the Verbal, Performance, and Total Scores.

REFERENCES

Bleuler, M. 1932. The delimitation of influences of environment and heredity on mental disposition. *Char. and Pers.* **1**: 286-300.

Cattell, R. B.; Blewett, D. B.; and Beloff, J. R. 1955. The inheritance of personality. *Amer. J. Hum. Genet.* **7**: 122-46.

Dahlberg, G. 1926. *Twin births and twins from a hereditary point of view.* Stockholm: Tidens Tryckeri.

Fuller, J. L. and Thompson, W. R. 1960. *Behavior Genetics.* New York: Wiley.

Galton, F. 1883. *Inquiry into Human Faculty.* London: Macmillan.

Guilford, J. P. 1956. *Fundamental Statistics in Psychology and Education.* New York: McGraw-Hill.

Hermann, L. and Hogben, L. 1933. The intellectual resemblance of twins. *Proc. Roy. Soc.* (Edinburgh) **53:** 105-29.

Holzinger, K. J. 1929. The relative effect of nature and nurture influences on twin differences. *J. Educ. Psychol.* **20:** 241-48.

Kohn, W. 1931. Vorfruchte aus einer psychologischen Reihenuntersuchung an Zwillingen, Geschwestern und nichtverwandten Schulkindern. *Archiv. für Rassen-und Gesellschaftsbiologie.* **25:** 62-73.

Kramer, E. and Lauterbach, C. E. 1928. Resemblances in the handwriting of twins and siblings. *J. Educ. Res.* **18:** 149-52.

Lauterbach, C. E. 1925. Studies in twin resemblance. *Genetics* **10:** 525-68.

Merriman, C. 1924. The intellectual resemblance of twins. *Psychol. Monogr.* **33** (5).

Newman, H. H.; Freeman, F. N.; and Holzinger, K. J. 1937. *Twins, A Study of Heredity and Environment.* Chicago: University of Chicago Press.

Ostlyngen, E. 1949. Possibilities and limitations of twin research as a means of solving problems of heredity and environment. *Acta Psychol.* **6:** 59-90.

Pearson, E. S. and Hartley, H. O. 1958. *Biometrika Tables for Statisticians.* Cambridge: Cambridge University Press.

Poll, H. 1930. Zwillinge in Dichtung und Wirklichkeit. *Zeitschrift für die gesamte Neurologie und Psychiatrie* **128:** 423-74.

Schulte, H. 1928. Psychiatrische Beitrage zur Zwillingssoziologie. *Psychiatrisch Neurologische Wochenschrift* **31.**

Stocks, P. 1930. A biometric investigation of twins and their brothers and sisters. *Ann. Eugen.* **4:** 49-107.

Stocks, P. and Karn, M. N. 1933. A biometric investigation of twins and their brothers and sisters. *Ann. Eugen.* **5:** 1-55.

Tallman, G. G. 1928. A comparative study of identical and non-identical twins with respect to intelligence resemblances. Pp. 83-86, in *27th Yearbook Nat. Soc. Stud. Educ.* (Pt. I). Bloomington, Ill.: Public School Pub. Co.

Thurstone, T. G.; Thurstone, L. L; and Strandskov, H H. 1953. *A Psychological study of twins.* I. *Distributions of absolute twin differences for identical and fraternal twins.* Rep. No. 4. Chapel Hill, N.C.: The Psychometric Laboratory, University of North Carolina.

Vandenberg, S. G. 1962. The hereditary abilities study: hereditary components in a psychological test battery. *Amer. J. Hum. Genet.* **14:** 220-37.

Vandenberg, S. G. and McGinty, P. 1964. *Hereditary Components in the Performance of Twins on the WISC.* Res. Rep. No. 4. Louisville, Ky.: Louisville Twin Study.

Wechsler, D. 1955. *Manual for WAIS.* New York: Psychological Corporation.

———. 1958. *The Measurement and Appraisal of Adult Intelligence.* 4th ed. Baltimore: Williams & Wilkins.

Wilson, P. T. 1934. A study of twins with special reference to heredity as a factor in the determination of differences in environment. *Hum. Biol.* **6:** 324-53.

Wingfield, A. H. and Sandiford, P. 1928. Twins and orphans. *J. Educ. Psychol.* **19:** 410-23.

PART III
Statistical Advances

Introduction

Since most of the multivariate statistical techniques proposed in this section are for data collected on twins, some problems with the twin method are considered here, rather than in the introduction to Part II which dealt with twin studies.

The comparison of identical and fraternal twin concordances or of twin differences has been used as a method in human genetics for about half a century without fundamental changes in methodology. Criticisms of the method have been made for almost as long. These can quickly be summarized. They are in increasing order of importance:

1. Assignment of twins into the two types, i.e. determination of twin zygocity is neither accurate nor objective, and at times may lead to circular reasoning.
2. The hereditary component in the fraternal twin differences consists not only of an additive component but includes also a genic component which is the effect of that combination of genes which is unique for each individual, and which is not transmitted to the individual's offspring. Estimation of the hereditary component of trait by the twin method does not permit the distinction between these two sources of hereditary variation.
3. The hereditary differences present in fraternal twins, but not in identical twins, are not the only source of the observable larger within-pair differences in fraternal compared with identical twins, but may be due in part to more differentiated parental treatment of fraternal twins.
4. Twins are different from single children; therefore no conclusions can be drawn which apply directly to the majority of man.

Before we consider these points briefly, it should be noted that the twin method is, in a sense, no more than a refined extension of common sense reasoning which is not just useful but inevitable as a first step toward genetic analysis in those areas of human variability which do not readily permit investigation by analysis of family pedigrees, or in which such studies would be prohibitively expensive.

Starting with the first point, let us now go over the criticisms briefly.

Point 1—The diagnosis whether a particular twin pair is fraternal or identical is, in principle, possible with complete objectivity and accuracy by mutual skin grafting between twins; but in practice 100 per cent accuracy is not necessary. What is required is a lack of bias such as can result from contamination of variables to be studied for hereditary effects by variables contributing to the zygocity diagnosis. The use of genetic markers, such as reactions to antisera, satisfy this demand. Addition of more and more variables such as, for instances, serum transferrins, urinary amino acids, fingerprint patterns, etc, can decrease (to as low a level as one desires and as one can justify financially) the probability that true fraternal pairs will not be recognized as such because they happen to be concordant on the tests used. Usually, however, it is not necessary to go further than a probability no greater than .05 for the joint concordance, if dizygous, in a given twin pair on the tests employed, to permit one to classify such a pair as identical. Each pair with any difference in one or more of the various tests is, of course, definitely fraternal. Inclusion of a few truly fraternal pairs in the group categorized as identical will (1) only tend to make it more difficult to find a statistically significant increase in the fraternal compared to the identical within-pair variance, and (2) will not affect noticeably calculations or conclusions unless we are dealing with a variable which can be measured exceedingly accurately. This is usually not the case with the phenomena investigated in twin studies.

Point 2—Separation of "genic" from "additive" hereditary components is indeed not possible in a twin study. Such an analysis requires the use of data on other sibs and on parents to allow an analysis to see whether or not certain phenomena are transmitted as unit entities, or whether they fall apart during segregation. There are no reasons, other than practical and economic ones, why such data cannot be collected in connection with a twin study. This is in effect what Cattell (1963, 1965) has suggested, although his proposed Multiple Abstract Variance Analysis is aimed primarily at traits under multifactoral inheritance.

Point 3—Besides hereditary differences, it seems likely that parental treatment and reaction of the twins to one another may produce within-pair differences in some or all pairs of twins. Whether fraternal twins are more subject to this than are identical twins is a moot question which is difficult to resolve. Arguments that intrauterine conditions affect fraternal twins less than identical twins have been summarized by Price (1950). I have discussed elsewhere (Vandenberg, 1968) the effects of parental treatment on twins and concluded that the evidence is not strong that there generally exists a systematic discrepancy which would lead to a greater role of parental influence in fraternal twin differences. This analysis was based on data obtained by Tienari (1966) and by Koch (1966) and on unpublished results from the Michigan and Louisville twin studies.

Point 4—That systematic differences of twins from single children have probably been exaggerated and may largely be limited to somewhat delayed language behavior in the preschool age range, with minor effects on later verbal ability, is another conclusion in the same paper by Vandenberg (1968).

Taking all these points into account, and remembering that groups studied will vary in genetic and cultural homogeneity, it may be wise to interpret published estimates of hereditary variance obtained by the twin method as tentative and to use them primarily as a means of rank-ordering variables with respect to the importance of hereditary components. In reading the papers that follow, this restriction should be kept in mind. An argument can be made that this restriction affects multivariate analyses considerably less than it does the analysis of single variables, because the restriction in genetic or cultural range is not likely to affect all variables equally or because parental treatment would be expected to have rather a general effect on all variables for a given twin pair, while such a general effect has not been found; but in principle the latter objection still holds, even if materially weakened—especially if such parental influences have been found in "child-development" type studies to work in opposite directions for a particular pair of variables.

In summary, it may be said that several criticisms of the twin method are probably based on exaggerated ideas of the uniqueness of being a twin, or of the difference producing effectiveness of parental treatment.

The statistical techniques for determining the significance of the increased fraternal within-pair differences have been considerably improved since the early studies of twins, and of late include methods to see whether the hereditary component in one variable is also present in one or more other variables. Most of these are presented in the next chapters.

In conclusion it seems that in spite of these advances, the basic assumptions of the twin method remain unchanged and even these refined methods cannot yield more than tentative indications of the importance of hereditary factors and where they may be most advantageously studied. Definitive genetic analyses will require data on sibs and parents as well.

References

Cattell, R. B. 1955. Research designs in psychological genetics with special reference to the multiple variance analysis method. *Amer. J. Hum. Genet.* **5**: 76-93.

———. 1965. Methodological and conceptual advances in evaluating hereditary and environmental influences and their interactions. In *Methods and Goals in Human Behavior Genetics,* ed. S. G. Vandenberg, pp. 95-130. New York: Academic Press.

Koch, Helen. 1966. *Twins and Twin Relations.* Chicago: University of Chicago Press.

Price, B. 1950. Primary biases in twin studies: a review of prenatal and natal difference producing factors in monozygotic pairs. *Amer. J. Hum. Genet.* **2**: 293-352.

Tienari, P. 1966. *On Intrapair Differences in Male Twins: With Special Reference to Dominance-Submissiveness.* Copenhagen: Munksgaard, ed. E. Nyman.

Vandenberg, S. G. 1968. In defense of twins and the twin method. In press. *Essays in honor of Professor Essen-Möller, Acta Psychiat. Scand. Suppl.*

14

R. DARRELL BOCK
The University of Chicago and
STEVEN G. VANDENBERG
The University of Colorado

COMPONENTS OF HERITABLE VARIATION IN MENTAL TEST SCORES

That a substantial component of individual differences in performance on composite tests of "intelligence" must be attributed to heritable factors now seems demonstrated beyond the point of any serious controversy (Erlenmeyer-Kimling and Jarvik, 1963). The studies of the resemblance of identical and fraternal twins reared together and apart (Newman, Freeman, and Holzinger, 1937), and that of adopted children and their natural and foster parents (Leahy, 1935), give especially convincing evidence that a purely environmental explanation of these individual differences is inadequate. A nondoctrinaire appraisal of the data now available leads one to the first principle of all genetic investigation: namely, that the variation of observable characteristics, whether physiological or behavioral, is the outcome of a lengthy sequence of interactions between the genetic material and the environment. In some characteristics the variation may be dominantly controlled by heritable genetic factors; in others, the variation contributed by nonheritable environmental factors may be more potent. In any case, the relative degree of genetic or environmental determination is a quantitative question which may be investigated empirically (by studies of familial resemblance, by twin studies, etc.) without prejudging the issue.

Information about the relative contribution of genetic and environmental factors has an important bearing on any behavioral study. It can and should influence the direction and methodology of the subsequent investigation of the behavior in question. If the variation is largely of heritable origin, the goals and methods of study will be those in the field of genetics. If nonheritable sources are dominant, conventional psychological investigations which ignore genetic factors may suffice. But in the great majority of cases, where both heritable and nonheritable variation are appreciable, procedures which take both sources into account will be necessary. The paradigm for this type of study will be the multifactor experiment in which the investigator controls one of these sources while observing the effects of the other. The present study is an example of the most familiar application of the paradigm—a comparison of monozygotic and dizygotic twin differences.

In this paper we employ data from the Louisville Twin Study* to assess the relative heritability in some cognitive, perceptual, and motor tasks. The data consists of scores on psychological tests similar to those found in omnibus intelligence tests. The most closely related previous work in this area is the Strandskov and Thurstone study (Strandskov, 1955) and Michigan Twin Study (Vandenberg, 1965). In particular, these studies included an analysis of heritability in scores on the Primary Mental Abilities Tests (Thurstone, 1941). They provided estimates of heritability ratios (dizygotic within-pair variance/monozygotic within-pair variance) for six of the primary mental abilities tests which were in agreement to the extent of showing significant ($p<.05$) heritable variation in the numerical, verbal, spatial, and word fluency tests, but not in the reasoning and memory tests. The negative results are not completely convincing, however, because like many commercially distributed tests, the PMA tests are short, highly speeded, and designed primarily for measuring individual differences among unrelated children. Considering how small a proportion of the variance between unrelated individuals is represented in dizygotic twin differences, the reliability of these tests may not be high enough to separate significantly the heritable variation from the measurement error. For any decisive comparison of heritabilities of different tests, we would prefer to use tests which are longer, have generous enough time limits to allow the subject to reach a stable level of responding, and have empirically demonstrated high reliability.

From this point of view, the third edition of the Differential Aptitude Tests (DAT) (Bennett, Seashore, and Wesman, 1959) seems especially well suited to the demands of a twin study. It has undergone an unusually thorough item development over a period of years. Subtests of the battery which showed unsatisfactory reliability in previous editions have been strengthened in the present edition. The number of items and time limits (quoted below in the description of the tests) are ample, and the reliability of certain of the tests is improved by the use of items with more than the conventional number (five) of multiple choice alternatives. The DAT is widely used in the vocational counseling of high school students, is considered sufficiently reliable for differential classifica-

* The data analyzed here were collected with support from USPHS grants HD 00843 and K 3-MH18382 to Dr. Vandenberg. The computations were supported by NSF grant GS-1025 to Dr. Bock. The authors wish to thank all the twins for their fine cooperation, as well as the various superintendents of schools, high-school principals, and counselors for their help and advice.

tion of individual students, and has been the subject of several reliability studies with large samples. The only possible objection to these tests for present purposes is that they are oriented toward practical abilities and are therefore more complex factorially than, say, the PMA tests. This does not pose a serious problem in the present context, however, for we have included in the analysis multivariate procedures which help identify independent factors in the heritable variation.

THE DIFFERENTIAL APTITUDE TESTS

The sample items from Form A of the DAT tests are shown in Figure 1. The *spatial test* (called "space relations" in the battery) is a variation of a test originally introduced by Thurstone (1938, pp. 36-37). In the present version, the subject is instructed to mark the space in the answer sheet corresponding to boxes which *can* be made from the pattern. More than one alternative may be correct; in the example, alternatives A, C and E are correct. The subjects were given 20 minutes to answer 40 items; some items are considerably more complex than that in the example.* Three cognitive operations appear to be required by the spatial test: 1) visual construction of the three-dimensional figure from the two-dimensional pattern; 2) holding the three-dimensional image in mind and matching it to the perspective drawings of the alternative objects; and 3) after locating a correct object, visualizing the rotation of the object in three-dimensional space and matching it with other objects.

The *numerical ability* test is not accurately represented by the sample items, which are merely exercises in simple mental arithmetic; many of the items in the body of the test require more advanced knowledge of topics such as: common and decimal fractions, placing the decimal point in multiplication and division, percentages, square root, cube root, ratios and proportions, factoring and cancellation, and the meaning of terms such as "list price," "discount" and "net price." The subjects had 20 minutes to complete 40 items.

The *abstract reasoning* test requires the subject to choose one of five alternatives which complete a series of four figures reading from left to right. In some of the series, two elements, such as shape and orientation, are varying at the same time. The changes of both elements must be extrapolated to identify the next member of the series. Although this test has no verbal content whatsoever, the publisher reports that it correlates about .60 with the verbal reasoning test. This degree of correlation is

* The standard time limits for the DAT tests were reduced approximately one-third in this study in order to allow other tests to be given at the same sitting.

1. Spatial Test

Example Y

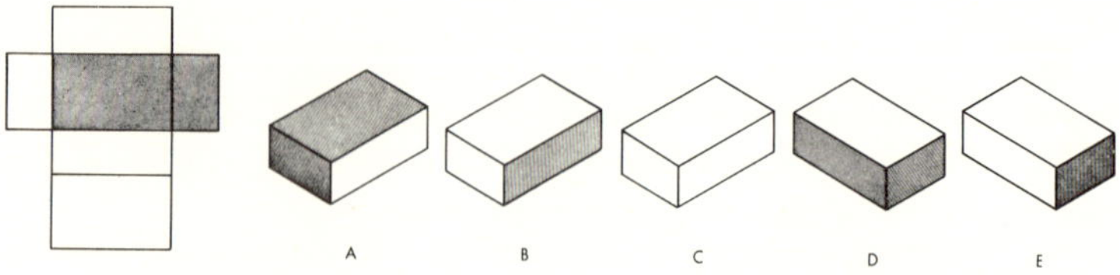

2. Numerical Test

EXAMPLE X

Add **13** A **14**
12 B **25**
— C **16**
D **59**
E **none of these**

In Example X, 25 is the correct answer, so the space under the letter for 25—B—has been filled in.

EXAMPLE Y

Subtract **30** A **15**
20 B **26**
— C **16**
D **8**
E **none of these**

In Example Y, the correct answer has not been given, so the space under the letter for "none of these"—E—has been blackened.

SAMPLE OF ANSWER SHEET

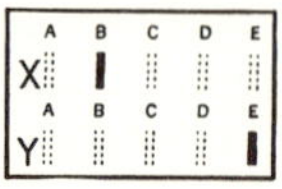

3. Abstract reasoning Test

PROBLEM FIGURES

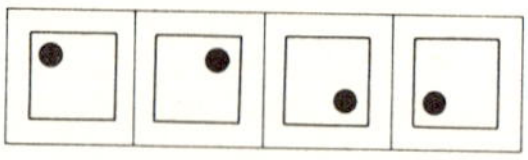

ANSWER FIGURES

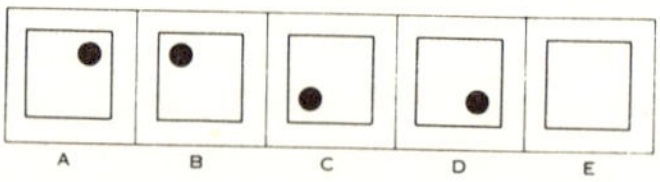

4. Verbal reasoning Test

EXAMPLE Z. .....is to night as breakfast is to.....

1. flow 2. gentle 3. supper 4. door
A. include B. morning C. enjoy D. corner

Sample items from the Differential Aptitude Tests 3rd Ed. (Bennett, G. K., Seashore, H. G. and Wesman, A. G.) New York: The Psychological Corp., 1959.

FIGURE 1

hard to understand since the verbal test, described below, requires an entirely different type of reasoning. We suspect that this correlation is the result of subject's transcoding the visual figures into verbal descriptions in order to induce the rule. Thus, in the sample item, one might proceed from left to right saying, "the dot is in the upper left, upper right, lower right, lower left, and, what comes next? . . . . . upper left!" Thus the test may depend upon subvocal use of language, but not necessarily the English language. Hence, the author's claim that this test is useful in assessing the ability of foreign language speaking students may be valid. Subjects were allowed 20 minutes to complete 50 of these items.

5. Mechanical reasoning Test

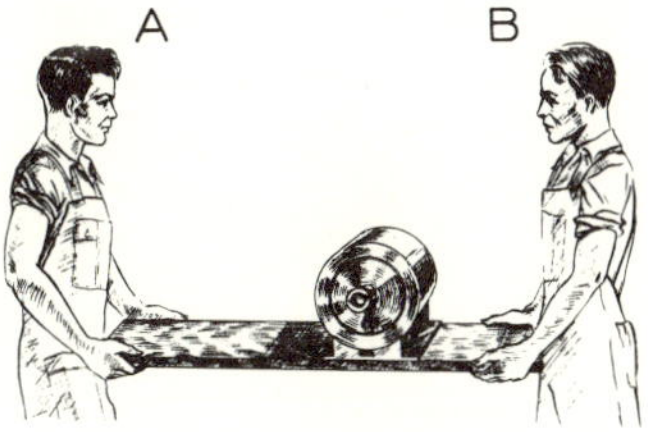

X

Which man has the heavier load?
(If equal, mark C.)

6. Clerical Speed Test

TEST ITEMS

V.	AB	AC	AD	AE	AF
W.	aA	aB	BA	Ba	Bb
X.	A7	7A	B7	7B	AB
Y.	Aa	Ba	bA	BA	bB
Z.	3A	3B	33	B3	BB

SAMPLE OF ANSWER SHEET

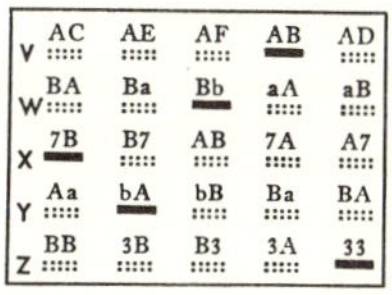

7. Spelling Test

EXAMPLES

W. man

X. gurl

Y. catt

Z. dog

SAMPLE OF ANSWER SHEET

8. Sentences (grammatical usage)

EXAMPLE

Ain't we / going to the / office / next week / at all.
A B C D E

SAMPLE OF ANSWER SHEET

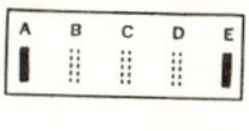

FIGURE 1 (Cont.)

The *verbal reasoning* test requires knowledge of synonyms as well as more factual knowledge. The items are in the form of analogies with two terms missing. The subject must supply these words from each of two lists of four words; thus each item has effectively sixteen alternatives. This text is a good predictor of academic achievement. Subjects had 20 minutes to complete 50 items.

The *mechanical reasoning* test uses a "true-false-other" format for response. In spite of its large number of items (68), this test is the least reliable in the battery (split-half reliability coefficient .85 for boys and .71 for girls). Many of the pictures in Form A seem somewhat ambiguous in relation to the question asked. This may contribute to unreliability. Most of the items require factual knowledge for a correct response, but some can perhaps be answered by visualizing mechanical motions, as in Thurstone's mechanical motions test. Subjects had 20 minutes to complete the 68 items.

The *clerical speed and accuracy* test is the simplest kind of perceptual-motor speed test. The subjects must find visually the underlined stimulus letter pair, compare it with the alternatives on the answer sheet, and mark the identical pair of letters. The subjects are given a three minute practice test (Part I) and a three minute test which is scored (Part II). The subject's task in this test probably involves manual dexterity as well as perceptual speed. It may be described as "eye-hand" coordination.

The *spelling* test is Part I of a "Language Usage" test. The subjects are required to identify printed words which are spelled incorrectly. This test is fairly highly correlated with Part II of the language usage test, which requires the *detection of grammatical errors* in printed material. Perceptual speed and reading speed as well as knowledge may be involved in both these tests. The subjects were allowed 8 minutes to respond to 100 words in the spelling test, and 20 minutes to respond to 50 sentences, each with five scorable sections, in the test of grammatical errors.

DESCRIPTION OF THE SUBJECTS

Data reported in this paper were obtained from 50 pairs of monozygotic and 25 pairs of dizygotic twin boys, and 56 pairs of monozygotic and 54 pairs of dizygotic twin girls. All subjects were normal students taken from grades 7 through 12 in schools in and around Louisville, Kentucky. The method of diagnosing zygosity, and other details of the Louisville Twin Study have been described by Vandenberg, Stafford, and Brown (see chapter 10).

ANALYSIS AND RESULTS

Part I: Univariate Analysis

In this part of the analysis we compute certain variance components for each of the eight tests separately. The symbolism for, and interpretation of, each of these components is as follows:

σ^2_F: the component of variance between twin pairs; i.e., the between-family component.

σ^2_H: the component of variance due to heritable effects in the test score, and of interaction of heritable and environmental effects.

σ^2_{EN}: the component of variance due to environmental effects.

σ^2_{ER}: the measurement error variance of the test.

With the exception of the measurement error variance (an estimate of which is supplied by the test publisher), we have estimated these components by the analysis of variance technique (see Graybill, 1961, Ch. 16). The form of the analysis is shown for dizygotic and monozygotic twins in Table 14-1. In this table, y_{M1i} represents the score of one twin of the i-th pair of monozygotic twins and y_{M2i} the score of the other twin of this pair. Which twin is designated 1 and which 2 is arbitrary and has no effect on the analysis. Similarly, y_{D1i} and y_{D2i} are the scores of twins in the i-th dizygotic pair. The total number of pairs is N_M and N_D respectively. The grand means for mono- and dizygotic twins are, respectively,

$$y_M = \Sigma_i^{N_M} (y_{M1i} + y_{M2i})/2N_M \quad \text{and}$$
$$y_D = \Sigma_i^{N_o} (y_{D1i} + y_{D2i})/2N_D \quad .$$

The formulas for estimating the variance components are derived by solving the equations for expected values of mean squares shown at the right in Table 14-1. The estimates, indicated by the "^", are expressed in terms of the sample mean squares as follows:

$$\hat{\sigma}^2_F = (ms_{BM} - ms_{WM})/2 \quad (1)$$
$$(\text{or } \hat{\sigma}^2_F = (ms_{BD} - ms_{WD})/2)$$
$$\hat{\sigma}^2_H = ms_{WD} - ms_{WM} \quad (2)$$
$$\hat{\sigma}^2_{EN} = ms_{WM} - \hat{\sigma}^2_{ER} \quad (3)$$
$$(\hat{\sigma}^2_{ER} \text{ estimated by the test publisher.})$$

Note that the measurement error variance cannot be estimated from the twin data, but is available in the test manual. Note also that the between-family component can be estimated from both the monozygotic and dizygotic scores. This suggests combining the estimates to obtain a single best estimate. We have not done this for the following reason: there is a consistent tendency in the data for the between-family estimate to be greater in the monozygotic data. We believe this occurs because at the extremes of the distribution of abilities more dizygotic than monozygotic twins are lost from the sample. At the lower end of the distribution there will be some point at which a subject will not have entered or will have dropped out of school. If the average ability of a monozygotic pair is above this point, then both twins are likely to be above and be found in school. If the average for a dizygotic pair is near this point, there is

TABLE 14-1
ANALYSIS OF VARIANCE OF TWIN DATA

Source of Variation	d.f.	Sum of Squares*	Mean Square	Expected Mean Square
Between MZ twin-pairs (families)	$N_M - 1$	$s_{BM} = \frac{1}{2} \sum_{i=1}^{N_M} (y_{M1i} + y_{M2i})^2 - 2N_M\bar{y}^2_M$	$m_{BM} = s_{BM}/(N_M - 1)$	$E(m_{BM}) = 2\sigma^2_F + \sigma^2_{EN} + \sigma^2_{ER}$
Within MZ twin-pairs	N_M	$s_{WM} = s_{TM} - s_{BM} = \frac{1}{2} \sum_{i=1}^{N_M} (y_{M1i} - y_{M2i})^2$	$m_{WM} = s_{WM}/N_M$	$E(m_{WM}) = \sigma^2_{EN} + \sigma^2_{ER}$
Total MZ	$2N_M - 1$	$s_{TM} = \sum_{i=1}^{N_M} (y_{M1i}^2 + y^2_{Mi}) - 2N_M\bar{y}^2_M$		
Between DZ twin-pairs (families)	$N_D - 1$	$s_{BD} = \frac{1}{2} \sum_{i=1}^{N_D} (y_{D1i} + y_{D2i})^2 - 2N_D\bar{y}^2_D$	$m_{BD} = s_{BD}/(N_D - 1)$	$E(m_{BD}) = 2\sigma^2_F + \sigma^2_H + \sigma^2_{EN} + \sigma^2_{ER}$
Within DZ twin-pairs	N_D	$s_{WD} = s_{TD} - s_{BD} = \frac{1}{2} \sum_{i=1}^{N_D} (y_{D1i} - y_{D2i}^2)^2$	$m_{WD} = s_{WD}/N_D$	$E(m_{WM}) = \sigma^2_H + \sigma^2_{EN} + \sigma^2_{ER}$
Total DZ	$2N_D - 1$	$s_{TD} = \sum_{i=1}^{N_D} (y^2_{D1i} + y^2_{D2i}) - 2N_D\bar{y}^2_D$		

* $\bar{y}_M = \sum_{i=1}^{N_M} (y_{M1i} + y_{M2i})/2N_M$; $\bar{y}_D = \sum_{i=1}^{N_D} (y_{D1i} + y_{D2i})/2N_D$.

greater probability that one twin will be below the point and will not be found in school. Since both twins must be in school in order to be included in the sample, the pair is lost. A similar effect may occur at high abilities if gifted children tend to be sent to private schools. We therefore consider the monozygotic data to give a more accurate estimate of between-family variation and have used ms_{BM} and ms_{WM} when estimating $\sigma_F{}^2$.

The same phenomenon may account for the widely different sample ratio of monozygotic to dizygotic male twin pairs as compared with the ratio for females. These ratios were 50/25 and 56/54 respectively. Aside from possibly reduced viability of male dizygotic twins, which may account for some of this discrepancy, it is also the case that in many communities the drop-out rate in high school is considerably higher for boys than for girls. Because of the greater variability of differences between dizygotic twins, and because the twin pair does not appear in the sample if data on either twin are missing, the school drop-out among the boys should remove more dizygotic than monozygotic male twins from the sample. The same reasoning should provoke a skeptical attitude toward any direct statistical comparison of boys and girls based on sample of high school students. Significant differences between the sexes may merely be the result of the differential drop-out rate.

Results of the Variance Components Analysis. The estimated variance components, calculated separately for males and females, are shown in Tables 14-2 and 14-3. The total variance, estimated by the sum of the separate variance components, is shown, and the variance components are expressed as a fraction of this total; i.e., in the form of intra-class correlation coefficients. At the right of the tables are shown the heritability ratios—ms_{WD}/ms_{WM}—and the probability under the hypothesis of no heritable variation (using the heritability ratio as an F statistic with N_D degrees of freedom in the numerator and N_M in the denominator; the dizygotic and monozygotic score differences are assumed to be normally and independently distributed).

On the basis of the heritability ratios in Tables 14-2 and 14-3, we may classify the eight DAT tests in three groups—those which give no evidence of heritable variation in either sex—those which show heritability in one sex but not the other—and those which clearly show heritability in both sexes. We will discuss these three groups separately.

1. The numerical reasoning test and the abstract reasoning test show no significant excess of dizygotic variation in these data. This does not appear to be the result of unreliability of the tests; the error variances are not larger in these tests than in some of the other tests which give evi-

TABLE 14-2
BOYS' DATA: VARIANCE COMPONENTS ANALYSIS
($N_M = 50, \quad N_D = 25$)

		Variance Components Relative to Total					
			Within Dizygotic Twin-Pairs			Heritability	
Test	Total Variance	Between[1] Families	Heritable	Environmental	Error[2]	Ratio	P
Spatial	618.9	.529	.354	.050	.066	4.05	.00001
Numerical	117.8	.788	.051	.094	.067	1.32	.20
Abstract	149.4	.626	.089	.209	.077	1.31	.21
Verbal	138.5	.799	.093	.051	.057	1.86	.031
Mechanical	171.2	.640	.000	.219	.140	.73	.80
Clerical	261.6	.631	.248	.069	.052	3.06	.0004
Spelling	1024.6	.658	.201	.097	.044	2.43	.004
Sentences	542.6	.737	.132	.083	.048	2.01	.018

[1] Estimated from monozygotic twin data.
[2] Measurement error estimates supplied by test publisher.

TABLE 14-3
GIRLS' DATA: VARIANCE COMPONENTS ANALYSIS
($N_M = 56, \quad N_D = 54$)

		Variance Components Relative to Total					
	Total	Between [1]	Within Dizygotic Twin-Pairs			Heritability	
Test	Variance	Families	Heritable	Environmental	Error[2]	Ratio	P
Spatial	455.0	.575	.138	.192	.096	1.48	.074
Numerical	83.0	.574	.113	.196	.116	1.36	.13
Abstract	148.7	.596	.024	.293	.087	1.06	.41
Verbal	116.8	.662	.212	.068	.058	2.68	.00017
Mechanical	165.9	.451	.297	.076	.176	2.18	.0024
Clerical	229.8	.526	.336	.072	.067	3.44	.000004
Spelling	982.7	.552	.280	.125	.043	2.67	.00018
Sentences	423.8	.682	.174	.078	.066	2.20	.0020

1 Estimated from monozygotic twin data.
2 Measurement error estimates supplied by test publisher.

dence of heritability. In the case of the numerical reasoning test, we would attribute this result to the rather narrowly specialized knowledge which the test requires. Since it is knowledge which is ordinarily acquired almost exclusively in the classroom, it will be held in common as much by dizygotic twins as by monozygotic, and no excess of dizygotic variation can be expected. If this interpretation is correct, we would expect to see a large component of between-family variation in this test (because differences in the effectiveness of schools in teaching this material will show up in this component). The data confirms this expectation; the numerical test has the second highest between-family component among boys and fourth highest among girls.

The abstract reasoning test shows a similar picture. Indeed, among girls, the error component is smaller and the between-family component larger than those of the numerical test. This result is harder to understand. Superficially, the test does not appear to require special knowledge. The task is novel and is not part of any instructional materials or other tests to which the subjects might have been exposed; reading ability is not required. We can only conjecture that success on this test requires some sort of sophistication about the meaning and use of symbolic conventions—in this case graphic symbols—and that this sophistication varies between families but tends to be common within families. This interpretation is consistent with the idea, expressed in our description of this test, that it is an indirect form of verbal test.

2. Two of the tests show differences in heritability between sexes. (This justifies our departure from the customary practice of pooling male and female data when analyzing twin data.) The heritability ratio for the mechanical reasoning test is clearly significant among girls, but is actually less than unity among boys. This is perhaps an indication that the test is not measuring the same trait in boys as in girls. For the boys it may be a test of knowledge which is acquired in experiences common to male members of a family—experiences like operating mechanical toys, building things out of wood, tinkering with automobiles, etc. This puts the variation between families rather than between siblings within families or, in particular, between dizygotic twins within families. The large between-family component for this test in the boys' data supports this interpretation. Among girls, on the other hand, experiential knowledge of mechanical principles is presumedly more limited. Thus girls must rely more on general skills and knowledge when responding to this test, and individual differences in the acquiring of these skills and this knowledge have more room to operate within families. Note that heritability is demonstrated among girls in spite of the relatively poor reliability of the

mechanical reasoning test. The error variance of this test is considerably larger than that of any of the other tests.

The spatial relations test shows a converse effect of sex. The heritability ratio for this test among boys is the largest anywhere in the data and is highly significant; the corresponding ratio for girls does not quite reach the .05 level. The explanation for this result is undoubtedly the well-known tendency for female subjects to perform poorly on tests which require the visualization of an object in three-dimensional space. On tests of this type, a large proportion of girls do not express the trait in sufficient strength to reveal any substantial degree of individual differences. Thus the evidence for heritability is only borderline. Among boys, on the other hand, the trait is fully expressed and shows purer heritable variation than any other test. Not only is the heritable variance component large, relative to the environmental and error components, but the between-family component is small. This means that the spatial visualization test is not especially sensitive to differences in social class, educational experience, etc., which are found between families. Thus, there is some justification for the widely accepted practice of including this type of item in tests which purport to measure innate general ability and intelligence. However, the present data suggest that at least in this age range, this practice is more relevant to boys than to girls.

3. The remaining four tests show clear evidence of heritable variation among both boys and girls. The results for the verbal reasoning test are especially interesting. As we would expect, this test shows the largest components of between-family variation. Since, in addition to their genetic similarity, members of the same family are actively sharing their vocabularies, they obviously tend to be similar in their knowledge of word meanings. This sharing must be especially intense between monozygotic twins, since they are known to interact more than dizygotic twins. The verbal environment of the monozygotic twins may therefore be more similar than it is for the dizygotic and, hence, lead to an overestimate of the heritable variance component. This is an example of an interaction between genetic and environmental variation: because monozygotic twins are more similar in ability and interests, they share more and develop more similar vocabularies. The genetic basis of this interaction may be of interest in its own right, but it has little bearing on the efficiency of acquisition of vocabulary, capacity of memory, probability of recall, or other cognitive factors which are of primary concern in this study.

The above remarks apply in large measure to the spelling and sentence tests, but must be qualified somewhat by the factor of clerical accuracy which appears to be present in these tests. The data indicate that skill in clerical speed and accuracy is highly heritable both for boys and for girls. It is hard to imagine this skill to be influenced by the more similar environments of monozygotic twins; thus, we would seem to be on safe ground in accepting it as heritable. The only question would be the relative degree of genetic determination of the perceptual as opposed to the motor component.

Part II: Multivariate Analysis

Methods. In the univariate analysis, we have detected heritable variation in five of the eight tests in both the male and female data. It does not necessarily follow, however, that we can detect five independent *dimensions* of variability in the heritable part of the eight test scores. We must allow the possibility that a smaller number of genetic sources of variation account for the major part of the heritability, and that any remaining sources are too weak to be distinguished from the environmental and error variation in these limited data. To pursue this question, we need a statistical test of the dimensionality of the heritable variation.

A procedure for this purpose based on a generalization of Fisherian discriminant analysis has been proposed by M. S. Bartlett (1951). It assumes that the components of variation in the data are multivariate normally distributed and may therefore be described in terms of their means and covariance matrices. Since the zero points of the scale of mental test scores are arbitrary, the means of the components may be set at a conventional value, say zero. Thus, only the component covariance matrices are of interest here. We will designate them as follows:

Σ_H: covariance matrix of the heritable components.

Σ_{EN}: covariance matrix of the environmental components.

Σ_{ER}: covariance matrix of the measurement errors.

In the present application these p x p matrices have eight rows and eight columns, each corresponding to one of the DAT tests. The hypothesis to be tested concerns the number of independent rows and columns, or "rank" of Σ_H. The sample quantities used for this test are the mean-product matrices designated M_{WM} and M_{WD} within monozygotic and dizygotic twin pairs, respectively. These matrices may be obtained in a multivariate analysis of variance analogous to the univariate analysis of Table 14-1. More straight-forwardly, however, their elements may be calculated from the between-twin difference for pairs of variables. For scores of tests j and k of the i-th monozygotic twin pair, let these differences be

$$d_{Mi}^{(j)} = y_{M1i}^{(j)} - y_{M2i}^{(j)}$$
$$d_{Mi}^{(k)} = y_{M1i}^{(k)} - y_{M2i}^{(k)} \qquad (4)$$

Let $d_{Di}^{(j)}$ and $d_{Di}^{(k)}$ be similar quantities for the i-th dizygotic twin pair. Then the (j, k) element of the mean product matrix within monozygotic twin pairs is,

$$m_M^{(j,k)} = \Sigma_i^{N_M} d_{Mi}^{(j)} d_{Mi}^{(k)} / 2N_M. \tag{5}$$

The mean-product matrix within dizygotic twin pairs is similarly calculated from the dizygotic differences.

Analogous to the univariate case, the expected values of the mean-product matrices are assumed to be,

$$E\,(M_{WD}) = \Sigma_H + \Sigma_{EN} + \Sigma_{ER}$$

and

$$E\,(_{WM}) = \Sigma_{EN} + \Sigma_{ER}.$$

Bartlett's procedure provides, first, a test of the hypothesis that $\Sigma_H = 0$. The statistic for this test may be calculated from the roots of the determinantal equation,

$$|M_{WD} - \lambda_l M_{WM}| = 0. \tag{6}$$

On the null hypothesis, the distribution of the quantity

$$\chi^2 = \{N_M + N_D - \frac{(N_D + p + 1)}{2}\} \sum_{l=1}^{p} \log\,(1 + \frac{N_D}{N_M} \lambda_l) \tag{7}$$

is closely approximated by the chi square distribution with pN_D degrees of freedom, where p is the number of variables. The significance of this χ^2 is evidence of heritable variation in one or more of the p tests. This over-all test is not of great interest, however, because Σ_H is seldom null. What is of interest is the residual of the chi square after 1, 2, up to s of the largest roots are deleted by summing from s + 1 to p in the above expression. Bartlett suggests using this residual as chi-square with (p — s) (N_D — s) degrees of freedom. If the residual chi square is not statistically significant, the data demonstrate no evidence of heritable variation in (p — s) dimensions.

As an index of the degree of heritable variation in the dimension corresponding to the root (λ_l), an intra-class correlation coefficient may be calculated:

$$r_l = \frac{\lambda_l - 1}{\lambda_l} \tag{8}$$

These coefficients represent sample values for the correlation of certain linear combinations of the variables with the heritable variation. These linear combinations are the canonical variates, and their sample variance is given by the corresponding root. In the sample, the set of weights for the linear compound, which defines the canonical variate, is given by the solution of the systems of homogeneous equations

$$(M_{WD} - \lambda_l M_{WM})\, x_l = 0.$$

(Computer routines which solve this system of equations are widely available.) If an estimate is needed of a subject's score on the canonical variate, the elements of the vector x provide the coefficients of the linear combination of that subject's test scores which best estimate his score for the canonical variate. Scores for different canonical variates with distinct roots are uncorrelated in the sample.

The combination of variables specified by x is usually called the *discriminant function.* The function corresponding to the largest root is that linear compound of the twin pair differences which has the largest possible variance; i.e., which maximizes the heritability ratio. In principal, this function could be used to diagnose zygosity when data are fallible, but to be of practical value for this purpose, the heritability ratio for the function would have to be far larger than is typically found for metric characters.

Usually, it is not possible to identify the source of heritable variation merely by inspecting the discriminant function. The weights in function are simultaneously maximizing dizygotic variation and minimizing monozygotic variation and are necessarily complicated. A better basis for interpretation is a principal component resolution of an estimate of Σ_H. Obtaining a suitable estimate of Σ_H presents a problem, however. Solving the equations of expectation of mean-product matrices, analogous to what was done in the univariate case, gives $M_{DW} - M_{MW}$ as the unbiased estimate. But this estimate may not be a proper covariance matrix in the sense that it is not positive definite or semi-definite. That is, some of the variances of some of the variables may be negative, or some of the principal components may have negative variances, neither of which are defined. A method of statistical estimation is needed which will constrain the estimate to be at least positive semi-definite (all principal component variances positive or, at least, zero).

Unfortunately, no such procedure is available at the present time. It may be conjectured that the maximum likelihood solutions for factor analysis can be generalized to estimate Σ_H and $\Sigma_{EN} + \Sigma_{ER}$ simultaneously from M_{DW} and M_{MW}, but no attempt in this direction has yet been made. As an alternative, we present an *algebraic,* rather than a statistical, solution for a positive semi-definite estimate of Σ_H.

Let

$$A = \Sigma_H + \Sigma_{EN} + \Sigma_{ER} \tag{9}$$

and

$$B = \Sigma_{EN} + \Sigma_{ER}. \tag{10}$$

Then for a symmetric positive definite B and a symmetric positive definite A it is possible to find a non-singular transformation T such that

$$T'AT = \Phi, \tag{11}$$

where Φ is diagonal with positive diagonal elements, and

$$T'BT = I, \tag{12}$$

where I is the p x p identity matrix. The columns of T are the solution of a system of homogeneous equations of the form:

$$(A - \phi_l B)\, t_l = 0, l = 1, 2, \ldots p,$$

and ϕ_l is a root of

$$|A - \phi B| = O.$$

From (10) and (12),

$$T'\Sigma_H T = \Phi - I;$$

thus,

$$\Sigma_H = (T^{-1})'(\Phi - I)T^{-1}.$$

We propose as an estimate of Σ_H, similar expression with sample values substituted for T and Φ, and setting to zero elements of $\Phi - I$ which correspond to insignificant dimensions of heritable variation (i.e., dimensions with insignificant canonical variance). The sample quantities for this purpose are obtained as part of the test of dimensionality described above. For elements in the columns of T are substituted the discriminant function coefficients $\underset{\sim}{x}_l$; for elements of ϕ are substituted the corresponding significant canonical variances λ_l ($l = 1,2, \ldots s$) (roots) and $p - s$ unities. Let these matrices be X and Λ^*, respectively. Then the estimate is

$$\Sigma_H\, (X^{-1})'(\Lambda^* - I)\, X^{-1}. \tag{13}$$

This estimate has the following properties. Because the elements of the diagonal matrix $(\Lambda^* - I)$ are non-negative, it can be expressed as the product of a matrix and its transpose and is therefore positive semi-definite. Its rank is s and its nullity is $p - s$. When all of the canonical variances are significant and $s = p$,

$$\begin{aligned}\Sigma_H &= (X^{-1})'(X'M_{DW}X - X'M_{MW}X)X^{-1} \\ &= M_{DW} - M_{MW};\end{aligned}$$

i.e., the proposed estimate is equal to the unbiased estimate.

Results of the Multivariate Analysis.[1] All of the sample information necessary for the preceding analysis is contained in the monozygotic and dizygotic difference covariance matrices. In Tables 14-4 and 14-5, we have shown this information in the form of correlations and standard deviations. The covariances can be recovered by multiplying the correlation coefficients of any two variables by the product of the standard deviations of these variables. The monozygotic difference correlations are potentially of interest as indicators of purely environmental factors which are common to more than one test. In the present study, however, interpretation of the environmental correlations is not at issue and, in fact, is not very rewarding because these correlations are rather small.

Since the dizygotic difference correlations reflect both heritable and environmental variation, we will not attempt to interpret them directly, but will proceed with the estimation of the heritable part. In Tables 14-6 and 14-7 are the results of the canonical analysis of the difference covariance matrices for boys and girls. According to Bartlett's criterion we have significantly resolved three dimensions of heritable variation in the boys' data and two dimensions in the girls' data. Thus a considerably simplified description of the heritable variation should result when Σ_H is estimated and examined. As expected, the discriminant function associated with each dimension in the canonical analysis is not readily interpretable (Tables 14-6 and 14-7).

We therefore proceed with the use of these functions to construct the estimated correlation matrix for the heritable part of the dizygotic differences. In this construction, we have taken the conservative course of retaining all discriminant functions and those which have a positive correlation with the heritable variation; i.e., which have canonical variance greater than unity (See Tables 14-6 and 14-7). Then from these functions, and the corresponding canonical variances reduced by unity (formula 13), we construct the covariance matrix for the heritable part and convert it to a correlation matrix. The elements of their correlation matrix for boys and girls, respectively, are shown in Tables 14-8 and 14-10.

There is a rather clear pattern of correlation in these matrices, especially for the girls' data where the effective dimensionality is smaller and the estimates more stable because of the larger sample size. The pattern may be summarized in terms of the characteristic root and vector resolutions of these matrices as in a principal component analysis (see Harmon, 1960). The results of this resolution are shown in Tables 14-9 and 14-11.

[1] Computer programming by Dr. Vidya Bhushan.

TABLE 14-4

BOYS' DATA: MONOZYGOTIC AND DIZYGOTIC DIFFERENCE CORRELATIONS AND STANDARD DEVIATIONS

Test	Standard Deviation	Monozygotic ($N_M = 50$) 1.	2.	3.	4.	5.	6.	7.	8.
1. Spatial	8.482	1.000							
2. Numerical	4.341	.117	1.000						
3. Abstract	6.536	.344	.016	1.000					
4. Verbal	3.868	.048	.179	.151	1.000				
5. Mechanical	7.847	.267	.302	.158	.246	1.000			
6. Clerical	5.623	.292	.250	.279	.139	.118	1.000		
7. Spelling	12.014	.104	.145	−.138	.204	.153	.288	1.000	
8. Sentences	8.417	.125	.381	.197	.161	.155	.457	.476	1.000

Test	Standard Deviation	Dizygotic ($N_D = 25$) 1.	2.	3.	4.	5.	6.	7.	8.
1. Spatial	17.068	1.000							
2. Numerical	4.994	−.041	1.000						
3. Abstract	7.479	.360	.058	1.000					
4. Verbal	5.274	.146	.412	.267	1.000				
5. Mechanical	6.710	.432	−.238	.079	−.189	1.000			
6. Clerical	9.829	.380	.053	.029	.180	.276	1.000		
7. Spelling	18.711	.201	.269	.300	.628	−.095	.187	1.000	
8. Sentences	11.947	.485	.193	.215	.488	−.015	.471	.449	1.000

TABLE 14-5
GIRLS' DATA: MONOZYGOTIC AND DIZYGOTIC DIFFERENCE CORRELATIONS AND STANDARD DEVIATIONS

Test	Standard Deviation	1.	2.	3.	4.	5.	6.	7.	8.
					Monozygotic ($N_M = 56$)				
1. Spatial	11.433	1.000							
2. Numerical	5.089	.016	1.000						
3. Abstract	7.517	.158	−.027	1.000			(symmetric)		
4. Verbal	3.836	.275	.153	.048	1.000				
5. Mechanical	6.461	−.119	−.174	.028	.124	1.000			
6. Clerical	5.628	.183	.180	−.117	−.020	.123	1.000		
7. Spelling	12.855	.109	.049	−.192	.266	.108	−.110	1.000	
8. Sentences	7.831	−.103	.237	.027	.023	−.012	−.004	.197	1.000
					Dizygotic ($N_D = 54$)				
1. Spatial	13.906	1.000							
2. Numerical	5.941	.261	1.000						
3. Abstract	7.753	.394	.420	1.000			(symmetric)		
4. Verbal	6.280	.291	.628	.415	1.000				
5. Mechanical	9.540	.403	.269	.480	.372	1.000			
6. Clerical	10.438	.225	.407	.127	.315	.046	1.000		
7. Spelling	20.991	.235	.599	.280	.668	.179	.263	1.000	
8. Sentences	11.615	.150	.499	.077	.621	.183	.270	.678	1.000

TABLE 14-6
BOYS' DATA: DISCRIMINANT FUNCTIONS, CANONICAL VARIANCES, AND BARTLETT'S CHI-SQUARE

	Functions							
Test	1.	2.	3.	4.	5.	6.	7.	8.
1. Spatial	.0862	−.0744	−.0191	−.0407	.0432	−.0163	.0083	−.0245
2. Numerical	−.1225	−.0372	−.0030	.0235	.2126	.0081	−.0716	.0545
3. Abstract	−.0956	−.0309	.0317	−.0658	−.0186	.1289	.0134	.0208
4. Verbal	.0040	−.1004	.0438	.1420	.0019	−.0469	.2052	.0082
5. Mechanical	.0241	.0575	−.0109	−.0430	−.0380	−.0310	.0154	.1081
6. Clerical	.0540	.1131	.1669	.0134	.0285	−.0146	−.0004	−.0245
7. Spelling	−.0585	−.0433	.0343	−.0405	−.0260	−.0305	−.0311	−.0037
8. Sentences	.0812	−.0267	−.0504	.0844	−.0569	.0448	−.0446	.0303
Variances	5.8547	4.5030	2.8494	2.0986	1.4067	0.9445	0.7060	0.3937
Chi-square	322.0	242.7	174.3	122.9	81.3	50.4	28.0	10.4
d.f.	200	168	138	110	84	60	38	18
P	<.0005	<.0005	.02	>.05	>.05	>.05	>.05	>.05
Correlation with heritable variation	.829	.777	.649	.522	.289	0	0	0

TABLE 14-7
GIRLS' DATA: DISCRIMINATION FUNCTIONS, CANONICAL VARIANCES, AND BARTLETT'S CHI-SQUARE

Test	Functions 1.	2.	3.	4.	5.	6.	7.	8.
1. Spatial	−.0218	.0487	−.0267	.0319	.0533	.0472	.0080	−.0050
2. Numerical	.0067	.0803	−.0246	−.0850	−.0072	−.0188	.1775	−.0048
3. Abstract	.0460	.0123	−.0247	−.0748	−.0142	.0115	−.0449	.0969
4. Verbal	.1144	−.0075	.0986	.0571	−.1936	.1283	−.0258	−.0543
5. Mechanical	−.0203	.1308	−.0498	.0212	−.0274	−.0740	−.0160	−.0329
6. Clerical	.1260	−.0999	−.1035	.0260	−.0063	−.0196	−.0191	−.0036
7. Spelling	.0408	−.0030	.0144	−.0499	.0439	−.0139	−.0239	−.0223
8. Sentences	.0320	.0175	.0513	.0963	.0289	−.0282	.0071	.0597
Variances	6.292	3.868	2.184	1.594	1.326	.752	.591	.449
Chi-square	608.7	455.2	333.2	244.2	171.1	106.5	63.7	28.4
d.f.	432	371	312	255	200	147	96	47
P	$<.0005$	.005	$>.05$	$>.05$	$>.05$	$>.05$	$>.05$	$>.05$
Correlation with heritable variation	.842	.742	.556	.373	.239	0	0	0

TABLE 14-8
BOYS' DATA: CORRELATIONS AND STANDARD DEVIATIONS OF THE HERITABLE VARIATION

Test	Standard Deviations	1.	2.	3.	4.	5.	6.	7.	8.
1. Spatial	14.823	1.000							
2. Numerical	3.148	−.150	1.000				(Symmetric)		
3. Abstract	4.098	.458	.245	1.000					
4. Verbal	3.946	.210	.683	.537	1.000				
5. Mechanical	3.848	.609	−.718	.143	−.540	1.000			
6. Clerical	8.097	.417	−.087	.336	.186	.487	1.000		
7. Spelling	14.738	.245	.486	.843	.866	−.333	.142	1.000	
8. Sentences	9.174	.669	.140	.271	.684	.043	.488	.459	1.000

TABLE 14-9
BOYS' DATA: CHARACTERISTIC ROOTS AND VECTORS OF THE HERITABLE VARIATION

Test	Vectors 1.	2.	3.	4.	5.
1. Spatial	.245	.470	.080	−.473	.447
2. Numerical	.304	−.402	.100	.214	.766
3. Abstract	.401	.139	−.650	.054	.035
4. Verbal	.500	−.177	.177	−.022	−.167
5. Mechanical	−.131	.590	−.195	.015	.166
6. Clerical	.196	.399	.191	.822	−.041
7. Spelling	.484	−.097	−.339	−.077	−.325
8. Sentences	.383	.223	.583	−.214	−.223
Root	3.529	2.570	.900	.623	.376

TABLE 14-10
GIRLS' DATA: CORRELATIONS AND STANDARD DEVIATIONS OF THE HERITABLE VARIATION

Test	Standard Deviation	1.	2.	3.	4.	5.	6.	7.	8.
1. Spatial	8.949	1.000							
2. Numerical	4.234	.526	1.000				(Symmetric)		
3. Abstract	4.979	.684	.819	1.000					
4. Verbal	5.136	.390	.918	.638	1.000				
5. Mechanical	7.426	.849	.530	.801	.491	1.000			
6. Clerical	8.820	.254	.573	.280	.467	.006	1.000		
7. Spelling	17.159	.342	.907	.569	.883	.274	.442	1.000	
8. Sentences	9.067	.333	.738	.305	.886	.276	.407	.866	1.000

TABLE 14-11
GIRLS' DATA: CHARACTERISTIC ROOTS AND VECTORS OF THE HERITABLE VARIATION

Test	Vectors				
	1.	2.	3.	4.	5.
1. Spatial	.300	.468	.204	−.542	.430
2. Numerical	.434	−.092	.054	.293	.095
3. Abstract	.364	.339	.091	.571	.069
4. Verbal	.414	−.194	−.207	.048	−.509
5. Mechanical	.296	.562	−.115	−.129	−.427
6. Clerical	.234	−.316	.859	−.113	−.188
7. Spelling	.389	−.296	−.218	.122	.561
8. Sentences	.353	−.341	−.324	−.497	−.092
Root	5.005	1.619	.747	.444	.185

Perhaps what is most to be noticed in these results is the large part of the heritable variance which can be accounted for by a general component having all elements positive (except for mechanical reasoning in boys, which has a small negative weight). This component accounts for 44.1 per cent of the boys' heritable variation and 62.6 per cent of that of girls'. The percentage for boys is smaller because relatively more of their heritable variation goes to the second component attributable to visual perceptual processes which are expressed more fully in boys than in girls (see below).

The fact that a strong general component appears in the heritable part of dizygotic twin differences clarifies the nature of the general factor obtained in conventional factor analyses of mental test data obtained from unrelated subjects. It has been argued that the general factor reflects primarily cultural and educational differences of subjects from different families and different social strata, because differences from these sources would be expected to affect test performance generally. According to this argument, the general factor should diminish as subjects of more homogeneous background are studied. The present analysis shows that this is not the case: the dizygotic twin pairs are the ultimate in subjects matched for background, and yet the differences within pairs between these twins are all positively correlated and, when corrected for environmental variation, exhibit a large general component.

Presumably, the variation of this component has its source in some feature or features common to each of the tests. In the absence of data on tests which might be contrasted with the DAT, it is difficult to say what these features might be. The DAT subtests do not all involve language, although the instructions do, but the tests do require the use of answer sheets and all make use of printed symbols of some kind. Each of the tests also requires long- or short-term memory, and, perhaps most important, the concentrated attention on a specific, detailed, visually presented task for a sustained period. Any or all of these common features of the tests, or the testing situation, may elicit individual differences in behavior which are ultimately under genetic control. The unmistakable evidence for a general component in the heritable variation in the test scores gives confidence in the genetic basis of a substantial component in test-taking behavior and encourages more detailed study of the perceptual and cognitive processes which make up this behavior.

Turning now to the second component we note that, both in boys and in girls, the signs of the coefficients contrast primarily the tests which employ graphics (pictures or diagrams) as stimuli with those which employ only literal symbols (numbers, letters and words). The mechanical reasoning and spatial relations tests define the positive pole of this component, and number and sentences define the negative pole in boys and girls, respectively. This component has the larger variance among boys, a finding which is consistent with the fact that spatial visualizing ability is better expressed among boys. In the boys' data also (but not the girls'), the clerical speed and sentences tests are on the positive, or visual, pole of the component. Both of these latter tests involve a visual search for details, and it appears that this facet of the tests is being picked up by the stronger expression of visual processes among boys.

As for the remaining components, we see in Table 14-11 that in the girls' data, the third component is clearly identified by the clerical speed test. This result seems clear enough to suggest that the perceptual motor feature of this test is reflecting heritable individual differences, even though the over-all statistical test shows only two significant components in the heritable variation. In the boys' data, it is the fourth component which is identified by this clerical speed test. These components are minor, however, and account for only 9.3 and 7.8 per cent of the respective heritable variation.

The third component in the boys' data has no obvious interpretation. It appears to be picking up correlation between the abstract reasoning and spelling tests, but these tests have so little in common that any relation between them, beyond that of the general factor, cannot be accounted for on the basis of what we presently know about these tests.

SUMMARY

Differential Aptitude Test (DAT) scores of 50 pairs of monozygotic and 25 pairs of dizygotic twin boys, and 56 pairs of monozygotic and 54 pairs of dizygotic twin girls were subjected to univariate and multivariate analysis in order to assess and characterize heritable variation in the data.

Subjects were obtained from grades 7 through 12 in schools in and around Louisville, Kentucky. The criterion for monozygosity was complete concordance of all important blood antigens. Components of variance due to a) measurement error; b) differences between twins attributable to environmental effects; c) differences attributable to heritable effects; and d) differences between twin pairs (i.e., between families) were estimated separately for each of the eight tests of the DAT battery. Data for boys showed significantly high heritability ratios (dizygotic variance/monozygotic variance) for all tests except numerical reasoning, abstract reasoning and mechanical reasoning. Data for girls showed significantly high heritability ratios for all tests except numerical reasoning, abstract reasoning and spatial relations (the latter fell just short of significance). The sex difference in heritability of the mechanical reasoning test and spatial relations suggests different strengths of expressions of these abilities in boys and girls scores.

The partition of variance between error, environmental and heritable parts for the most part confirms our expectations about the role of formal training, cultural differences, and inherited ability in performance on the DAT tests. Where formal training and cultural differences are important in performance, heritable variance tends to be suppressed. The exception to this generalization is the abstract reasoning test, for which the absence of heritable variance is difficult to explain. The tests which show the largest components of heritable variance are spatial relations in the boys' and clerical speed and accuracy in the girls'.

A multivariate analysis of these data based on an estimate of the covariance matrix for heritable variation was carried out. Statistically significant heritable variation was found in only three dimensions in the boys' data and two dimensions in the girls' data. The most striking aspect of this analysis was the appearance of a general component accounting for a large part of the heritable variation in both the boys' and girls' data. This component was interpreted as reflecting heritable individual differences in general test-taking behavior, including use of symbols, memory, and sustained attention to a visually presented task.

Of the remaining components, one was clearly identified with tests which involved pictures or graphic symbols as contrasted with those that used only numbers, letters, and words. This component accounted for substantially more of the heritable variation of boys than of girls, and reflects a stronger expression of visual processes in boys than in girls. Another component was identified with the clerical speed test both in the boys' and girls' data.

References

Bartlett, M. S. 1951. The goodness of fit of a single hypothetical discriminant function in the case of several groups. *Ann. Eugen.* **16:** 199-214.

Bennett, G. K.; Seashore, H. G., and Wesmann, A. G. 1959. *Differential Aptitude Tests,* 3rd edition. New York: The Psychological Corporation.

Erlenmeyer-Kimling, L. and Jarvik, L. F. 1963. Genetics and intelligence: a review. *Science* **142:** 1477-79.

Graybill, F. A. 1961. *An Introduction to Linear Statistical Models,* Vol. I. New York: McGraw-Hill.

Harmon, H. H. 1960 *Modern Factor Analysis.* Chicago: The University of Chicago Press.

Leahy, Alice M. 1935. Nature-nurture and intelligence. *Genet. Psychol. Monog.* **17:** 236-308.

Newman, H. H.; Freeman, F. N.; and Holzinger, K. J. 1937. *Twins: A Study of Heredity and Environment.* Chicago: The University of Chicago Press.

Strandskov, H. H. 1955. Some aspects of the genetics and evaluation of man's behavioral characteristics. *Eugen. Quart.* 2: 152-61.

Thurstone, L. L. 1941. *The Primary Mental Abilities Tests.* Chicago: Science Research Associates.

———. 1938. *Primary Mental Abilities.* Chicago: The University of Chicago Press.

Vandenberg, S. G. 1965. Multivariate analysis of twin differences. In *Methods and goals in human behavior genetics,* Ed. S. G. Vandenberg, pp. 29-43. New York: Academic Press.

Vandenberg, S. G.; Stafford, R. E. and Brown, A. 1968. The Louisville Twin Study. In *Progress in human behavior genetics,* Ed. S. G. Vandenberg, pp. 153–204. Baltimore: The Johns Hopkins Press.

15

JOHN C. LOEHLIN
University of Texas and
STEVEN G. VANDENBERG
University of Colorado

GENETIC AND ENVIRONMENTAL COMPONENTS IN THE COVARIATION OF COGNITIVE ABILITIES: AN ADDITIVE MODEL

For a number of years, geneticists and psychologists have concerned themselves with methods for analyzing the variance of physical or behavioral traits in a population into components reflecting genetic and environmental influences, and various interactions between and among them (Broadhurst and Jinks, 1961; Burt and Howard, 1956; Cattell, 1960; Fisher, 1918; Mather, 1949; Wright, 1952). Only fairly recently, at least among psychologists, has it become recognized that trait *co*variation (covariances, correlations) can also be subjected to such analysis; a number of recent proposals have been made concerning methods of doing this (Kempthorne and Osborne, 1961; Nichols, 1964; Loehlin, 1965; Vandenberg, 1965; Meredith [see chap. 18, this volume]).

There are some attractive features for the psychologist in working with trait covariation rather than with single traits. First, there is a shift of emphasis away from "heredity versus environment" towards a consideration of the structuring of genetic and environmental influences. And second, while most behavior traits that have been empirically studied appear to reflect both genetic and environmental influence (Fuller and Thompson, 1960; Vandenberg, 1966a), the possibility remains that some *associations* among traits may result solely from common genes or common environment.

We would like to express our gratitude to the many twins whose cooperation made this research possible. We are also indebted in many ways to officials of schools in Michigan at Ann Arbor, Ypsilanti, Dearborn, and Detroit; and in Louisville and Jefferson County, Kentucky; and in a number of communities in Southern Indiana. Data were collected and analyzed in Michigan under grants from McGregor Fund of Detroit and grants M-1045 and RG 5527 of the National Institutes of Health, and in Louisville under grants K3—MH 18382, M 6203, MH 07033, MD 00843 of the National Institutes of Health and grant GB 466 of the National Science Foundation.

In addition, we are grateful to the Computation Centers of the Universities of Nebraska and Texas for computer time for the analyses described in this paper.

The present paper will investigate trait covariation in the realm of cognitive abilities, using data from Thurstone's Primary Mental Abilities test (PMA) administered to two samples of identical and fraternal twin pairs. A simple additive model of the relationship between heredity and environment is employed. One of the writers has elsewhere reported an analysis of some of the same data based on a multiplicative model (Vandenberg, 1965). The relationship between the results of these two approaches will be discussed in a later section of this paper.

Cognitive abilities represent a particularly interesting area for investigation of genetic and environmental components of covariation. Such an investigation may, for example, clarify the question of whether mental abilities are basically general or specific, an issue debated by Spearman (1904), Thomson (1920), Thurstone (1934), Tryon (1935), Cattell (1943), and Guilford (1956), among others. If separation of the covariation of cognitive tests into genetic and environmental portions can be carried out, subsequent factoring of these components should be illuminating. Such a factoring might reveal, for example, a single general factor in the genetic covariation and a number of independent environmental factors; or a general environmental factor tending to pull together relatively independent genetic factors; or an identical factor structure in both realms; or even a different multiple factor structure in each. Any one of these outcomes would have relevance for the generality-specificity question.

The twin method is, of course, not without flaws as a technique for separating genetic and environmental components of variance or covariance, since it leaves together certain sources of variance which would ideally be kept distinct: for example, in the usual application of the twin method (a) any correlated hereditary and environmental influences are grouped with heredity, (b) any gene-environment interaction is grouped with environment, (c) no distinction is made between the prenatal biochemical environment and the postnatal stimulus environment, and (d) genic and additive effects cannot be separated. Not all of these limitations are inherent in every application of the twin method—such as a longitudinal study of twins from birth—but they are characteristic of the method in its usual cross-sectional form. Still, even the conventional twin method does achieve *some* degree of separation of genetic and environmental effects, and if interest is chiefly focused on the pattern of relationships rather than their absolute magnitude, it should be able to provide at least some useful approximations to the true state of affairs.

METHOD

Samples

The first twin sample has been described in detail elsewhere (Sutton, Vandenberg, and Clark, 1962). Briefly, it consists of 82 pairs of like-sexed twins of high school age from four Michigan cities: Ann Arbor, Ypsilanti, Dearborn, and Detroit. Of these, 45 pairs were classified as identical and 37 as fraternal, on the basis of blood typing, eye color, and physical appearance.

The second sample consists of 116 pairs of like-sexed twins of high school age from the metropolitan area around Louisville, Kentucky, 78 identical and 38 fraternal pairs, classified by blood typing (Vandenberg, Stafford, and Brown, Chapter 10, this volume).

The two samples, broken down by zygosity and sex, are described in Table 15-1. Note that dizygotic twins appear to be somewhat undersampled, particularly the males. We believe that in most of the schools from which the twins were recruited a complete list of twins was obtained. There seemed to be no obvious bias in the few refusals to cooperate. The difference, then, would appear to suggest that dizygotic twins, especially males, were less likely to be found in school together at these ages. If the more divergent pairs are the ones most often not in school together, the dizygotic undersampling will tend to lead to an underestimation of genetic effects; however, the focus of the present investigation is on the relationships among effects rather than on their absolute magnitude, so this is not in itself a major problem. The dizygotic sex bias is potentially more serious in view of sex differences to be expected on some of the PMA scales. However, the most critical analyses in the present study are

Table 15-1
Twin Samples by Zygosity and Sex, Number of Like-sexed Twin Pairs

Group	Male	Female	Combined
Monozygotic			
Michigan	24	21	45
Louisville	40	38	78
Combined Samples	64	59	123
Dizygotic			
Michigan	14	23	37
Louisville	15	23	38
Combined Samples	29	46	75
All twins	82	116	198

based on within-pair covariance, and sex differences do not present problems here, since only like-sexed pairs are used.

TESTS

The testing procedures have been described in detail elsewhere (Sutton, Vandenberg, and Clark, 1962; Vandenberg, Stafford, and Brown, 1968). In both studies, Thurstone's PMA Tests (Thurstone and Thurstone, 1941) were administered as part of a large battery of psychological tests. In the first study anthropometric and biochemical measures were also obtained. In the Michigan study, 17 subtests of the PMA were administered, three each for the factors of Numerical Ability (N), Verbal Comprehension (V), Spatial Ability (S), Word Fluency (W), and Reasoning Ability (R); and two for Memory (M). At the time of the Louisville study, the Memory subtests were not available and hence not included. In the present paper, therefore, only the five factors N, V, S, W, and R will be considered, each represented by three subtests in both samples.

RESULTS

Means of the PMA scales and subscales for various breakdowns of the total sample are shown in Table 15-2. Analyses of variance indicated that interactions among zygosity, sample, and sex were generally negligible, permitting direct interpretation of the main effects.[1]

It will be seen from Table 15-2 that there was no substantial difference in the mean level of performance between identical and fraternal twin groups on any PMA scale or subscale. There were, however, significant differences for the other classifications, the Michigan sample scoring higher on Number and Word Fluency, and the girls scoring higher on Word Fluency and Reasoning and lower on Space. The sex differences are generally similar to those reported by other investigators using this test in non-twin groups. For example, Herzberg and Lepkin (1954), with 705 17-year-old Pittsburgh high school students, found corresponding significant differences on the same three scales, plus a difference in favor of girls on Verbal Comprehension.

Despite the presence of the differences noted, it was decided to combine the two regional samples and the two sexes to achieve a larger sample for further analyses. Within-pair variance, the main focus of in-

[1] No interactions involving main scales were significant, and of 60 interactions involving subscales only two reached significance at the .05 level, a result probably safely attributable to sampling fluctuation.

TABLE 15-2
MEANS OF PMA SCALES AND SUBSCALES BY ZYGOSITY, SAMPLE AND SEX

Scale	MZ	DZ	Mich.	Lvle.	Male	Female
1. Addition	19.1	20.3	22.4**	17.5	19.8	19.3
2. Multiplication	40.0	42.0	43.3**	38.9	40.3	41.2
3. 3-higher	46.3	47.3	48.7*	45.2	49.1**	44.6
Number	106.0	110.2	114.6**	102.5	109.3	106.0
4. Sentences	17.9	18.6	19.4**	17.3	18.2	18.1
5. Vocabulary	27.8	28.2	28.7	27.5	27.8	28.1
6. Completion	27.5	27.3	28.2	26.9	28.4*	26.6
Verbal	73.3	74.1	76.1	71.9	74.5	72.9
7. Flags	40.5	40.3	39.8	40.9	43.2**	37.9
8. Figures	37.9	38.5	38.0	38.3	41.1**	35.5
9. Cards Space	34.3	33.3	33.3	34.3	36.0**	32.0
Space	114.4	113.0	111.4	115.7	121.8**	106.6
10. First letters	35.3	36.1	38.0**	33.9	33.2	37.7**
11. 4-letter words	11.1	11.0	11.2	10.9	10.1	11.9**
12. Suffixes	10.4	10.7	11.5**	9.8	9.4	11.5**
Word fluency	57.2	58.0	60.8**	55.2	53.1	61.4**
13. Letter series	15.5	15.5	15.3	15.6	14.7	16.2*
14. Letter groupings	14.5	14.3	13.7	14.9**	13.7	15.1**
15. Pedigrees	23.4	24.5	24.6	23.2	21.7	25.6**
Reasoning	53.4	54.6	53.7	54.0	50.3	56.9**

Note: Total scores were computed only for subjects completing all three subtests, therefore subtest means may not sum exactly to total means.

* Significantly higher at .05 level.
** Significantly higher at .01 level.

terest, should not be appreciably affected by such differences. The variance between pairs should tend to be somewhat inflated by sex and regional differences, but this variance is of secondary interest in any case.

The principal results to be reported are based on intercorrelations among twin sums and twin differences on PMA scales and subscales for the monozygotic (MZ) and dizygotic (DZ) groups, for the total sample. (The basic variance-covariance matrices are contained in Tables A through H in the Appendix.) The correlations among twin sums (or means) are based on between-pairs covariance and should largely reflect between-family influences, both genetic and environmental. The correlations among twin differences are based on within-pair covariance and should reflect within-families influences. As has been noted elsewhere (Vandenberg, 1965), correlations among MZ twin differences are of unique psychological interest, since any such correlations must be due to common environmental influences on the traits (in some broad sense of the term environment). Correlations among DZ differences, on the other hand, may reflect both common genetic and common environmental influences on the traits in question; the nature of the discrepancies between MZ- and DZ-difference correlations may, therefore, serve to shed some light on genetic effects on the traits.

The intercorrelations of MZ and DZ pair sums and differences on the 15 PMA subscales were factored (principal axes factoring with unities in the diagonal), and the first five factors in each case rotated by the Varimax method (Kaiser, 1958). The rotated factor loadings are shown in Tables 15-3 and 15-4. Loadings of .50 or greater are underlined.

The main question of interest in interpreting these data is whether the structuring of the subtests into Thurstone's five primary mental abilities

TABLE 15-3
ROTATED FACTOR LOADINGS FROM BETWEEN-PAIR COVARIATION*

	a. Dizygotic Twins					b. Monozygotic Twins				
	I	II	III	IV	V	I	II	III	IV	V
1. Addition	70	34	40	18	20	84	20	24	14	10
2. Multiplication	89	18	16	19	19	83	26	08	24	10
3. 3-higher	62	41	35	14	40	71	23	23	08	42
4. Sentences	29	75	22	29	22	29	80	12	22	22
5. Vocabulary	35	76	20	33	26	21	83	06	31	24
6. Completion	22	82	24	20	20	16	85	17	17	21
7. Flags	27	31	80	11	19	26	05	89	07	10
8. Figures	13	08	94	14	07	05	11	93	05	05
9. Cards	15	28	88	13	11	12	12	91	09	13
10. First letters	39	36	08	75	09	16	28	15	82	20
11. 4-letter words	−05	17	19	79	35	07	06	−01	89	21
12. Suffixes	38	35	19	67	15	22	34	08	76	06
13. Letter series	20	36	26	23	78	37	35	12	19	70
14. Letter grouping	42	27	03	29	74	12	22	07	23	82
15. Pedigrees	10	63	29	27	48	18	56	12	18	59

* Note: Decimal points omitted in this and subsequent tables.

TABLE 15-4
ROTATED FACTOR LOADINGS FROM WITHIN-PAIR COVARIATION

	a. Dizygotic Twins					b. Monozygotic Twins				
	I	II	III	IV	V	I	II	III	IV	V
1. Addition	82	02	−02	32	02	79	11	08	−00	24
2. Multiplication	75	22	08	−04	33	80	17	−07	03	−11
3. 3-higher	71	18	36	15	04	17	−13	−01	03	81
4. Sentences	39	74	08	05	−26	04	59	−19	18	−26
5. Vocabulary	21	70	02	37	08	09	77	05	19	−12
6. Completion	−08	70	06	51	13	08	65	17	09	−24
7. Flags	15	18	86	03	00	12	−06	69	03	06
8. Figures	08	05	86	23	02	−07	16	65	−14	−21
9. Cards	04	−02	71	08	49	−06	00	59	31	17
10. First-letters	08	15	37	78	02	03	11	13	71	−24
11. 4-letter words	27	−07	02	81	27	−03	29	−24	61	09
12. Suffixes	12	30	11	77	07	03	05	13	68	07
13. Letter series	23	21	15	27	77	22	48	−00	35	12
14. Letter grouping	36	−02	41	19	33	−28	59	06	−10	46
15. Pedigrees	07	76	14	−12	33	14	51	02	−00	12

factors will appear in the analyses based on various sources of covariation. If we look first at Table 15-3, based on covariation between-pair sums, it is clear that Thurstone's five factors show up here. With the exception of one of the tests assigned to Reasoning, which appears to load at least as heavily on the Verbal Comprehension factor, the high loading for each test is on its designated factor. The general tendency for the tests to be positively correlated is reflected in the generally positive loadings elsewhere in the matrix. No very striking difference is evident between the MZ and DZ loadings. To some extent this is to be expected, since both matrices should reflect both environmental and genetic sources of covariation. One might expect, however, to find some quantitative difference, since the between-pair sum matrix contains all the genetic variation of the MZ twins, but only part of the genetic variation of the DZ twins. We will return to this point later.

Looking next at Table 15-4, it appears that at least four of Thurstone's five factors may be extracted from either MZ or DZ within-pair correlations. The Reasoning factor has been rather badly split apart in the DZ data, but the other four factors are tolerably clear. In the MZ data, Reasoning appears to have merged with Verbal Comprehension, and the fifth factor has split one of the Number tests off from the others, but Number, Verbal, Space, and Word Fluency factors are still fairly clearly identifiable.

One conclusion can immediately be drawn from the MZ data: since any covariation among identical twin differences is necessarily environmentally determined, it follows that—except perhaps for Reasoning—the factor structure of Thurstone's PMA battery reflects the structuring of the environmental influences that have been brought to bear on the development of cognitive abilities in these subjects.

Does the PMA also reflect the structure of genetic influences? For an answer to this, the DZ data do not suffice, since they reflect both genetic and environmental effects. What is needed is DZ data from which the environmental effects reflected in the MZ data have been removed. The approach used in the present study was to subtract the MZ variance-covariance matrix from the DZ variance-covariance matrix, and analyze the resulting difference matrix. Since the environmental effects on within-pair differences may well be somewhat greater for fraternal twins than for identical twins, this procedure may not eliminate environmental effects entirely, but it should markedly reduce them. There is, however, one problem in applying this method to empirical data. A substantial part of the variance of the difference scores is likely to be error variance resulting from test unreliability. If error variances are in fact precisely equal in the MZ and DZ groups, the subtraction of the variance-covariance matrices in effect corrects for attenuation any correlations derived from the resulting difference matrix. But if for some variable the MZ error of measurement happens to exceed the DZ error of measurement, overcorrection, which can lead to absurd correlations, will result; even if this does not occur, the corrected correlations will not be readily comparable with the uncorrected *r*s on which the analyses in Tables 15-3 and 15-4 are based. One simple solution to this difficulty is to subtract only reliable MZ variance from the DZ variances. For this purpose, a rough estimate of reliability was obtained from the highest correlation involving each subtest; this will tend to be an underestimate of reliability, but with the present data not a very gross one. The variance-covariance matrix of MZ differences, with the variances thus corrected, was subtracted from the variance-covariance matrix of DZ differences. The resulting matrix was converted to correlations and factored and rotated as before. The rotated factor loadings are shown in Table 15-5. It will be observed that the loadings are somewhat irregular, but on the whole display the same structure observed in the other matrices—once more, with a fragmented Reasoning factor. Granted our assumptions, then, it appears that the basic factor structure of the test battery reflects a characteristic patterning of genetic as well as of environmental influences on cognitive trait development.

The results so far have focused on the loading of the subtests on the PMA factors; possible differences in the relationships among these factors have not been examined. One way to do this is to take the grouping of the subtests as given, and to use the scores on the five regular PMA scales [2] to compute variance-covariance and correlation matrices as before.

[2] The total scale scores are simple sums of the subtest scores.

TABLE 15-5
ROTATED FACTOR LOADINGS FROM WITHIN-FAMILY GENETIC COVARIATION

	DZ-MZ (corr.) [a]				
	I	II	III	IV	V
1. Addition	<u>82</u>	06	01	33	−06
2. Multiplication	<u>70</u>	31	15	−07	40
3. 3-higher	<u>59</u>	48	37	25	11
4. Sentences	31	<u>81</u>	12	07	−21
5. Vocabulary	18	<u>60</u>	04	47	−09
6. Completion	−12	47	−01	<u>72</u>	22
7. Flags	06	31	<u>76</u>	08	07
8. Figures	08	04	<u>83</u>	23	07
9. Cards	05	−04	<u>68</u>	02	50
10. First-letters	14	03	<u>51</u>	<u>68</u>	−09
11. 4-letter words	44	−30	14	<u>69</u>	13
12. Suffixes	16	19	09	<u>84</u>	11
13. Letter series	21	−03	23	22	<u>83</u>
14. Letter grouping	46	−13	<u>57</u>	08	17
15. Pedigrees	−13	<u>71</u>	11	00	47

[a] For explanation, see text.

Tables 15-6, 15-7, and 15-8 show the correlation matrices based on twin pair sums, pair differences, and the corrected DZ-MZ difference, respectively. The reliabilities used in correcting the MZ variances of the five scales were estimated by way of the intercorrelations among their subtests, and should therefore be somewhat more accurate than those used in the preceding analysis. Each table shows the original correlations above

TABLE 15-6
PMA SCALE CORRELATIONS FROM BETWEEN-PAIR COVARIATION
(CORRELATIONS ABOVE DIAGONAL, FIRST-FACTOR RESIDUALS BELOW)

	a. Dizygotic Twins						b. Monozygotic Twins				
	N	V	S	W	R		N	V	S	W	R
N	*	73	55	63	69	N	*	57	39	44	58
V	−04	*	54	71	79	V	−07	*	25	54	71
S	05	−00	*	40	49	S	13	−03	*	14	24
W	00	02	−04	*	66	W	−02	06	−05	*	50
R	−02	01	−01	02	*	R	−04	05	−03	03	*

TABLE 15-7
PMA SCALE CORRELATIONS FROM WITHIN-PAIR COVARIATION
(CORRELATIONS ABOVE DIAGONAL, FIRST-FACTOR- RESIDUALS BELOW)

	a. Dizygotic Twins						b. Monozygotic Twins				
	N	V	S	W	R		N	V	S	W	R
N	*	38	38	39	49	N	*	−01	00	−01	15
V	−02	*	18	51	56	V	−01	*	06	33	48
S	05	−13	*	42	41	S	00	−00	*	10	06
W	−04	09	08	*	39	W	−01	07	05	*	27
R	01	09	03	−12	*	R	15	01	−03	−07	*

TABLE 15-8
PMA SCALE CORRELATIONS FROM WITHIN-FAMILY GENETIC COVARIATION
(CORRELATIONS ABOVE DIAGONAL, FIRST-FACTOR RESIDUALS BELOW)

	DZ-MZ (corr.)				
	N	V	S	W	R
N	*	51	48	50	54
V	05	*	20	48	36
S	−01	−14	*	48	48
W	−05	10	07	*	32
R	02	01	10	−11	*

the diagonal, and below the diagonal the residuals after extracting a general factor by means of a Spearman formula (DuBois, 1965, p. 462). It will be observed that the one factor accounts for the bulk of the covariation in each case. The general factor loadings are assembled together in Table 15-9.

The contrast of greatest interest is that between the purely environmental effects (Table 15-7b) and the matrix representing mostly genetic effects (Table 15-8), a contrast which may be observed either in the correlations themselves or in the corresponding factor loadings in Table 15-9. Briefly, (a) the "environmental" factors tend to be less intercorrelated than the "genetic" ones, and (b) the correlations in the former case appear to reflect a common factor that mainly involves verbal performance (Verbal Comprehension, Reasoning, Word Fluency), while in the latter case, the correlations suggest a rather uniform second-order factor general to all tests.

By contrast, the correlations based on twin sums (Table 15-6) show a very similar pattern in the MZ and DZ data, with the correlations and factor loadings based on the DZ data consistently somewhat higher. Since, as noted earlier, the MZ between-pair data contain all the genetic variance and the DZ between-pair data only part of it, one might, on the basis of the within-pair findings, have expected a difference in the opposite direction. Interpretation of the between-pair data in both instances is complicated by the fact that the total variance within the DZ group in the present sample materially exceeds that within the MZ group. The difference is not large enough to reach statistical significance, and hence may be attributable to chance differences in sampling. We can merely note that if the DZ group, for whatever reason, actually does include more genetic variability than the MZ group, this fact could account for the anomalies noted.

TABLE 15-9
GENERAL-FACTOR LOADINGS FROM PMA SCALE CORRELATIONS
(BASED ON TABLES 15-6 TO 15-9)

	Between-pair		Within-pair		"Genetic"
	DZ	MZ	DZ	MZ	DZ-MZ
Number	84	78	64	00	81
Verbal	91	82	62	61	57
Space	59	34	51	11	60
Word Fluency	75	59	68	44	68
Reasoning	85	80	75	78	63

The somewhat higher levels of correlation in the between-pair DZ data compared to the within-pair DZ data are quite reasonable: age, sex, socio-economic, and regional differences are contributing to the former and not to the latter, and will enhance correlations. In addition, if assortative mating occurs for these abilities—if like marries like—then within-family genetic variation will tend to be reduced; furthermore, a greater degree of parental similarity probably implies reduced within-family environmental variation as well. Both would further accentuate the observed discrepancy.

Thurstone's Chicago Study

For purposes of comparison, a parallel analysis was made of data from a twin study by Thurstone, Thurstone, and Strandskov (1953, 1955).[3] The data were only available for the five main PMA scales, not for the subscales, so it was possible to check only on the second part of our findings. The Thurstone study involved 48 identical and 55 fraternal twin pairs from Chicago high schools. The test versions and scoring procedures differed in some minor respects from those of the Michigan and Louisville studies, but one would expect at least general comparability of results.

The Chicago data were analyzed in exactly the same fashion as the Michigan-Louisville data, and Tables 15-10 to 15-13 present these results, paralleling Tables 15-6 to 15-9. Since Chicago subscale data were not available, the Michigan-Louisville reliabilities were used in correcting the variances, although this procedure is obviously somewhat hazardous.

Comparison of the results brings out several points of interest. The somewhat lower correlations in the between-pair data (Table 15-10, com-

[3] We are grateful to Drs. Thelma G. Thurstone, H. H. Strandskov, and Thomas Jeffrey for providing additional information about these data.

Table 15-10
Chicago Sample: Between-pair Covariation

	a. DZ Twins						b. MZ Twins				
	N	V	S	W	R		N	V	S	W	R
N	*	38	05	22	28	N	*	46	16	64	56
V	08	*	35	62	60	V	02	*	29	60	50
S	−08	00	*	25	37	S	−10	02	*	35	44
W	−00	02	01	*	49	W	08	02	00	*	62
R	02	−10	09	−02	*	R	03	−06	11	−09	*

pared to Table 15-6) suggest that the Chicago twin pairs may have been sampled from a less wide range of the population than the Michigan and Louisville twins.

Total variability is somewhat more comparable between the Chicago MZ and DZ groups than it is in the Michigan-Louisville data, and this has improved matters: the Chicago between-pair MZ correlations tend on

TABLE 15-11
CHICAGO SAMPLE: WITHIN-PAIR COVARIATION

	a. DZ Twins						b. MZ Twins				
	N	V	S	W	R		N	V	S	W	R
N	*	36	28	57	55	N	*	17	17	15	21
V	−01	*	31	45	39	V	−06	*	−06	24	40
S	−04	05	*	42	33	S	15	−08	*	07	02
W	00	−02	02	*	64	W	−01	08	06	*	08
R	05	−02	−02	00	*	R	−01	19	00	−07	*

TABLE 15-12
CHICAGO SAMPLE: WITHIN-FAMILY GENETIC CORRELATION

	DZ-MZ (Corr.)				
	N	V	S	W	R
N	*	37	27	67	54
V	04	*	39	46	23
S	−09	12	*	48	39
W	01	−03	−04	*	78
R	06	−12	01	09	*

TABLE 15-13
CHICAGO SAMPLE: GENERAL-FACTOR LOADINGS
(BASED ON TABLES 15-10 THROUGH 15-12)

	Between-pair		Within-pair		"Genetic"
	DZ	MZ	DZ	MZ	DZ-MZ
Number	33	65	67	49	67
Verbal	92	68	55	48	50
Space	37	40	47	04	53
Word Fluency	66	87	85	33	98
Reasoning	77	82	75	44	71

the whole to run somewhat higher than the DZ correlations, as expected, rather than the reverse, as in the Michigan-Louisville data.

Let us look now at the within-pair correlation matrices, both the purely environmental and the largely genetic (Tables 15-7b and 15-11b; Tables 15-8 and 15-12; and the corresponding factors in Tables 15-9 and 15-13). From these we may conclude: (a) the within-family correlations are of the same order of magnitude in the two sets of data; (b) in both sets of data there is less environmental than genetic intercorrelation; (c) there is a difference in the second-order environmental factor, in that Number tends to go along with the verbal tests in the Chicago data, although Space remains independent; and (d) the second-order genetic factor, while still general to all the tests, does not load them as evenly in the Chicago data as in the Michigan-Louisville data. That faulty reliability corrections may be partly to blame is suggested by the suspiciously high loading of Word Fluency.

On the whole, then, the Chicago results tend to support the findings of the present study, except that a question is raised concerning the status of the Numerical Ability factor. What looks in the Michigan-Louisville environmental data like a second-order verbal factor might better be described in the Chicago data as a general educational factor, with Numerical Ability joining the three verbal tests.

It may be worth noting that the Numerical factor has proved erratic in other respects, showing relatively high heritability in the Michigan twin data (Vandenberg, 1962), moderate heritability in the Chicago data (Thurstone, *et al.,* 1955) and close to zero heritability in a study by Blewett in Great Britain (1954)—despite fairly good agreement among the three studies concerning all of the other factors except Reasoning.

DISCUSSION

Summarizing our results, and neglecting a few complications, we can say that (1) the same cognitive ability factors have been found in environmental and in genetic covariation, but that (2) the relationship among these factors differs in the two cases, taking the form of a verbal (educational?) second-order factor in the environmental covariation, and a general factor in the genetic.

In connection with the first of these observations—the similar dimensions found in the genetic and environmental components of cognitive abilities—a few additional comments might be made. First, the case for similar dimensionality in the two realms is not as strong as it would be had the same results emerged from factoring a randomly selected group of cognitive tests. The three subtests of each factor were after all selected in Thurstone's analyses for their similar correlations with other tests, and might conceivably have been selected simultaneously on quite distinct genetic and environmental dimensions. In this case they would still be expected to hang together in either kind of data, even though for different reasons. This is, of course, hypothetical: a simpler interpretation of the present data is the one suggested earlier—that there are rather similar dimensions in both the genetic and environmental influences which act on cognitive development. It is perhaps worth noting a somewhat parallel finding for personality traits, in a study using a different method (Loehlin, 1965). Separate factoring of groups of personality inventory items of relatively high and low heritability yielded rather similar factors in both groups of items. On general grounds findings such as these are perhaps not unreasonable. Presumably the development of cultural institutions is to some extent influenced by the human biological tendencies they control or exploit. A sex factor, for example, might emerge either from purely sociological or from purely biological data. The case is perhaps less obvious for cognitive traits, but it is at least conceivable that the biological capacities of the human organism have historically had some bearing on what society has tended to recognize, name, and educate as a unit.

The second main finding of this study, and perhaps the more interesting one, is the difference between the second-order factors in the environmental and the genetic components of covariation. In the environmental data the "quantitative" factors, Number and Space, are essentially independent from each other and from the remaining factors—Verbal Comprehension, Word Fluency, and Reasoning. The most plausible basis for the interrelationship of the latter three factors would appear to be their common connection with language and verbal behavior. It is easy to conceive of environmental variables—such as amount of early interaction with adults or number of books in the home—that might have broad effects on verbal development, and clearly language behavior is a major focus of the educational process. The association of Numerical Ability with these factors in the Chicago data would seem most consonant with the last of these.

The second-order factor emerging from the largely genetic component of covariation, on the other hand, involves all five of the primary abilities. This factor may well reflect some general potential for intellectual functioning of the sort Spearman had in view when he proposed his general intellective factor "g." It is highly unlikely that this factor represents even a major effect of a single gene, in view of the continuous distribution of intelligence measures in human populations, and the results of the animal studies of Tryon (1940), Thompson (1954), and others. However, the factor could conceivably represent a single critical parameter of neural functioning or of cognitive development, under multiple gene control. The data of the present study of course offer little clue to the nature of such a hypothetical neural or biochemical variable, if it exists.

The results of the present investigation also bear some relevance to Cattell's hypothesis of two kinds of intelligence, "fluid" and "crystallized" (Cattell, 1943, 1963). This hypothesis holds that two "general intelligence" factors exist, strongly correlated but functionally distinct, the one reflecting innate ability, the other the effects of educational and cultural processes. These factors are said to be best measured by nonverbal and verbal tests, respectively. In some ways the present study offers more direct support for such a notion than the data Cattell himself presents, although it should be noted that the hereditary factor in this study is a quite general one, and by no means represented only or chiefly in nonverbal tests.

The present results may also be regarded as at least compatible with the findings of Nichols (1965), who examined the heritability of both general and specific abilities in the National Merit Scholarship Qualifying Test with a large twin sample. His method of analysis did not lead to a separate assessment of hereditary and environmental components, but he found evidence for the heritability both of total score and of subtest residual scores with total score partialed out. Because of the high intercorrelation of the subtests the residual scores yielded rather erratic results, but on the whole the data suggested that they were under a substantial degree of separate genetic control. Nichols' total score may perhaps be regarded as comparable to our general factors, his specifics to our subtest groupings.

Finally it is appropriate to compare the present results with those obtained for some of the same data by Vandenberg (1965), using a multiplicative model of the relation between heredity and environment, and analyzing total variation rather than common variation. The evaluation of environmental effects in both analyses was based on the within-pair MZ variance-covariance matrix, but hereditary effects were represented in the earlier study not by a difference matrix, as here, but by finding the matrix which multiplicatively transforms the matrix of MZ differences into that of DZ differences. In the earlier study, four significant latent roots were found in both the "heredity" and "environment" matrices, and interpreted as possibly corresponding approximately to the Number, Verbal, Space, and Word Fluency factors.

Actually, the differences between the results of the previous study and the present one appear to reflect more the difference between analyzing total variation and analyzing common variation than they do the difference between additive and multiplicative models. If specific variance is included in a factor analysis, the number of dimensions obtained tends to approach the number of tests. If common variance alone is analyzed, fewer and broader factors tend to emerge. In this view, one might expect correspondence between the previous study and the first analyses of the present study (those based on subtests), since in both these cases variance specific to the five main scales is included. And, indeed, the results do not differ markedly here. It is only in the second part of the present analysis, with variance specific to the main factors absent, that the present investigation yields general factors not found in the earlier one.

The question of the ultimate usefulness of additive and multiplicative models in this area is thus left open by the present results. For now, all we can suggest is that more experience with both kinds of models is desirable.

References

Blewett, D. B. 1954. An experimental study of the inheritance of intelligence. *J. Ment. Sci.* **100:** 922-33.

Broadhurst, P. L. and Jinks, J. L. 1961. Biometrical genetics and behavior: reanalysis of published data. *Psychol. Bull.* **58:** 337-62.

Burt, C. and Howard, Margaret. 1956. The multifactorial theory of inheritance and its application to intelligence. *Brit. J. Stat. Psychol.* **9:** 95-131.

Cattell, R. B. 1943. The measurement of adult intelligence. *Psychol. Bull.* **40:** 153-93.

———. 1960. The multiple abstract variance analysis equations and solutions. *Psychol. Rev.* **67:** 353-72.

———. 1963. Theory of fluid and crystallized intelligence: a critical experiment. *J. Educ. Psychol.* **54:** 1-22.

DuBois, P. H. 1965. *An Introduction to Psychological Statistics.* New York: Harper.

Fisher, R. A. 1918. The correlations between relatives on the supposition of Mendelian inheritance. *Trans. Roy. Soc.* (Edinburgh) **52:** 399-433.

Fuller, J. L. and Thompson, W. R. 1960. *Behavior Genetics.* New York: Wiley.

Guilford, J. P. 1956. The structure of intellect. *Psychol. Bull.* **53:** 267-93.

Herzberg, F. and Lepkin, M. 1954. A study of sex differences on the Primary Mental Abilities test. *Educ. Psychol. Meas.* **14:** 687-89.

Kaiser, H. F. 1958. The varimax criterion for analytic rotation in factor analysis. *Psychometrika* **23:** 187-200.

Kempthorne, O. and Osborne, R. H. 1961. The interpretation of twin data. *Amer. J. Hum. Genet.* **13:** 320-39.

Loehlin, J. C. 1965. A heredity-environment analysis of personality inventory data. In *Methods and goals in human behavior genetics,* ed. S. G. Vandenberg, pp. 163-68. New York: Academic Press.

Mather, K. 1949. *Biometrical Genetics.* New York: Dover.

Nichols, R. C. 1954. The National Merit Twin Study. In *Methods and goals in human behavior genetics,* ed. S. G. Vandenberg, pp. 231-42. New York: Academic Press.

———. 1964. "A statistical model for twin research." Personal Communication.

Spearman, C. 1904. 'General intelligence,' objectively determined and measured. *Amer. J. Psychol.* **15:** 201-93.

Sutton, H. E.; Vandenberg, S. G.; and Clark, P. J. 1962. The hereditary abilities study: selection of twins, diagnosis of zygosity and program of measurements. *Amer. J. Hum. Genet.* **14:** 52-63.

Thompson, W. R. 1954. The inheritance and development of intelligence. *Proc. Assoc. Res. Nerv. Ment. Dis.* **33:** 309-31.

Thomson, G. H. 1920. General versus group factors in mental activities. *Psychol., Rev.* **27:** 173-90.

Thurstone, L. L. 1934. The vectors of mind. *Psychol. Rev.* **41:** 1-32.

Thurstone, L. L. and Thurstone, Thelma G. 1941. *The Chicago Tests of Primary Mental Abilities.* Chicago: Science Research Associates.

Thurstone, Thelma G.; Thurstone, L. L.; and Strandskov, H. H. 1953. *A psychological study of twins. 1. Distributions of absolute twin differences for identical and fraternal twins.* Report No. 4, Psychometric Laboratory. Chapel Hill, N.C.: University of North Carolina.

———. 1955. *Scores of one hundred and twenty-five pairs of twins on fifty-nine tests.* Report No. 12, Psychometric Laboratory. Chapel Hill, N.C.: University of North Carolina.

Tryon, R. C. 1935. A theory of *psychological* components—an alternative to 'mathematical factors.' *Psychol. Rev.* **42:** 425-54.

———. 1940. Genetic differences in maze-learning ability in rats. *39th Yearbk. Nat. Soc. Study of Educ.* Part. 1, pp. 111-19.

Vandenberg, S. G. 1962. The hereditary abilities study: hereditary components in a psychological test battery. *Amer. J. Hum. Genet.* **14:** 330-37.

———. 1965. Innate abilities: one or many? *Acta Genet. Med. Gemell.* **14:** 41-47.

———. 1966. Contributions of twin research to psychology. *Psychol. Bull.* **66:** 327-52.

———, Stafford, R. E., and Brown, Anne. 1968. The Louisville Twin Study. In *Progress in Human Behavior Genetics,* ed. S. G. Vandenberg, pp. 153–204. Baltimore: The Johns Hopkins Press.

Wright, S. 1952. The genetics of quantitative variability. In *Quantitative Inheritance,* ed. E. C. R. Reeve and C. H. Waddington, pp. 5-41. London: H.M.S.O.

APPENDIX

Tables A–H

Basic variance-covariance matrices
for MZ and DZ twins

TABLE A
VARIANCES AND COVARIANCES, PMA SCALES: MZ PAIR SUMS

	N	V	S	W	R
N	3603.41	1521.73	1449.47	889.60	1034.20
V		2047.49	712.91	837.76	959.43
S			4059.80	304.36	453.67
W				1161.32	497.75
R					937.22

TABLE B
VARIANCES AND COVARIANCES, PMA SCALES: DZ PAIR SUMS

	N	V	S	W	R
N	3943.48	2160.52	2248.69	1282.17	1425.28
V		2161.06	1683.94	1062.12	1199.44
S			4595.79	877.62	1084.55
W				1064.25	682.36
R					1064.74

TABLE C
VARIANCES AND COVARIANCES, PMA SCALES: MZ PAIR DIFFERENCES

	N	V	S	W	R
N	372.74	−1.48	1.36	−2.57	30.04
V		161.25	18.13	58.41	69.88
S			449.28	28.26	13.66
W				196.90	43.60
R					126.89

TABLE D
VARIANCES AND COVARIANCES, PMA SCALES: DZ PAIR DIFFERENCES

	N	V	S	W	R
N	1183.74	242.25	464.66	307.51	226.77
V		325.00	111.14	198.69	132.65
S				478.80	114.34
W			1110.50	313.88	183.25
R					177.84

TABLE E
VARIANCES AND COVARIANCES, PMA SUBSCALES: MZ PAIR SUMS

	1	2	3	4	5	6	7	8	9	10	11	12	13	14	15
1. Addit.	186.87	233.77	274.20	88.91	115.00	82.08	158.75	99.79	111.22	97.06	25.76	51.82	67.55	28.90	94.37
2. Mult.		567.85	473.11	164.73	229.54	161.82	180.95	100.80	142.42	208.16	64.02	96.93	113.85	58.13	168.76
3. 3-Hi.			902.84	202.66	301.78	225.44	316.61	201.24	257.38	233.65	58.65	101.89	195.30	98.91	272.65
4. Sent.				182.43	214.63	156.59	83.58	85.63	88.50	134.82	37.80	66.71	80.22	45.21	151.62
5. Vocab.					394.14	255.46	98.28	88.89	108.45	224.68	69.16	109.83	113.44	70.06	235.86
6. Compl.						251.72	112.08	92.62	117.09	149.26	38.98	70.50	87.58	50.37	171,61
7. Flags							619.55	466.07	436.14	119.11	25.42	49.92	77.95	30.31	119.58
8. Fig.								589.60	406.70	103.09	12.08	37.39	47.68	28.44	94.09
9. Cards									504.75	133.74	16.91	50.21	67.73	41.86	97.04
10. 1st L.										406.48	125.73	138.21	92.60	59.99	174.70
11. 4-LW.											74.13	49.63	29.61	22.26	47.82
12. Suff.												98.58	42.83	23.54	71.58
13. L. Ser.													101.91	43.68	127.55
14. L. Grp.														47.84	67.26
15. Ped.															313.39

TABLE F
VARIANCES AND COVARIANCES, PMA SUBSCALES: DZ PAIR SUMS

	1	2	3	4	5	6	7	8	9	10	11	12	13	14	15
1. Addit.	178.96	261.02	315.45	114.07	199.58	111.52	214.68	170.51	173.57	134.45	36.24	76.75	89.72	51.14	140.57
2. Mult.		636.52	565.04	187.89	320.73	167.44	264.49	211.82	213.88	269.76	51.26	132.94	140.26	106.49	193.12
3. 3-Hi.			918.54	300.30	485.20	283.47	525.74	351.73	417.86	356.86	92.65	162.06	252.88	149.16	343.52
4. Sent.				191.02	255.44	146.58	204.87	124.53	158.67	165.95	51.97	80.24	97.98	58.19	174.16
5. Vocab.					457.05	254.76	295.01	189.32	249.88	276.71	84.29	146.44	175.66	94.76	287.05
6. Compl.						199.42	196.30	134.78	173.22	163.85	43.33	78.23	102.95	55.09	176.95
7. Flags							641.25	485.71	478.28	210.47	61.07	116.67	166.90	71.33	235.40
8. Fig.								608.72	483.32	120.39	58.76	81.37	113.01	37.33	182.51
9. Cards									533.67	155.86	53.52	101.51	136.08	48.44	209.86
10. 1st L.										386.11	85.92	148.50	117.44	77.55	183.08
11. 4-LW.											54.72	36.91	42.27	25.52	67.17
12. Suff.												97.02	70.85	38.28	101.93
13. L. Ser.													144.84	64.93	157.14
14. L. Grp.														47.79	69.88
15. Ped.															311.52

Table G
Variances and Covariances, PMA Subscales: MZ Pair Differences

	1	2	3	4	5	6	7	8	9	10	11	12	13	14	15
1. Addit.	42.72	24.22	16.71	1.78	5.10	1.00	3.53	.68	2.71	1.55	−.80	2.624	7.58	.36	9.26
2. Mult.		86.91	4.14	5.21	15.32	8.62	6.23	−8.55	−3.66	5.74	3.98	1.544	7.26	−1.25	6.31
3. 3-Hi.			159.12	−9.66	−12.37	−8.70	5.08	−11.07	7.67	−13.03	3.23	−1.408	3.53	3.77	−.21
4. Sent.				19.66	13.50	4.93	−4.91	.96	−.69	8.55	4.74	1.591	6.61	2.48	6.56
5. Vocab.					46.44	17.10	−.52	4.52	9.67	12.25	8.97	4.955	12.81	7.92	12.17
6. Compl.						26.12	2.12	8.71	2.67	10.42	4.82	2.097	7.85	3.23	9.96
7. Flags							139.28	23.00	20.78	9.70	−4.10	3.710	1.22	2.01	2.17
8. Fig.								119.18	15.26	4.69	−2.59	.615	2.31	−.04	5.87
9. Cards									74.49	11.15	1.93	5.914	3.26	1.29	−2.73
10. 1st L.										90.89	12.02	13.433	11.16	.95	4.51
11. 4-LW.											22.83	4.698	6.95	1.04	5.80
12. Suff.												20.581	3.67	1.19	5.30
13. L. Ser.													27.64	3.05	5.80
14. L. Grp.														14.86	4.83
15. Ped.															56.39

TABLE H
VARIANCES AND COVARIANCES, PMA SUBSCALES: DZ PAIR DIFFERENCES

	1	2	3	4	5	6	7	8	9	10	11	12	13	14	15
1. Addit.	95.70	80.94	83.74	19.38	21.08	13.41	25.39	17.08	15.89	30.11	29.84	20.84	14.59	11.98	5.30
2. Mult.		237.55	118.84	26.97	45.75	10.67	66.88	39.69	34.20	29.53	24.84	19.90	33.02	22.71	42.82
3. 3-Hi.			259.57	43.08	34.19	26.49	86.43	86.64	46.06	67.95	31.89	31.25	34.92	34.25	30.56
4. Sent.				37.90	26.63	23.28	19.08	9.59	.50	18.56	3.66	6.80	4.98	4.81	21.49
5. Vocab.					62.97	32.61	19.83	17.77	13.17	41.61	14.02	26.79	13.58	8.23	28.04
6. Compl.						59.09	27.35	17.64	10.38	45.11	16.95	26.19	18.61	3.75	26.81
7. Flags							215.97	131.77	96.58	68.58	7.21	19.61	18.71	22.40	26.47
8. Fig.								202.91	94.40	81.71	21.08	30.66	21.14	23.36	21.81
9. Cards									139.27	46.08	20.94	9.60	27.40	23.20	20.10
10. 1st L.										160.18	50.77	44.88	22.82	23.39	11.12
11. 4-LW.											40.22	22.43	14.26	7.36	2.41
12. Suff.												38.60	10.65	8.95	12.37
13. L. Ser.													29.31	10.27	12.55
14. L. Grp.														24.52	5.40
15. Ped.															65.46

16

RONALD S. WILSON
University of Louisville School of Medicine

AUTONOMIC RESEARCH WITH TWINS: METHODS OF ANALYSIS

The expansion of interest in behavior genetics has touched off studies of central and autonomic nervous system functioning in twins, since these physiological measures are expected to be heavily influenced by genetic factors. While the results seem generally confirmatory (e.g., Dustman and Beck, 1965), there is an indigenous problem in this research area of choosing the statistical method that is best suited to the data and that will provide a test of concordance for the twins.

Some statistical methods based on analysis of variance (ANOVA) have been adapted for use with autonomic data (Wilson, 1967), and they provide a test of individual differences in autonomic patterning and responsiveness. The purpose of the present paper is to illustrate how the same ANOVA methods may be used to measure the similarity of autonomic reaction patterns within twin pairs, as a first step toward delineating the role of genetic and environmental factors in autonomic functioning.

The use of ANOVA to determine the components of variance associated with genetic and environmental factors is well established. Kempthorne and Osborne (1961) have presented a comprehensive model for analysis of twin data, in which some twelve separate sources of variance are identified that would affect the covariance for monozygotic (MZ) and dizygotic (DZ) twins. Cattell (1953, 1960) has worked out a more elaborate model in which the nature-nurture ratio for a given trait is determined from data obtained on groups where the pair members differ in specified ways along a dimension of genetic and environmental similarity (e.g., MZ twins raised apart, unrelated children raised together, etc.).

The methods presented herein differ from the above methods principally in focusing upon a statistical test of within-pair covariance rather than estimating the expected values for the separate components of variance. These tests signify whether the presumed link between genetic similarity and concordance of autonomic reactions is actually present. If the

---

Preparation of this paper was supported by Public Health Service Research Grants MH 12290 and HD 00843 from the National Institutes of Health.

results are positive, then it may be informative to proceed further and estimate the expected values.

A more complete appraisal of the role of genetic factors may be gained by supplementing the MZ and DZ twin groups with a group of sibling pairs, a group of half siblings, and a group of unrelated children formed into pairs. After ANOVA has been performed for each group, a further test may be made to determine whether within-pair concordance in autonomic functioning increases progressively across the groups, from unrelated children to MZ twins.

Before presenting the analysis-of-variance model certain requirements about the data will be briefly mentioned—they are discussed in more detail elsewhere (Wilson, 1967). The sample of twins should be reasonably good-sized (say 20 pairs) and the data should be in quantitative form and normally distributed. It is assumed that the experimental procedure and scoring method are identical for all Ss, and that several trials are given to assure the stability of the data.

The proposed model is designated by Winer (1962, pp. 335-37) as a Case I three-factor design, where one factor is random and two factors are fixed. The model provides for repeated measures on the two fixed factors that are common to all Ss in the experiment. The third factor orders Ss into different groups; in this case each twin pair constitutes a separate group, with $K = 2$. One important feature of the design is that all interactions can be tested for significance, and as the illustration will show, the evidence for within-pair similarity in autonomic patterning is contained in the interactions.

For the single variable case, assume that E is interested in beat-to-beat changes in heart rate produced by shock. He measures beat-to-beat (or R-R) time in milliseconds for 5 beats preceding and 15 beats following shock. His experiment is also designed to study adaptation, so 10 trials are selected to span the duration of the experiment. The design is shown in Figure 16-1.

The sources of variance and the comparisons of major interest to E are presented in Table 16-1. Aside from the F ratios, the variance estimates may be combined to yield intraclass correlation coefficients as well. These coefficients express the degree of similarity in autonomic functioning within the twin pairs. Haggard (1958) provides an informative discussion of intraclass correlation and analysis of variance; see also Winer (1962, Chap. 4) and Wilson (1967).

Each source of variance designated as Error in Table 16-1 is a measure of within-pair variability, and each is used to test the significance of the observed between-pair variance for a particular aspect of autonomic func-

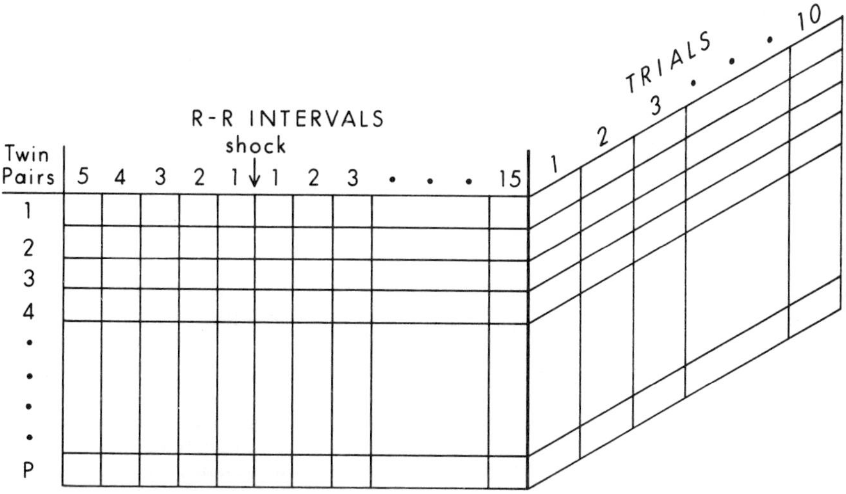

FIG. 16-1. DESIGN FOR *ANOVA* OF CARDIAC RESPONSE TO SHOCK IN A SAMPLE OF TWINS.

tioning. The larger the F ratio of the between-pair variance to the within-pair variance, and hence the larger the intraclass correlation coefficient, the better the evidence for twin concordance, as noted below.

1. *Pairs.* The test reveals whether there are significant between-pair differences in average R-R time, or sustained heart rate, and the correlation coefficient for within-pair covariance in average heart rate is given by

$$R_I = \frac{MS_{Pairs} - MS_{Error\ (Pairs)}}{MS_{Pairs} + MS_{Error\ (Pairs)}}, \text{ or } \frac{F - 1}{F + 1}. \tag{1}$$

The other main effects reveal the consequence of the experimental treatment for the sample as a whole, ignoring pair membership. The test for R-R Intervals signifies whether a typical cardiac response curve is induced by shock, while the test for Trials shows whether there is a significant drift in average heart rate during the experiment.

2. *Pairs × R-R Intervals.* This interaction signifies whether each twin pair displays a distinctive cardiac response curve. It also serves as the error term for the test of R-R Intervals, so the investigator can judge the magnitude of the stable, sample-wide response to shock as contrasted to the idiosyncratic response displayed by each pair.

TABLE 16-1
ANALYSIS OF VARIANCE OF CARDIAC RESPONSE TO SHOCK IN TWINS

Source	Tested Against	Results Signify Whether There Is:
Between Subjects		
Pairs	Error (Pairs)	Difference in average R-R time between pairs
Error (Pairs); within Pairs		
Within Subjects		
R-R Intervals	Pairs × R-R Intervals	Common cardiac response to shock
Pairs × R-R Intervals	Error 1	Differential cardiac response for each pair
Error 1: within Pairs × R-R Intervals		
Trials	Pairs × Trials	Common change in average R-R time over trials
Pairs × Trials	Error 2	Differential change in average R-R time over trials
Error 2: within Pairs × Trials		
R-R Intervals × Trials	Pairs × R-R Intervals × Trials	Common change in cardiac response curve over trials
Pairs × R-R Intervals × Trials	Error 3	Differential change in cardiac response curve over trials
Error 3: within Pairs × R-R Intervals × Trials		

The interaction itself is tested against a measure of within-pair variability in the response curve. The more substantial the F ratio, the better the evidence that each twin pair generates a stable but distinctive curve that distinguishes it from other pairs. The intraclass correlation expresses the extent to which the two members of each pair display homogeneous cardiac response curves, and is given by

$$R_I = \frac{MS_{\text{Pairs x R-R Intervals}} - MS_{\text{Error 1}}}{MS_{\text{Pairs x R-R Intervals}} + MS_{\text{Error 1}}}, \text{ or } \frac{F - 1}{F + 1}. \quad (2)$$

3. *Pairs* × *Trials.* If there are differential changes in average heart rate for the twin pairs over successive trials, it will be shown in this interaction. A significant F ratio might lead one to inspect the records to see if those pairs that displayed a high rate initially then showed a steady

decrement over trials, while the pairs that started at a lower rate remained constant throughout.

4. *R-R Intervals* × *Trials.* This interaction ignores pair membership and reveals whether the cardiac response curve for the sample showed changes during the experiment. In the case of adaptation, one might expect a flattening of the response curve over successive trials.

5. *Pairs* × *R-R Intervals* × *Trials.* The triple interaction represents the trial-to-trial variation in the cardiac response curve for each pair. It serves as the error term for the preceding interaction and the ratio so formed indicates whether the sample-wide changes in the response curve exceed the idiosyncratic changes for each pair.

The triple interaction is tested against a measure of within-pair variance and a significant F ratio would point to a distinctive heart-rate curve for each pair on a given trial, but the form of the curve would change over trials. Probably this would represent differential adaptation effects across the trials.

Advantages and Limitations

The model as presented here can be employed for any variable where quantitative scoring is carried out over successive intervals, and where the scores are distributed in approximately normal fashion. The magnitude and phasing of the response are preserved by interval scoring, and estimates of within-pair response consistency are available from the variance table.

One caution is in order, however. Designs that provide for repeated measures on each S assume that the correlations are equal between all the measures. For autonomic variables scored over sequential intervals this assumption is likely to be violated, and some adjustment should be made in evaluating the F tests that involve repeated measures (Winer, 1962).

One recommended adjustment is to reduce the degrees of freedom for the F test in proportion to the degree of violation. The data matrix can be tested for deviation from constant variance and covariance, and the reduction factor (epsilon) can be computed from the elements of the matrix (Box, 1950, 1954; Greenhouse and Geisser, 1959).

Unfortunately the computations are laborious and likely to be ignored unless a computer is available. Some empirical results have been obtained, however, for heart-rate curves on a sizable sample ($N = 111$), and these results showed that the reduction factor did not fall below 0.60 for any matrix (Wilson and Bartels, 1968).

Based on these results, the following rule of thumb is suggested: For all comparisons involving repeated measures, reduce the degrees of freedom by one-half when establishing the significance level of the F ratio. By setting epsilon equal to 0.50, the investigator is provided reasonable protection against a Type I error even when the variance-covariance matrix is deviant.

THE MULTIVARIATE CASE

When several autonomic variables are simultaneously recorded the investigator may wish to explore the possibility of response patterning among the vairables, and the further possibility of distinctive and reliable patterns that characterize each pair of twins. The same model may be used for the multivariate case (Fig. 16-2), although certain modifications are required for the data entries.

For this application, separate variables are substituted in place of the Trials dimension. The Intervals dimension is retained and E must specify the intervals in such a way that a score is available on each variable at each interval. One will probably define his intervals on a time base, scoring each variable at successive one- or two-second intervals. This system preserves the sequential changes that appear simultaneously in all

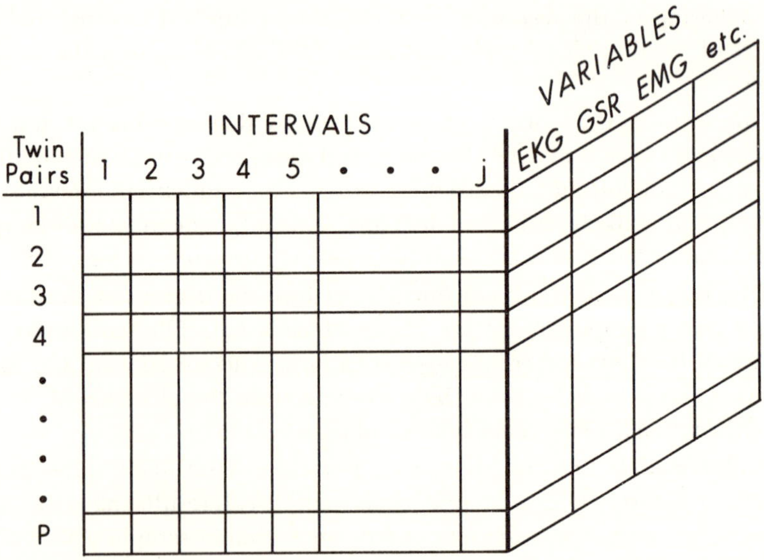

FIG. 16-2. DESIGN FOR ANOVA OF MULTIVARIATE AUTONOMIC REACTIONS IN A SAMPLE OF TWINS.

variables, and the evidence for autonomic patterning comes from the response curves generated by the separate variables.

Since trials are omitted as a treatment dimension, the data entries should be the mean of t trials to stabilize the data. One may choose to organize the experiment into several periods with t trials per period, and the results of the analysis for each period may be compared to determine whether progressive changes are evident throughout the experiment.

Inclusion of separate variables raises the problem of the widely different scoring units for each variable. The solution is to transform the raw scores to standard scores for each variable, using the sample mean and sigma. In the ANOVA there will be no main effect for variables, but all other main effects and interactions are preserved intact.

A second problem concerns the interpretation of scores that have been pooled across the separate variables. Theoretically if the variables are positively intercorrelated, so that twins high on heart rate are also high on skin conductance, muscle action potential, etc., a combined score might be meaningfully interpreted as a measure of overall autonomic tone. A method for estimating the average intercorrelation among variables will be illustrated shortly, but until this is available one will wish to proceed cautiously when interpreting these pooled scores.

Given these restraints, the ANOVA is performed just as in the univariate case, and the summary table is shown in Table 16-2. The comparisons of major interest are elaborated below.

1. *Pairs.* This source of variance is derived from the total scores for each twin pair, summed over variables and intervals. A significant F ratio suggests a positive intercorrelation between variables and significant between-pair differences in overall autonomic tone. However the final interpretation will also be influenced by the size of other variance components; specifically, the interactions of pairs $\times$ variables, and of pairs $\times$ variables $\times$ intervals.

2. *Variables.* No test here since all data entries are converted to standard scores.

3. *Intervals.* The mean score for each interval is obtained by adding across the separate variables, so a systematic effect would appear here only if the response curves for the separate variables displayed the same direction and magnitude of change. In view of the evidence for directional fractionation of response (e.g., Lacey, Kagan, Lacey and Moss, 1963) it seems improbable that the separate variables would generate congruent response curves. However the nature of the experimental task and the method of scoring will have some bearing on the form of the curves, and one's interpretation would be guided accordingly.

TABLE 16-2
SUMMARY TABLE FOR ANOVA OF MULTIVARIATE AUTONOMIC DATA RECORDED WITH TWINS

Source	Tested Against	Results Signify Whether There Is:
Between Subjects		
Pairs	Error (Pairs)	Between-pair differences in overall autonomic tone (see text).
Error (Pairs): within Pairs		
Within Subjects		
Variables*	Pairs × Variables	
Pairs × Variables	Error 1	Between-pair differences in basal autonomic profile.
Error 1: within Pairs × Variables		
Intervals	Pairs × Intervals	See text.
Pairs × Intervals	Error 2	See text.
Error 2: within Pairs × Intervals		
Variables × Intervals	Pairs × Variables × Intervals	Divergence among response curves for separate variables.
Pairs × Variables × Intervals	Error 3	Between-pair differences in autonomic reaction patterns.
Error 3: within Pairs × Variables × Intervals		

* When standard scores are used there will be no main effect for Variables.

4. *Variables × Intervals.* It is more probable that each variable would generate a distinctive response curve of its own, and the variance associated with the distinctive response curves is contained in this interaction. The larger the F ratio, the better the evidence for directional fractionation of response that is exhibited consistently by all twin pairs. It signifies that the experimental procedure evokes a relatively constant autonomic reaction pattern in the twins even though the curves for the separate variables do not follow a parallel course.

5. *Pairs × Variables.* Each member of the pair is represented by his mean score (averaged across intervals) on each variable, and the interaction appraises for each pair the change in average score from one variable to the next.

Since the average scores represent sustained autonomic activity for each variable, the term *basal autonomic profile* has been suggested as a descriptive title for this vector of scores (Wilson, 1967). The interaction is a test of between-pair discrepancy and within-pair concordance in autonomic profile. Profile concordance within each pair can be expressed by an intraclass correlation, where

$$R_I = \frac{MS_{\text{Pairs x Variables}} - MS_{\text{Error 1}}}{MS_{\text{Pairs x Variables}} + MS_{\text{Error 1}}}, \text{ or } \frac{F - 1}{F + 1}. \quad (3)$$

The investigator will in general expect a substantial mean square and F ratio for this interaction. If the F test is nonsignificant, two possible conclusions should be considered: (a) each individual displays a basal autonomic profile that is stable for him, but within the twin pairs the members differ as much from each other as they do from the other twin pairs; or (b) the correlations between variables are so high that each pair is principally identified by its total score for all variables combined. The profile for each pair approximates a flat line, and between-pair differences are apparent only in profile elevation, not in profile contour. (If this conclusion is tenable, the original test for Pairs as a main effect should have produced a highly significant F ratio.)

A close estimate of the average intercorrelation between variables can be obtained from the variance table. The dimension of pairs is temporarily ignored and the data are analyzed to reveal the extent to which a given S retains the same rank order across variables. The intraclass correlation is calculated as

$$R_I = \frac{\dfrac{SS_{\text{Pairs}} + SS_{\text{Error (Pairs)}}}{df_{\text{Pairs}} + df_{\text{Error (Pairs)}}} - \dfrac{SS_{\text{Pairs x Var}} + SS_{\text{Error 1}}}{df_{\text{Pairs x Var}} + df_{\text{Error 1}}}}{\dfrac{SS_{\text{Pairs}} + SS_{\text{Error (Pairs)}}}{df_{\text{Pairs}} + df_{\text{Error (Pairs)}}} + \dfrac{(V - 1)\,(SS_{\text{Pairs x Var}} + SS_{\text{Error 1}})}{df_{\text{Pairs x Var}} + df_{\text{Error 1}}}} \quad (4)$$

where $V =$ the number of variables.

The larger this coefficient, the clearer the evidence that individual Ss maintain the same relative position across variables; the Ss are distinguished by differences in profile elevation. Now if the two Ss in each twin pair generate profiles that are equally elevated, the concordance within each pair is revealed in the F ratio and R_I coefficient for the main effect of pairs; and if the twins also duplicate each other in the contour of their basal autonomic profile, the concordance will be represented by the F ratio and R_I coefficient for the pairs $\times$ variables interaction. One advantage of this design is that it allows one to pull apart the sources of variance and determine from the data the interpretation that is most appropriate.

6. *Pairs $\times$ Intervals.* The problem of interpretation here is the same as for the main effect of intervals. If a significant interaction were obtained,

it would suggest that each twin pair generated a relatively uniform response curve for the separate variables, but there was considerable discrepancy between pairs in the form of the curve. The prospects seem guarded at best for a uniform response curve over the separate variables, and consequently the interaction is expected to be small.

7. *Pairs × Variables × Intervals.* If the average response curve is plotted for each variable, and the curve for each twin pair is then projected onto the graph, how closely do the individual curves fit? This is the test for idiosyncratic response patterning, to determine the extent to which each twin pair is distinguished by patterned autonomic reactions that differ in form and magnitude from the pattern observed in other twins. The larger the interaction, the better the evidence that each pair is identified by an idiosyncratic autonomic response pattern, and the coefficient of within-pair concordance can be calculated in the manner given previously.

Assuming this interaction is significant, E will want to establish whether idiosyncratic patterning is a more prominent factor in his results than the sample-wide response pattern shared by all the twin pairs. The common response pattern is appraised in the test of variables × intervals, for which the triple interaction serves as the error term, and if this test is also significant the investigator should proceed to estimate the expected values for both the common and idiosyncratic sources of variance.

Applications

The preceding design may be used with any sample of subject pairs that one wishes to employ, on the assumption that the pairs so selected are a random sample from a population of such pairs. As mentioned earlier, a more comprehensive appraisal of the role of genetic factors in autonomic functioning may be obtained by performing the same experiment with other groups where the pairs are composed of siblings, half siblings and unrelated children. The prediction is that within-pair concordance in autonomic functioning will increase as genetic similarity and common experience increase, and therefore one needs some method to evaluate the change in concordance from group to group.

COMPARISONS BETWEEN GROUPS

Two methods have often been employed for such comparisons. First, the error terms representing the within-pair variance for two groups (usually MZ and DZ twins) can be formed into an F ratio, with the denominator being the within-pair variance for the group expected to display greater concordance. Other two-group comparisons would follow the same pattern, and significant results in the predicted direction would support the hypothesis.

This method has been widely used for comparing the similarity of MZ and DZ twins on single variables, and the variance estimates are sometimes processed to produce a heritability index—see Vandenberg (1966) for a discussion of the method. The chief limitations are to be found in the assumption of equal score distribution in the groups, the exclusion of between-pair variance from the calculations, and the restriction of comparing only two groups at a time.

The second method is an extension of the first and makes use of the intraclass correlation coefficients obtained for each group, which take into account the between-pair and within-pair variance. The coefficients are transformed to z values by the formula

$$z = \tfrac{1}{2} \log_e \frac{1 + (K - 1)R}{1 - R}, \text{ or } \tfrac{1}{2} \log_e F, \tag{5}$$

where K is the number of members in the class; for twin pairs this is 2. Many statistical texts contain a transformation table for r and z that can be used when K is 2.

The z values for any two groups can be compared in reference to the expected distribution of differences for two independent groups, which according to Haggard (1958) is given by

$$\begin{aligned} \sigma^2_{z_1 - z_2} &= \frac{K_1}{2(C_1 - 2)(K_1 - 1)} + \frac{K_2}{2(C_2 - 2)(K_2 - 1)}, \\ &= \frac{1}{C_1 - 2} + \frac{1}{C_2 - 2}, \end{aligned} \tag{6}$$

where C = the number of classes, or twin pairs.

The test of significance takes the form

$$\text{Critical Ratio} = \frac{z_1 - z_2}{\sqrt{\sigma^2_{z_1 - z_2}}} \tag{7}$$

and the table of the normal curve may be used to determine the p value for the critical ratio.

This method circumvents some of the difficulties of dealing with estimates of within-pair variance alone, but it is still limited to the two-group comparison. What would be preferable is a technique for multigroup comparisons.

MULTI-GROUP CONCORDANCE ANALYSIS

If the ideal statistical test could be designed for these data, it would be one in which the within-pair concordance in autonomic functioning was determined separately for each group, and then tested for change in relative magnitude across the groups. The latter qualification is not easily met by ANOVA because of the restrictive condition of subject pairs, but the covariance within pairs suggests a form of regression analysis in which the slope coefficients for the several groups would be tested for homogeneity.

The procedures for regression analysis involving independent groups are readily available (Hays, 1963; McNemar, 1962), but the assumption of a clearly defined independent variable and dependent variable is not satisfied by these data, where the two members of each pair are regarded as indistinguishable. Therefore, it appears that the between-group comparisons must make use of correlation coefficients that represent the within-pair concordance for each group.

In brief, the procedure to be outlined below is a generalization to the multi-group case of the two-group comparison given earlier. It makes use of the intraclass correlation coefficient (transformed to z) obtained in each group for a particular aspect of autonomic functioning, and evaluates whether the separate group coefficients are significantly different from each other. Further, it provides for more specific comparisons among the groups by dividing the between-group sum of squares into separate components.

In the illustration to be given, it will be assumed that ANOVA has been carried out for five separate groups and intraclass correlations have been computed for each group. With reference to specific aspects of autonomic functioning, we wish to determine (a) whether the groups differ in the degree of within-pair concordance they display, and (b) whether the concordance increases progressively from unrelated children to MZ twins.

Some hypothetical data and the appropriate calculations are shown in Table 16-3. The summary statistics at the bottom show that the hypothesis of a common population value can be rejected; the z values differ significantly from each other.

The further question is whether the within-pair concordance increases progressively as genetic and environmental similarity increases. The hypothetical z values fall in this order, and one can apply a set of linear coefficients to the z values to determine whether a progressive increment in within-pair concordance is manifest over the groups.

The use of specialized coefficients to make comparisons between groups

TABLE 16-3
HYPOTHETICAL DATA FOR ANOVA OF TRANSFORMED CORRELATION COEFFICIENTS

Group	Number of Pairs (P_i)	Intraclass Correlation	z_i	$\sigma^2_{error} = \left(\frac{1}{P_i - 2}\right)$	$(z_i - \bar{z}_{weighted})$
MZ Twins	30	.82	1.157	.0357	.542
DZ Twins	28	.71	.887	.0385	.272
Full Sibs	44	.56	.633	.0238	.018
Half Sibs	33	.42	.448	.0323	−.167
Unrelated	50	.24	.245	.0208	−.370

$$\bar{z}_{weighted} = \frac{(P_1 - 2)\, z_1 + \ldots + (P_5 - 2)\, z_5}{(P_1 - 2) + \ldots + (P_5 - 2)} = \frac{107.692}{175} = .615$$

$$\text{SS Betw. Groups} = \sum_1^g (z_i - \bar{z}_{wtd})^2 = (.542)^2 + \ldots + (-.370)^2 = .5329$$

$$df = g - 1 = 4$$

$$\text{MS Pooled Error} = \frac{1}{g}\sum_1^g \left(\frac{1}{P_i - 2}\right) = \frac{1}{5}(.0357 + \ldots + .0208) = .0302$$

$$df = \Sigma P_i - 2g = 175$$

Source	df	MS	F	p
Between Groups	4	.1332	4.41	<.01
Error (Pooled)	175	.0302		

is covered in Winer (1962, p. 65 ff.). The computations are performed in the same way as fitting a set of orthogonal polynomials to a set of treatment means, but in the present illustration there is no implication of curve fitting nor that the groups are equally spaced along some quantitative dimension. The choice of comparisons to be made should be settled before the analysis begins and should be plausibly linked to the logic of the experiment, so that capitalizing upon chance differences is minimized. The comparisons do not have to be orthogonal; the more important consideration is the rationale that underlies each comparison.

Technically the groups must contain an equal number of pairs before the specialized coefficients can be employed. However, the unusual circumstance here of dealing with a single z value for each group rather than sums of squares and cross-products introduces less of an error with unequal pairs, and if one has an irremediable problem with the size of his groups the comparisons can still furnish an approximate answer.

The application of the specialized coefficients to the z values is illustrated in Table 16-4. The first comparison is the one of principal interest since it evaluates whether there is a progressive increment across groups in within-pair concordance. The remaining orthogonal comparisons are included to exhaust the between-group sum of squares and reveal the discrepancy that arises when the groups have unequal pairs.

It is apparent from Table 16-4 that the linear comparison accounts for practically all the between-group variance. With one degree of freedom for each component, the F test for the linear comparison is

$$F_{\text{Linear}} = \frac{MS_{\text{Linear}}}{MS_{\text{Error}}} = \frac{.5121}{.0302} = 16.96, \tag{8}$$

TABLE 16-4
APPLICATION OF ORTHOGONAL COEFFICIENTS TO HYPOTHETICAL DATA*

Group	z_i	Linear v_i	Linear v_iz_i	Quad v_i	Quad v_iz_i	Cubic v_i	Cubic v_iz_i	Quartic v_i	Quartic v_iz_i
MZ Twins	1.157	2	2.314	−2	−2.314	1	1.157	−1	−1.157
DZ Twins	.887	1	.887	1	.887	−2	−1.774	4	3.548
Full Sibs	.633	0	0	2	1.266	0	0	−6	−3.798
Half Sibs	.448	−1	−.448	1	.448	2	.896	4	1.792
Unrelated	.245	−2	− .490	−2	− .490	−1	− .245	−1	− .245
Σv_i^2			2.263		− .203		.034		.140
Σv_iz_i		10		14		10		70	

$$\text{Component sum of squares} = \frac{(\Sigma v_iz_i)^2}{\Sigma v_i^2}$$

$$\text{Linear: } \frac{(2.263)^2}{10} = \frac{5.1212}{10} = .5121$$

$$\text{Quadratic: } \frac{(-.203)^2}{14} = \frac{.0412}{14} = .0029$$

$$\text{Cubic: } \frac{(.034)^2}{10} = \frac{.0012}{10} = .0001$$

$$\text{Quartic: } \frac{(.140)^2}{70} = \frac{.0196}{70} = .0003$$

Sum of component SS = .5154
Between-group SS from Table 16-3 = .5329
Difference = .0175

* Each vector of orthogonal coefficients is identified by its conventional name solely for descriptive purposes; curve-fitting is not implied.

which for 1 and 175 degrees of freedom gives $p < .001$. None of the other components approaches significance, so it may be said that there is a highly significant progression across groups in the size of the coefficient that represents within-pair concordance in autonomic functioning.

As noted at the bottom of Table 16-4, the sum of the component SS is smaller than the between-group SS computed in Table 16-3. The discrepancy arises from the difference between the weighted $\bar{z}$ used in Table 16-3 and the simple average $\bar{z}$ implicitly contained in the polynomial calculations. The weighted $\bar{z}$ compensates for differences in group size while the simple $\bar{z}$ does not, and when the two $\bar{z}$'s differ it can be shown that

$$\sum_{1}^{g} (z_i - z_{wtd})^2 = \sum_{1}^{g} (z_i - z_{simpl})^2 + g(z_{wtd} - z_{simpl})^2, \quad (9)$$

or

$$\sum_{1}^{g} (z_i - .615)^2 = \sum_{1}^{g} (z_i - .674)^2 + 5(.615 - .674)^2,$$

$$.5329 = .5159 + 5(.00348)$$

so

$$.5329 = .5333 \text{ within rounding error.}$$

The simple average is inherent in the method of extracting orthogonal components of variance, so when unequal group size produces a discrepancy between the simple and the weighted mean, the sum of the component SS will underestimate the between-group SS in proportion to the discrepancy. Consequently the tests for individual components will be imprecise, but the error is in the conservative direction and if one cannot obtain equal representation in all groups, the comparisons can be performed to furnish an approximate answer.

As a final note, this analysis may be applied to the several intraclass correlation coefficients that are based on independent sources of within-pair covariance for each group. It is both plausible and desirable to make between-group comparisons for the extent of within-pair concordance in sustained autonomic activity, autonomic profile and autonomic responsiveness, where these measures are obtained as described in this paper and elsewhere (Wilson, 1967). At this point it is still largely an open question whether there is increasing concordance of autonomic functioning as genetic and environmental factors coincide, and whether the separate dimensions of autonomic activity display the same trend in concordance. Provided the assumptions of ANOVA are respected, the methods outlined herein provide a means of securing an answer to the question.

References

Box, G. E. P. 1950. Problems in the analysis of growth and wear curves. *Biometrics,* **6:** 362-89.

———. 1954. Some theorems on quadratic forms applied in the study of analysis of variance problems, II. Effects of inequality of variance and of correlation between errors in the two-way classification. *Ann. Math. Stat.* **25:** 484-98.

Cattell, R. B. 1953. Research designs in psychological genetics with special reference to the multiple variance analysis method. *Amer. J. Hum. Genet.* **5:** 76-93.

———. 1960. The multiple abstract variance analysis equations and solutions: For nature-nurture research on continuous variables. *Psychol. Rev.* **67:** 353-72.

Dustman, R. E. and Beck, E. C. 1965. The visually evoked potential in twins. *Electroencephal. and Clin. Neurophysiol.* **19:** 570-75.

Greenhouse, S. W. and Geisser, S. 1959. On methods in the analysis of profile data. *Psychometrika* **24:** 95-111.

Haggard, E. A. 1958. *Intraclass Correlation and the Analysis of Variance.* New York: Dryden Press.

Hays, W. L. 1963. *Statistics for Psychologists.* New York: Holt, Rinehart and Winston.

Kempthorne, O. and Osborne, R. H. 1961. The interpretation of twin data. *Amer. J. Hum. Genet.* **13:** 320-39.

Lacey, J. I., Kagan, J., Lacey, B. C., and Moss, H. A. 1963. The visceral level: Situational determinants and behavioral correlates of autonomic response patterns. In *Expression of the Emotions in Man,* ed. P. Knapp, pp. 161-96. New York: International Universities Press.

McNemar, Q. 1962. *Psychological Statistics,* 3rd ed. New York: Wiley.

Vandenberg, S. G. 1966. Contributions of twin research to psychology. *Psychol. Bull.* **66:** 327-52.

Wilson, R. S. 1967. Analysis of autonomic reaction patterns. *Psychophysiology,* **4:** 125-42.

Wilson, R. S. and Bartels, B. L. 1968. The cardiac response: Stability of principal components and individual differences. *J. Compar. Physiol. Psychol.,* **65:** 132-39.

Winer, B. J. 1962. *Statistical Principles in Experimental Design.* New York: McGraw- Hill.

17

GLENN E. ROUDABUSH
American Institutes for Research
Palo Alto, California

ANALYZING DYADIC RELATIONSHIPS

A dyadic relationship is a relationship between pairs of entities, and a dyadic score is one obtained by combination of the scores belonging to two different entities, most often a simple difference between them. A dyadic score is dyadic by virtue of this double source, and the relative contribution of each member of the pair is lost. According to Sears (1951), a dyadic score (or unit) is "one that describes the combined actions of two or more persons" and for Cronbach (1958) "the term *dyadic* . . . may be applied to any study which compares descriptions of, statements about, or actions by two persons." In studies of interpersonal perception, etc., the scores for two persons are generally combined into an overall index of similarity or dissimilarity of the two entities involved —by person intercorrelations or by the Euclidean distance index—and inferences are then drawn from comparisons of these overall indices. There are serious methodological problems concerned with this procedure, but they need not be considered here. Cronbach (1958) has given a rather complete review of them. Our attention will be restricted to simple dyadic scores and the relations between them, which I take to be typical of twin studies.

We shall begin by considering the simple difference score. In this case we have two data matrices, say X_1 and X_2, each with N rows and p columns. The columns of these two matrices correspond in that they each represent the same p variates arranged in the same order. The rows of X_1 and X_2 also correspond in that they represent corresponding first and second members of pairs of entities to be compared. These might be N twin pairs with the first member of each pair assigned to X_1 and the second member assigned to the corresponding row of X_2. The matrix of difference scores, Y, is obtained by subtracting X_2 from X_1:

$$Y = X_1 - X_2 \tag{1}$$

Equation 1 can be written in supermatrix form. First form the supermatrix $X = (X_1X_2)$ by appending X_2 to X_1 on the right. Then form the supermatrix $E' = (I—I)$ by appending the negative of a p by p identity matrix to a p by p identity matrix on the right. The matrix of difference scores is then:

$$Y = XE = [X_1\ X_2]\begin{bmatrix} I \\ -I \end{bmatrix} = X_1 - X_2 \tag{2}$$

The covariance matrix of difference scores is:

$$\frac{Y'Y}{N} - \frac{Y'11'Y}{N^2} = \frac{E'X'XE}{N} - \frac{E'X'11'XE}{N^2}$$
$$= E'\left[\frac{X'X}{N} - \frac{X'11'X}{N^2}\right]E \tag{3}$$

where 1 is the unit vector. Notice that the expression in brackets on the right of (3) is the covariance matrix of the supermatrix X, which can be partitioned into four parts:

$$\text{Cov}\,(X) = C = \begin{bmatrix} C_{11}\ C_{12} \\ C_{21}\ C_{22} \end{bmatrix} \tag{4}$$

where C_{11} is the covariance matrix of the p variables for the first pair members in X_1, C_{22} is the covariance matrix of the p variables for the second pair members in X_2, and C_{12} and C_{21} (one is the transpose of the other) are "cross" covariance matrices between the pairs. The covariance matrix of difference scores is then:

$$\text{Cov}(Y) = E'CE = [I\ -I]\begin{bmatrix} C_{11}\ C_{12} \\ C_{21}\ C_{22} \end{bmatrix}\begin{bmatrix} I \\ -I \end{bmatrix} \tag{5}$$

Multiplying out the right side of (5) gives:

$$\text{Cov}(Y) = C_{11} + C_{22} - C_{12} - C_{21} \tag{6}$$

The covariance matrix of difference scores is the sum of the covariance matrices of first pair members and second pair members less the covariance that members of pairs share. If the second member of a pair is always identical to the first (with respect to the p measured variables), then the covariances of differences are all zero, as are the differences themselves. If members of pairs differ only by independent errors of measurement, then the covariances of difference scores are all error covariances, and the correlation matrix of difference scores derived from it will not differ significantly from the identity matrix. Vandenberg (1965) published some correlation matrices of this kind computed from monozygous twin difference scores on some cognitive tests. In these matrices about 5 per cent of the correlations are significant at the 5 per cent level.

Such matrices are residuals which represent whatever is dissimilar between members of pairs of persons. In the case of twin studies, two such matrices can be obtained, one for identical twin differences and one for fraternal twin differences. The identical twins differ with respect to errors of measurement, and, for some sets of variables at least, they differ with respect to the experiences they have had. Fraternal twins differ (to an unknown extent) genetically as well. It is appropriate, then, to attempt to assess the nature and extent of the genetic differences existing in the fraternal twin matrix. Vandenberg (1965; 1966) has done just this.

In place of considering the dissimilarity between pairs of entities, it is possible instead to consider their similarity. This similarity is represented in the off-diagonal submatrices of (4) or in the corresponding correlation matrix:

$$r = \begin{bmatrix} r_{11}\, r_{12} \\ r_{21}\, r_{22} \end{bmatrix} \tag{7}$$

Equation 7 suggests the use of canonical correlations for comparison of similarity between pair members by obtaining the eigenvalues and eigenvectors of the matrix $r_{11}^{-1}\, r_{12}\, r_{22}^{-1}\, r_{21}$. This approach would provide transformations for the (standardized) X_1 and X_2 matrices to transform them into orthogonal variates that are maximally correlated. A comparison of these for different groups of pairs would then be possible. I would like to suggest a somewhat different procedure, however, which accomplishes about the same thing, but does it one step at a time, thus providing more information with which to work.

Suppose our purpose is to assess the similarity between identical twins on the one hand and fraternal twins on the other and then compare these similarities. We have the same measures of some kind of performance on each group of twins, and we are looking for any underlying genetic factors contributing to these measures of performance. We can suppose that all of the persons in our total sample are members of some general population, the characteristics of which (in respect to our set of variables) could be independently assessed. Our total sample is selected in that each member is a twin. Each of the two groups of twins is also selected in that one group contains only monozygous twins, the other only dizygous. Each of these two groups is further subdivided into two matched subgroups by assigning one pair member (randomly) to one group and the other to the other group. This provides us with four data matrices, X_{m1}, X_{m2}, X_{d1}, and X_{d2}, where the first subscript indicates monozygous or dizygous, and the second subscript indicates first or second pair members. The first step in the procedure is to perform a principal axes decomposition or factoring of the four data matrices separately, obtaining the basic structure (Horst, 1963) of each. That is, let:

$$\begin{aligned} X_{m1} &= P_{m1}\Delta_{m1}Q'_{m1} \\ X_{m2} &= P_{m2}\Delta_{m2}Q'_{m2} \\ X_{d1} &= P_{d1}\Delta_{d1}Q'_{d1} \\ X_{d2} &= P_{d2}\Delta_{d2}Q'_{d2} \end{aligned} \tag{8}$$

Here the P_{ij} are vertical orthonormal factor score (or component) matrices, the matrices $Q_{ij}\Delta_{ij}$ are factor loading matrices containing the correlations between the orthogonal factors and the original variables, and the diagonal matrices Δ_{ij} contain the square roots of the factor variances arranged in descending order along the main diagonals. At this point one could examine the elements of the diagonal matrices, Δ_{ij}, to decide if any factors can be discarded, or this can be left until later.

The next step is to form the products of factor scores for each of the two groups of twins separately, $P'_{m1}P_{m2}$ and $P'_{d1}P_{d2}$, and obtain the basic structure of these, that is, let:

$$\begin{aligned} P'_{m1}P_{m2} &= V_m\delta_m U'_m \\ P'_{d1}P_{d2} &= V_d\delta_d U'_d \end{aligned} \tag{9}$$

Here V_m and U_m are orthonormal transformation matrices which will rotate P_{m1} and P_{m2} (respectively) to maximal congruence and V_d and U_d will similarly transform P_{d1} and P_{d2}. The elements of δ_m and δ_d are the corresponding coefficients of congruence which are analogous to canonical correlations. To see that this is so, form the supermatrix $(P_{m1}P_{m2})$ by appending P_{m2} to P_{m1} on the right. Then obtain the minor product moment of this matrix:

$$\begin{bmatrix} P'_{m1} \\ P'_{m2} \end{bmatrix} [P_{m1}\ P_{m2}] = \begin{bmatrix} I & P'_{m1}P_{m2} \\ P'_{m2}P_{m1} & I \end{bmatrix} \tag{10}$$

The elements of this matrix are correlations between factor scores except for a correction for origin. As such, they are coefficients of congruence (Tucker, 1957) for factor scores. From (10) proceed as in canonical correlations to find:

$$\begin{aligned} P'_{m2}P_{m1}P'_{m1}P_{m2} &= V_m\delta_m{}^2V'_m \\ U_m &= P'_{m2}P_{m1}V_m\delta_m{}^{-1} \end{aligned} \tag{11}$$

From (8) note that:

$$P_{m1} = X_{m1}Q_{m1}\Delta_{m1}{}^{-1} \tag{12}$$

Substituting equations (12) into the first equation in (9):

$$\Delta_{m1}^{-1}Q'_{m1}X'_{m1}X_{m2}Q_{m2}\Delta_{m2}^{-1} = V_m\delta_m U'_m \tag{13}$$

If we started our analysis with properly standardized scores, that is, if we factored the correlation matrices from r_m and r_d, then the product $X'_{m1}X_{m2}$ is the submatrix r_{12} from r_m. If we factored covariance matrices, then this product is C_{12} from the supermatrix C_m. For standardized scores (13) becomes:

$$\Delta_{m1}^{-1}Q'_{m1}r_{12}Q_{m2}\Delta_{m2}^{-1} = V_m\delta_m U'_m \tag{14}$$

That is, we need not actually compute the factor score matrices.

The elements of δ_m and δ_d are measures of the similarity of monozygous and dizygous twins, respectively, on certain orthogonal factors. The four factor loading matrices can be obtained by applying the appropriate rotation to the original principal axes loadings in (8):

$$\begin{aligned} F_{m1} &= Q_{m1}\Delta_{m1}V_m \\ F_{m2} &= Q_{m2}\Delta_{m2}U_m \\ F_{d1} &= Q_{d1}\Delta_{d1}V_d \\ F_{d2} &= Q_{d2}\Delta_{d2}U_d \end{aligned} \tag{15}$$

The two monozygous twin factor loading matrices should look very much alike as should the two dizygous matrices, though perhaps not as much. The monozygous and dizygous matrices may still be quite dissimilar since nothing has yet been done to make them similar. It may, however, be possible to interpret at least some factors as being the same in all four matrices. If so, the analysis might be completed at this point, noting that the matrix $(\delta_m^2 - \delta_d^2)$ gives a measure of the proportion of the variance of identified factors attributable to genetic (and perhaps experiential) differences in the group of fraternal twins. If few factors can be identified across twin groups, then a number of things can be done. I will describe only one of them.

I take the two (hopefully quite similar) monozygous twin factor loading matrices as primary and find two orthonormal transformation matrices which will rotate F_{d1} toward F_{m1} and F_{d2} toward F_{m2}, respectively. This can be done in the following way. Form the products $F'_{m1}F_{d1}F'_{d1}F_{m1}$ and $F'_{m2}F_{d2}F'_{d2}F_{m2}$ and find the basic structure of each of these:

$$\begin{aligned} F'_{m1}F_{d1}F'_{d1}F_{m1} &= G_1D_1G'_1 \\ F'm_2F_{d2}F'_{d2}D_{m2} &= G_2D_2G'_2 \end{aligned} \tag{16}$$

Then define:

$$\begin{aligned} H_1 &= F'_{d1}F_{m1}G_1D_1^{-1/2} \\ H_2 &= F'_{d2}F_{m2}G_2D_2^{-1/2} \end{aligned} \tag{17}$$

Here H_1 projects F_{d1} into the space of F_{m1}, and the columns of $F_{d1}H_1$ can be directly compared with those of F_{m1}. Similarly, H_2 projects F_{d2} into the space of F_{m2}. H_1 and H_2 are orthonormal matrices. Since we have rotated the factors of the two dizygous factor loading matrices, the congruence coefficients appearing in δ_d can no longer be meaningfully compared with those in δ_m. From (9) we have:

$$[V'_d P'_{d1}]\,[P_{d2} U_d] = \delta_d \tag{18}$$

Pre- and postmultiplying (18) by H_1' and H_2', respectively, gives:

$$\phi_d = S[H_1' V_d' P_{d1}']\,[P_{d2}\ U_d H_1] = H_1' \delta_d H_2 \tag{19}$$

The elements of ϕ_d are the coefficients of congruence between the dizygous twin pairs in the new (rotated) factor space. Note that ϕ_d is not diagonal. The elements in the diagonal of ϕ_d can be compared with those in δ_m, and the elements of $(\delta_m^2 - D_{\phi d}^2)$ are a measure of the proportion of factor variance attributable to genetic differences in the fraternal twins.

One further step is possible. That is to choose one of the factor loading matrices, say F_{m1}, and compute:

$$F_g = F_{m1}\,(\delta_m^2 - D_{\phi d}^2)^{1/2} \tag{20}$$

This matrix is the factor loadings of the fraternal twin genetic differences. The sum of squared elements of each row gives a measure of the variance (or proportion of variance depending upon the original metric) of each of the original p variables that is attributable to these genetic differences so that "heritability" coefficients can be obtained. Finally, a rotation of this matrix to "simple structure" might lead to the identification of genetic factors.

This suggested procedure has been programmed for an IBM 7090 computer in the FORTRAN language, and an example problem will be presented here. The data for the analysis were kindly provided by Dr. Vandenberg and consist of scores on 15 of Thurstone and Thurstone's (1941) Primary Mental Abilities (PMA) tests for 77 monozygous twin pairs and 38 dizygous twin pairs. These data have been previously analyzed by Vandenberg (1965) using another procedure. The 15 tests were originally constructed to measure five factors—number, verbal, spatial, word fluency, and reasoning. In the present analyses, only four of these factors emerged.

The analysis begins with the two supermatrices, r_m and r_d, of intercorrelations for monozygous and dizygous twins, respectively. These are shown in Tables 17-1 through 6. Table 17-1 gives the intercorrelations for the first members of monozygous twin pairs, Table 17-2 gives the

TABLE 17-1

INTERCORRELATIONS OF 15 PMA TESTS FOR THE FIRST MEMBERS OF THE MONOZYGOUS TWIN PAIRS (r_{11} FOR MZ). BASED ON 77 CASES

		1	2	3	4	5	6	7	8
1	Addition	1.000	.614	.642	.446	.400	.334	.230	.101
2	Multiplication	.614	1.000	.566	.458	.455	.396	.191	.125
3	Three higher	.642	.566	1.000	.467	.447	.466	.405	.319
4	Sentences	.446	.458	.467	1.000	.797	.660	.229	.284
5	Vocabulary	.400	.455	.447	.797	1.000	.806	.291	.287
6	Completion	.334	.396	.466	.660	.806	1.000	.296	.369
7	Flags	.230	.191	.405	.229	.291	.296	1.000	.601
8	Figures	.101	.125	.319	.284	.287	.369	.601	1.000
9	Cards	.181	.239	.340	.318	.317	.366	.658	.701
10	First letters	.289	.451	.353	.399	.458	.410	.070	.145
11	Four letter words	.168	.324	.133	.228	.369	.266	−.046	.054
12	Suffixes	.119	.233	.203	.254	.331	.289	.091	.087
13	Letter series	.543	.403	.622	.589	.659	.614	.324	.337
14	Letter grouping	.258	.293	.431	.404	.476	.418	.255	.173
15	Pedigrees	.291	.396	.465	.616	.675	.599	.299	.326
	Means	16.818	39.908	47.041	17.408	28.234	27.013	28.539	25.534
	Stand. Dev.	5.929	12.273	16.363	6.948	10.193	8.758	13.715	13.592

TABLE 17-1

INTERCORRELATIONS OF 15 PMA TESTS FOR THE FIRST MEMBERS OF THE MONOZYGOUS TWIN PAIRS (r_{11} FOR MZ). BASED ON 77 CASES

		9	10	11	12	13	14	15
1	Addition	.181	.289	.168	.119	.543	.258	.291
2	Multiplication	.239	.451	.324	.233	.403	.293	.396
3	Three higher	.340	.353	.133	.203	.622	.431	.465
4	Sentences	.318	.399	.228	.254	.589	.404	.616
5	Vocabulary	.317	.458	.369	.331	.659	.476	.675
6	Completion	.366	.410	.266	.289	.614	.418	.599
7	Flags	.658	.070	−.046	.091	.324	.255	.299
8	Figures	.701	.145	.054	.087	.337	.173	.326
9	Cards	1.000	.231	.156	.200	.358	.287	.249
10	First letters	.231	1.000	.698	.501	.382	.270	.359
11	Four letter words	.156	.698	1.000	.525	.193	.311	.190
12	Suffixes	.200	.501	.525	1.000	.230	.193	.293
13	Letter series	.358	.382	.193	.230	1.000	.476	.682
14	Letter grouping	.287	.270	.311	.193	.476	1.000	.456
15	Pedigrees	.249	.359	.190	.293	.682	.456	1.000
	Means	22.123	33.740	11.618	9.105	15.844	15.247	23.870
	Stand. Dev.	10.860	10.996	4.904	4.644	5.976	3.833	9.569

TABLE 17-2
INTERCORRELATIONS OF 15 PMA TESTS FOR THE FIRST MEMBERS OF THE MONOZYGOUS TWIN PAIRS (r_{11} FOR MZ). BASED ON 77 CASES

		1	2	3	4	5	6	7	8
1	Addition	1.000	.635	.589	.455	.421	.424	.324	.191
2	Multiplication	.635	1.000	.527	.581	.507	.424	.234	.047
3	Three higher	.589	.527	1.000	.352	.421	.418	.314	.138
4	Sentences	.455	.581	.352	1.000	.782	.724	.258	.142
5	Vocabulary	.421	.507	.421	.782	1.000	.801	.216	.101
6	Completion	.424	.424	.418	.724	.801	1.000	.296	.150
7	Flags	.324	.234	.314	.258	.216	.296	1.000	.664
8	Figures	.191	.047	.138	.142	.101	.150	.664	1.000
9	Cards	.190	.170	.389	.157	.236	.259	.724	.665
10	First letters	.313	.434	.230	.506	.499	.487	.347	.180
11	Four letter words	.151	.261	.121	.325	.378	.253	.125	−.008
12	Suffixes	.317	.477	.305	.514	.588	.526	.228	.029
13	Letter series	.441	.478	.499	.545	.575	.529	.356	.105
14	Letter grouping	.240	.157	.326	.354	.401	.414	.335	.163
15	Pedigrees	.468	.330	.475	.569	.637	.640	.339	.234
	Means	17.724	38.342	45.158	17.247	27.566	27.195	27.685	25.458
	Stand. Dev.	6.332	12.032	17.016	6.477	10.274	8.977	11.789	11.636

TABLE 17-2
INTERCORRELATIONS OF 15 PMA TESTS FOR THE SECOND MEMBERS OF THE MONOZYGOUS TWIN PAIRS (r_{11} FOR MZ). BASED ON 77 CASES

	9	10	11	12	13	14	15
1 Addition	.190	.313	.151	.317	.441	.240	.468
2 Multiplication	.170	.434	.261	.477	.478	.157	.330
3 Three higher	.389	.230	.121	.305	.499	.326	.475
4 Sentences	.157	.506	.325	.514	.545	.354	.569
5 Vocabulary	.236	.499	.378	.588	.575	.401	.637
6 Completion	.259	.487	.253	.526	.529	.414	.640
7 Flags	.724	.347	.125	.228	.356	.335	.339
8 Figures	.665	.180	−.008	.029	.105	.163	.234
9 Cards	1.000	.257	.050	.207	.263	.421	.274
10 First letters	.257	1.000	.578	.703	.470	.373	.454
11 Four letter words	.050	.578	1.000	.558	.441	.214	.263
12 Suffixes	.207	.703	.558	1.000	.438	.205	.400
13 Letter series	.263	.470	.441	.438	1.000	.517	.534
14 Letter grouping	.421	.373	.214	.205	.517	1.000	.486
15 Pedigrees	.274	.454	.263	.400	.534	.486	1.000
Means	21.507	34.273	10.597	10.653	15.597	14.829	23.052
Stand. Dev.	10.533	10.893	4.916	4.958	5.257	3.761	9.222

TABLE 17-3

CORRELATIONS OF 15 PMA TESTS ACROSS FIRST AND SECOND MEMBERS OF THE MONOZYGOUS TWIN PAIRS (r_{11} FOR MZ). BASED ON 77 CASES

		1	2	3	4	5	6	7	8
1	Addition	.513	.502	.556	.462	.361	.340	.241	.063
2	Multiplication	.504	.749	.507	.502	.403	.383	.253	.152
3	Three higher	.530	.599	.725	.449	.516	.487	.332	.169
4	Sentences	.383	.472	.470	.793	.651	.677	.325	.246
5	Vocabulary	.311	.360	.466	.706	.744	.691	.264	.231
6	Completion	.327	.308	.480	.615	.665	.808	.243	.166
7	Flags	.256	.180	.373	.212	.316	.324	.566	.599
8	Figures	.088	.058	.306	.187	.181	.240	.477	.556
9	Cards	.203	.215	.361	.273	.275	.325	.615	.603
10	First letters	.236	.308	.276	.404	.323	.320	.197	.226
11	Four letter words	.132	.235	.047	.241	.222	.216	.247	.227
12	Suffixes	−.071	.172	.039	.303	.334	.362	.104	.123
13	Letter series	.343	.312	.647	.493	.524	.499	.313	.179
14	Letter grouping	.283	.278	.333	.340	.313	.434	.275	.095
15	Pedigrees	.371	.399	.540	.533	.587	.524	.204	.128

Table 17-3

Correlations of 15 PMA Tests Across First and Second Members of Monozygous Twin Pairs (r_{12} for MZ). Based on 77 Cases

		9	10	11	12	13	14	15
1	Addition	.060	.261	.349	.392	.394	.191	.234
2	Multiplication	.120	.478	.270	.450	.375	.168	.300
3	Three higher	.308	.341	.256	.506	.493	.342	.462
4	Sentences	.246	.423	.224	.444	.439	.325	.550
5	Vocabulary	.245	.507	.350	.446	.457	.327	.559
6	Completion	.268	.424	.227	.385	.420	.347	.557
7	Flags	.628	.120	.042	.188	.326	.241	.390
8	Figures	.685	.130	−.000	.037	.130	.261	.323
9	Cards	.697	.260	.158	.220	.301	.412	.361
10	First letters	.214	.632	.479	.495	.346	.265	.348
11	Four letter words	.168	.573	.574	.439	.172	.182	.185
12	Suffixes	.086	.480	.352	.534	.193	.107	.173
13	Letter series	.261	.343	.243	.376	.512	.397	.577
14	Letter grouping	.273	.411	.316	.285	.425	.475	.471
15	Pedigrees	.137	.415	.229	.380	.566	.310	.670

TABLE 17-4

INTERCORRELATIONS OF 15 PMA TESTS FOR THE FIRST MEMBERS OF THE DIZYGOUS TWINS PAIRS (r_{11} FOR DZ). BASED ON 38 CASES

		1	2	3	4	5	6	7	8
1	Addition	1.000	.706	.723	.545	.673	.629	.617	.506
2	Multiplication	.706	1.000	.667	.431	.616	.442	.339	.280
3	Three higher	.723	.667	1.000	.712	.772	.707	.599	.444
4	Sentences	.545	.431	.712	1.000	.859	.692	.580	.497
5	Vocabulary	.673	.616	.772	.859	1.000	.831	.564	.560
6	Completion	.629	.442	.707	.692	.831	1.000	.597	.484
7	Flags	.617	.339	.599	.580	.564	.597	1.000	.601
8	Figures	.506	.280	.444	.497	.560	.484	.601	1.000
9	Cards	.598	.393	.516	.560	.622	.554	.687	.755
10	First letters	.474	.413	.578	.548	.658	.670	.542	.356
11	Four letter words	.489	.269	.385	.422	.466	.486	.398	.371
12	Suffixes	.658	.420	.620	.549	.745	.775	.685	.470
13	Letter series	.680	.425	.684	.682	.734	.735	.616	.488
14	Letter grouping	.633	.596	.823	.727	.772	.658	.621	.468
15	Pedigrees	.274	.182	.467	.538	.608	.677	.436	.371
	Means	18.789	38.811	44.784	16.895	26.447	26.158	26.250	23.778
	Stand. Dev.	7.506	15.050	17.490	6.546	11.959	8.900	13.179	11.784

TABLE 17-4

INTERCORRELATIONS OF 15 PMA TESTS FOR THE FIRST MEMBERS OF THE DIZYGOUS TWINS PAIRS (r_{11} FOR DZ). BASED ON 38 CASES

	9	10	11	12	13	14	15
1 Addition	.598	.474	.489	.658	.680	.633	.274
2 Multiplication	.393	.413	.269	.420	.425	.596	.182
3 Three higher	.516	.578	.385	.620	.684	.823	.467
4 Sentences	.560	.548	.422	.549	.682	.727	.538
5 Vocabulary	.622	.658	.466	.745	.734	.772	.608
6 Completion	.554	.670	.486	.775	.735	.658	.677
7 Flags	.687	.542	.398	.685	.616	.621	.436
8 Figures	.755	.356	.371	.470	.488	.468	.371
9 Cards	1.000	.490	.405	.614	.559	.622	.423
10 First letters	.490	1.000	.587	.770	.456	.540	.425
11 Four letter words	.405	.587	1.000	.422	.500	.425	.272
12 Suffixes	.614	.770	.422	1.000	.585	.619	.490
13 Letter series	.559	.456	.500	.585	1.000	.729	.669
14 Letter grouping	.622	.540	.425	.619	.729	1.000	.483
15 Pedigrees	.423	.425	.272	.490	.669	.483	1.000
Means	19.514	33.895	10.921	9.459	15.105	15.026	23.081
Stand. Dev.	11.323	11.517	4.534	5.173	7.675	4.505	9.739

TABLE 17-5

INTERCORRELATIONS OF 15 PMA TESTS FOR THE SECOND MEMBERS OF THE DIZYGOUS TWIN PAIRS (r_{22} FOR DZ). BASED ON 38 CASES

		1	2	3	4	5	6	7	8
1	Addition	1.000	.741	.588	.681	.709	.633	.359	.304
2	Multiplication	.741	1.000	.687	.595	.631	.498	.400	.348
3	Three higher	.588	.687	1.000	.492	.544	.541	.572	.417
4	Sentences	.681	.595	.492	1.000	.755	.669	.362	.175
5	Vocabulary	.709	.631	.544	.755	1.000	.767	.428	.339
6	Completion	.633	.498	.541	.669	.767	1.000	.363	.372
7	Flags	.359	.400	.572	.362	.428	.363	1.000	.712
8	Figures	.304	.348	.417	.175	.339	.372	.712	1.000
9	Cards	.258	.330	.517	.301	.443	.493	.680	.704
10	First letters	.475	.535	.468	.542	.503	.500	.259	.390
11	Four letter words	.247	.328	.348	.400	.354	.348	.246	.513
12	Suffixes	.504	.450	.478	.529	.657	.605	.300	.443
13	Letter series	.380	.457	.723	.345	.547	.499	.517	.501
14	Letter grouping	.469	.562	.596	.420	.487	.408	.432	.461
15	Pedigrees	.407	.508	.547	.552	.724	.600	.469	.356
	Means	17.158	38.579	42.417	17.211	27.026	26.342	28.737	26.278
	Stand. Dev.	6.828	14.854	16.657	6.063	9.865	7.771	12.416	13.458

TABLE 17-5
INTERCORRELATIONS OF 15 PMA TESTS FOR THE SECOND MEMBERS OF THE DIZYGOUS TWIN PAIRS (r_{22} FOR DZ). BASED ON 38 CASES

		9	10	11	12	13	14	15
1	Addition	.258	.475	.247	.504	.380	.469	.407
2	Multiplication	.330	.535	.328	.450	.457	.562	.508
3	Three higher	.517	.468	.348	.478	.723	.596	.547
4	Sentences	.301	.542	.400	.529	.345	.420	.552
5	Vocabulary	.443	.503	.354	.657	.547	.487	.724
6	Completion	.493	.500	.348	.605	.499	.408	.600
7	Flags	.680	.259	.246	.300	.517	.432	.469
8	Figures	.704	.390	.513	.443	.501	.461	.356
9	Cards	1.000	.225	.389	.493	.653	.443	.544
10	First letters	.225	1.000	.524	.648	.407	.423	.430
11	Four letter words	.389	.524	1.000	.626	.464	.344	.341
12	Suffixes	.493	.648	.626	1.000	.560	.388	.651
13	Letter series	.653	.407	.464	.560	1.000	.632	.736
14	Letter grouping	.443	.423	.344	.388	.632	1.000	.559
15	Pedigrees	.544	.430	.341	.651	.736	.559	1.000
	Means	21.324	33.816	10.342	10.263	15.711	14.459	22.711
	Stand. Dev.	11.766	11.003	4.669	5.461	7.078	4.080	9.676

TABLE 17-6

CORRELATIONS OF 15 PMA TESTS ACROSS FIRST AND SECOND MEMBERS OF DIZYGOUS TWIN PAIRS (r_{12} FOR DZ). BASED ON 38 CASES

		1	2	3	4	5	6	7	8
1	Addition	.388	.479	.497	.346	.588	.450	.412	.375
2	Multiplication	.325	.356	.319	.219	.405	.407	.020	.054
3	Three higher	.378	.532	.589	.234	.549	.486	.257	.223
4	Sentences	.435	.428	.471	.524	.720	.627	.337	.246
5	Vocabulary	.467	.411	.459	.457	.725	.698	.320	.232
6	Completion	.518	.476	.440	.388	.596	.570	.353	.214
7	Flags	.557	.559	.647	.560	.568	.497	.419	.224
8	Figures	.539	.463	.387	.321	.542	.432	.203	.243
9	Cards	.519	.463	.449	.437	.504	.599	.262	.307
10	First letters	.390	.470	.475	.501	.516	.505	.183	.128
11	Four letter words	.242	.387	.346	.406	.447	.324	.443	.309
12	Suffixes	.403	.362	.362	.381	.524	.501	.255	.198
13	Letter series	.362	.418	.487	.259	.570	.388	.445	.287
14	Letter grouping	.481	.539	.564	.390	.614	.532	.352	.187
15	Pedigrees	.448	.267	.367	.252	.473	.413	.185	.127

Table 17-6

Correlations of 15 PMA Tests Across First and Second Members of Dizygous Twin Pairs (r_{12} for DZ). Based on 38 Cases

		9	10	11	12	13	14	15
1	Addition	.341	.247	.004	.301	.461	.418	.414
2	Multiplication	−.007	.195	−.082	.137	.115	.034	.030
3	Three higher	.390	.224	.100	.320	.563	.459	.408
4	Sentences	.494	.339	.349	.481	.665	.552	.588
5	Vocabulary	.425	.334	.156	.463	.572	.428	.578
6	Completion	.332	.312	−.016	.309	.501	.535	.556
7	Flags	.409	.280	.195	.468	.570	.519	.515
8	Figures	.387	−.041	.123	.306	.471	.375	.414
9	Cards	.534	.220	.187	.415	.542	.250	.461
10	First letters	.228	.443	.118	.358	.344	.403	.508
11	Four letter words	.229	.180	.083	.046	.312	.276	.417
12	Suffixes	.285	.371	−.020	.395	.369	.363	.513
13	Letter series	.351	.279	.129	.391	.680	.476	.607
14	Letter grouping	.384	.242	.165	.342	.550	.381	.507
15	Pedigrees	.187	.246	.113	.448	.595	.367	.641

intercorrelations for the second members, and Table 17-3 gives the correlations across pairs. In Table 17-3 the correlations for the same test across twin pairs are underlined. Tables 17-4, 5, and 6 give the corresponding information for the dizygous twins. Comparing Table 17-3 with Table 17-6, one can see that the correlations between monozygous twins are generally higher than those between dizygous twins. Tables 17-7 through 10 show the factor loading matrices represented in Equations (8)—the $Q\Delta$ matrices. At this point all but five factors were discarded since we knew that the battery was meant to contain just this number. These tables, however, present evidence of only four factors for the two monozygous matrices and three for the dizygous.

Tables 17-11 and 12 give what I have called the factor score similarities for MZ and DZ twins, respectively. These are the matrices $P_{m1}'P_{m2}$ and $P_{d1}'P_{d2}$ and represent the extent to which the factor score vectors are similar (Tucker's coefficient of congruence). The largest element in each column of these matrices is underlined. Tables 17-13 and 14 show the basic structure of these two matrices as represented in Equations (9). The

TABLE 17-7
FACTOR LOADINGS ON 5 FACTORS OF THE 15 PMA TESTS FROM FIRST MEMBERS OF MONOZYGOUS TWIN PAIRS ($Q_{m1}\Delta_{m1}$)

		I	II	III	IV	V	Communality
1	Addition	.598	.067	−.464	.495	.072	.828
2	Multiplication	.645	.207	−.234	.444	.128	.727
3	Three higher	.726	−.118	−.293	.355	−.127	.769
4	Sentences	.779	.058	−.183	−.272	.223	.767
5	Vocabulary	.841	.113	−.073	−.350	.135	.866
6	Completion	.789	−.004	−.032	−.338	.172	.767
7	Flags	.482	−.667	.212	.182	−.068	.760
8	Figures	.483	−.631	.379	.016	.142	.796
9	Cards	.547	−.544	.440	.171	.008	.818
10	First letters	.608	.490	.347	.175	.088	.769
11	Four letter words	.439	.593	.499	.143	−.113	.827
12	Suffixes	.437	.411	.495	.047	.042	.609
13	Letter series	.802	−.065	−.234	−.108	−.072	.719
14	Letter grouping	.597	.031	−.033	−.150	−.746	.937
15	Pedigrees	.748	−.006	−.127	−.361	−.021	.706
	Factor Variance (Eigenvalues)						
		6.313	1.983	1.465	1.160	.745	
	Cumulative Proportion of Total Variance Accounted for						
		.421	.553	.651	.728	.778	

TABLE 17-8
FACTOR LOADINGS ON 5 FACTORS OF THE 15 PMA TESTS FROM SECOND MEMBERS OF MONOZYGOUS TWIN PAIRS ($Q_{m2}\Delta_{m2}$)

		I	II	III	IV	V	Communality
1	Addition	.641	−.008	−.499	−.308	.066	.759
2	Multiplication	.666	.203	−.333	−.452	−.016	.800
3	Three higher	.629	−.117	−.492	−.134	.312	.767
4	Sentences	.786	.238	−.074	.126	−.333	.807
5	Vocabulary	.819	.249	−.029	.229	−.255	.851
6	Completion	.792	.140	−.068	.299	−.309	.836
7	Flags	.533	−.693	.137	−.155	−.033	.808
8	Figures	.315	−.785	.144	−.112	−.285	.830
9	Cards	.475	−.755	.123	−.036	.039	.814
10	First letters	.709	.144	.464	−.184	.014	.773
11	Four letter words	.486	.330	.569	−.223	.272	.793
12	Suffixes	.690	.303	.345	−.269	−.088	.767
13	Letter series	.757	.084	−.017	.100	.358	.719
14	Letter grouping	.568	−.195	.092	.515	.441	.829
15	Pedigrees	.750	−.010	−.098	.333	−.035	.684
	Factor Variance (Eigenvalues)						
		6.449	2.127	1.344	1.055	.863	
	Cumulative Proportion of Total Variance Accounted for						
		.430	.572	.661	.732	.789	

elements of δ_m and δ_d can be considered canonical correlations. Those for the MZ twins are considerably larger than the corresponding ones for the DZ twins. Tables 17-15 through 18 give the rotated factor loadings represented in Equation (15).

The next step in the procedure is to force congruence between the monozygous and dizygous factors. This part of the procedure is displayed in Tables 17-19 through 22 for the first MZ and DZ matrices and in Tables 17-23 through 26 for the second MZ and DZ matrices. Table 17-19 shows the orthonormal transformation matrix H_1 for rotating F_{d1} toward F_{m1}. Table 17-20 shows the DZ factors rotated by H_1. Tables 17-21 and 22 show the factor loading similarities before and after this rotation, respectively. Tables 17-23 through 26 give the corresponding information for the second MZ and DZ pairs. Table 17-27 shows the first and second DZ twin congruences in the rotated space $H_1' \delta_d H_2$. The diagonal elements in this matrix should be compared with the elements of δ_d in Table 17-14. This table also shows the squares of the diagonal elements labeled $D_{\phi_d}{}^2$, and the differences between these and the squares of the MZ congruence

TABLE 17-9
FACTOR LOADINGS ON 5 FACTORS OF THE 15 PMA TESTS FROM FIRST MEMBERS OF DIZYGOUS TWIN PAIRS ($Q_{d1}\Delta_{d1}$)

		I	II	III	IV	V	Communality
1	Addition	.799	.358	−.230	.030	−.065	.825
2	Multiplication	.621	.682	−.034	−.053	.038	.855
3	Three higher	.848	.301	.126	−.122	.022	.841
4	Sentences	.817	−.036	.174	−.166	−.122	.742
5	Vocabulary	.915	.053	.175	−.066	.016	.875
6	Completion	.867	−.131	.250	.063	.081	.842
7	Flags	.771	−.185	−.294	−.028	.172	.745
8	Figures	.656	−.287	−.531	−.204	−.075	.842
9	Cards	.759	−.198	−.458	−.125	.058	.844
10	First letters	.735	−.067	.113	.530	.239	.896
11	Four letter words	.585	−.090	−.087	.580	−.536	.982
12	Suffixes	.822	−.091	.003	.250	.401	.907
13	Letter series	.831	−.090	.159	−.175	−.281	.834
14	Letter grouping	.849	.176	.068	−.166	−.059	.787
15	Pedigrees	.634	−.452	.437	−.228	−.017	.850
	Factor Variance (Eigenvalues)						
		8.972	1.126	1.012	.901	.654	
	Cumulative Proportion of Total Variance Accounted for						
		.598	.732	.741	.801	.844	

coefficients labeled $(\delta^2_m - D^2_{\phi d})$. The latter can be interpreted as the proportion of factor variance attributable to genetic differences in the fraternal twins.

Table 17-28 shows the genetic differences factor loading matrix, F_g, obtained by Equation (20). The communalities from this matrix can be interpreted as the proportion of the variance of each test attributable to genetic differences in the fraternal twins. In order to obtain an estimate of the proportion of variance in the tests that is inherited, we must multiply these communalities by the inverse of the proportion of genes that is shared by fraternal twins. If we take this proportion to be 0.5, then the communalities should be multiplied by 2. This is shown in the column labeled $h^2 \times 2$. The last column labeled Prop. is the proportion of the common factor variance of the tests attributable to inheritance (under this assumption). The last row in this table gives the proportion of the factor variance for each factor attributable to inheritance. Since neither the test nor the factor proportions of inherited variance approach unity, we may conclude that either (1) the abilities measured by

TABLE 17-10
FACTOR LOADING ON 5 FACTORS OF THE 15 PMA TESTS FROM SECOND MEMBERS OF DIZYGOUS TWIN PAIRS ($Q_{d2}\Delta_{d2}$)

		I	II	III	IV	V	Communality
1	Addition	.723	.427	.240	.210	.195	.845
2	Multiplication	.749	.272	.241	.355	−.061	.823
3	Three higher	.790	−.059	.273	.183	−.208	.779
4	Sentences	.728	.464	.020	−.071	.195	.789
5	Vocabulary	.831	.306	.077	−.261	.121	.873
6	Completion	.773	.236	.003	−.286	.223	.785
7	Flags	.644	−.486	.302	.081	.320	.851
8	Figures	.629	−.584	−.128	.238	.299	.899
9	Cards	.681	−.559	.047	−.213	.168	.852
10	First letters	.672	.243	−.402	.294	−.050	.761
11	Four letter words	.581	−.131	−.658	.186	−.034	.823
12	Suffixes	.769	.090	−.451	−.170	−.015	.832
13	Letter series	.778	−.308	.040	−.156	−.400	.886
14	Letter grouping	.702	−.115	.195	.212	−.397	.747
15	Pedigrees	.786	−.020	.044	−.423	−.254	.864
	Factor Variance (Eigenvalues)						
		7.898	1.708	1.146	.866	.794	
	Cumulative Proportion of Total Variance Accounted for						
		.527	.640	.717	.776	.829	

TABLE 17-11
SIMILARITY COEFFICIENTS FOR MONOZYGOUS FACTOR SCORES ($P_{m1}' P_{m2}$)

	1	2	3	4	5
1	.860	−.046	−.019	.010	−.053
2	.005	.591	.261	−.190	.011
3	−.029	−.379	.587	.008	−.120
4	.041	−.135	−.013	−.584	.224
5	−.017	.051	−.101	−.060	−.419

TABLE 17-12
SIMILARITY COEFFICIENTS FOR DIZYGOUS FACTOR SCORES ($P_{d1}'P_{d2}$)

	1	2	3	4	5
1	.696	.114	.279	−.205	−.105
2	−.183	.093	.049	.228	.185
3	−.037	.046	−.140	−.109	−.285
4	−.041	.103	.048	.156	.168
5	−.049	.285	−.143	−.015	−.028

TABLE 17-13
BASIC STRUCTURE OF THE MONOZYGOUS TWIN FACTOR SCORE SIMILARITIES MATRIX
(V_m, U_m, and δ_m)

	The Matrix V_m for Rotating First Twin Factor Loadings				
	1	2	3	4	5
1	.994	.036	−.000	.103	.023
2	−.093	.577	.441	.679	.050
3	−.019	−.812	.322	.483	−.060
4	.059	.029	.808	−.521	−.267
5	.003	.074	−.222	.152	−.960
	The Matrix U_m for Rotating Second Twin Factor Loadings				
	1	2	3	4	5
1	.990	.077	.044	.089	.066
2	−.117	.866	.027	.476	.094
3	−.065	−.447	.471	.734	.188
4	−.008	−.183	−.804	.279	.492
5	−.045	.104	.358	−.386	.843
	Elements of the Diagonal Matrix δ_m = Canonical Correlations				
	.865	.747	.671	.613	.414

TABLE 17-14
BASIC STRUCTURE OF THE DIZYGOUS TWIN FACTOR SCORE SIMILARITIES MATRIX
(V_d, U_d, and δ_d).

	The Matrix V_d for Rotating First Twin Factor Loadings				
	1	2	3	4	5
1	.939	.236	−.179	−.159	.070
2	−.309	.486	−.257	−.597	.495
3	.037	−.687	−.200	−.666	−.209
4	−.128	.451	−.285	−.128	−.826
5	−.068	−.180	−.884	.398	.155
	The Matrix U_d for Rotating Second Twin Factor Loadings				
	1	2	3	4	5
1	.861	.191	−.045	.091	−.460
2	.057	.076	−.969	−.043	.224
3	.295	.492	.225	−.453	.645
4	−.344	.445	−.090	−.595	−.568
5	−.224	.720	.016	.657	.009
	Elements of the Diagonal Matrix δ_d = Canonical Correlations				
	.832	.474	.345	.096	.018

TABLE 17-15
ROTATED FIRST MONOZYGOUS TWIN FACTOR LOADING MATRIX (F_{m1})

	I	II	III	IV	V
1	.626	.457	.265	−.364	−.156
2	.653	.355	.347	−.118	−.202
3	.758	.197	.169	−.351	.056
4	.757	.219	−.302	.208	−.110
5	.807	.154	−.287	.331	−.007
6	.765	.055	−.324	.265	−.056
7	.547	−.540	−.064	−.406	−.019
8	.533	−.644	−.175	−.182	−.184
9	.596	−.646	.038	−.188	−.095
10	.563	.035	.450	.485	−.114
11	.379	−.051	.563	.597	.080
12	.389	−.144	.369	.546	−.052
13	.801	.173	−.175	−.029	.127
14	.580	.007	.047	.031	.774
15	.725	.115	−.330	.197	.140

TABLE 17-16
ROTATED SECOND MONOZYGOUS TWIN FACTOR LOADING MATRIX (F_{m2})

	I	II	III	IV	V
1	.667	.328	.064	−.424	−.148
2	.661	.457	.236	−.208	−.236
3	.655	.224	.012	−.519	.136
4	.769	.242	−.215	.293	−.158
5	.794	.223	−.247	.332	−.031
6	.784	.126	−.345	.290	−.060
7	.603	−.596	.182	−.212	−.108
8	.408	−.729	.048	−.161	−.321
9	.550	−.662	.101	−.252	−.001
10	.656	.006	.407	.416	.069
11	.395	.138	.575	.451	.289
12	.632	.201	.386	.418	−.067
13	.724	.157	.075	−.016	.406
14	.555	−.215	−.194	−.002	.661
15	.749	.028	−.293	.096	.165

TABLE 17-17
ROTATED FIRST DIZYGOUS TWIN FACTOR LOADING MATRIX (F_{d1})

	I	II	III	IV	V
1	.632	.546	−.140	−.218	.246
2	.376	.471	−.298	−.462	.438
3	.722	.201	−.239	−.374	.286
4	.815	.003	−.016	−.251	.122
5	.857	.088	−.207	−.279	.111
6	.851	−.017	−.261	−.202	−.096
7	.762	.250	−.175	.256	.073
8	.717	.301	.187	.417	.171
9	.769	.331	−.010	.342	.163
10	.632	.259	−.500	−.125	−.406
11	.536	.512	.244	−.269	−.548
12	.741	.188	−.550	.049	−.132
13	.855	.015	.141	−.273	.081
14	.771	.175	−.111	−.288	.260
15	.781	−.471	−.005	−.099	−.085

TABLE 17-18
ROTATED SECOND DIZYGOUS TWIN FACTOR LOADING MATRIX (F_{d2})

	I	II	III	IV	V
1	.602	.522	−.409	−.058	.200
2	.623	.396	−.277	−.304	.331
3	.741	.212	.063	−.295	.306
4	.640	.293	−.468	.208	.176
5	.819	.191	−.291	.262	.114
6	.729	.201	−.234	.376	.136
7	.517	.502	.507	.104	.254
8	.322	.334	.492	.195	.636
9	.604	.137	.544	.302	.286
10	.384	.044	−.384	.025	.681
11	.242	−.164	−.065	.224	.826
12	.596	−.154	−.208	.362	.528
13	.807	−.212	.280	−.104	.317
14	.671	.030	.098	−.406	.347
15	.891	−.201	.028	.137	.100

TABLE 17-19
THE TRANSFORMATION MATRIX, H_1, FOR ROTATING THE FIRST DZ TWIN FACTOR LOADINGS TOWARD THE FIRST MZ FACTOR LOADINGS

	1	2	3	4	5
1	.943	−.210	−.019	.249	.069
2	.209	.055	.523	−.778	.271
3	−.167	−.299	−.207	.112	.910
4	−.185	−.869	.386	.065	−.239
5	.074	−.330	−.731	−.561	−.192

TABLE 17-20
THE FIRST DZ TWIN FACTOR LOADINGS ROTATED BY H_1

	I	II	III	IV	V
1	.792	.047	.039	−.435	.069
2	.620	.293	−.197	−.582	−.091
3	.853	.162	−.213	−.187	−.078
4	.827	.013	−.197	.115	.079
5	.920	.093	−.116	.042	−.060
6	.872	.105	.020	.237	−.117
7	.758	−.340	.198	−.048	−.114
8	.643	−.608	.141	−.104	.168
9	.745	−.491	.173	−.136	.021
10	.726	.273	.475	.120	−.233
11	.580	.257	.503	.053	.567
12	.811	.020	.313	.055	−.384
13	.842	−.010	−.203	.155	.241
14	.854	.046	−.202	−.121	.018
15	.651	−.074	−.236	.602	−.038

TABLE 17-21
FACTOR LOADING SIMILARITY COEFFICIENTS BETWEEN THE FIRST MZ TWIN FACTORS AND THE FIRST DZ TWIN FACTORS BEFORE ROTATION

	1	2	3	4	5
1	.980	.512	−.484	−.489	.264
2	−.087	−.109	−.168	−.808	.115
3	−.002	.649	−.325	−.205	−.273
4	.205	−.047	−.264	−.279	−.704
5	.100	−.177	.105	−.256	.046

Table 17-22
Factor Loading Similarity Coefficients between the First MZ Twin Factors and the First DZ Twin Factors after Rotation

	1	2	3	4	5
1	.985	−.028	−.032	−.021	−.002
2	−.021	.821	−.468	−.096	−.018
3	.078	.423	.606	−.485	.028
4	.195	.447	.421	.512	−.008
5	.087	.125	−.288	.195	.141

Table 17-23
The Transformation Matrix, H_2, for Rotating the Second DZ Twin Factor Loadings toward the Second MZ Factor Loadings

	1	2	3	4	5
1	.869	.038	−.327	−.175	.327
2	.184	−.262	−.433	.640	−.549
3	−.082	−.954	−.005	−.168	.236
4	.096	−.002	.415	.721	.546
5	−.442	.143	−.731	.110	.488

Table 17-24
The Second DZ Twin Factor Loadings Rotated by H_2

	I	II	III	IV	V
1	.735	.248	−.299	.234	−.315
2	.754	.138	−.258	−.065	−.406
3	.785	−.130	−.233	−.251	−.170
4	.746	.369	−.119	.285	−.034
5	.846	.242	−.156	.204	.182
6	.785	.178	−.068	.296	.212
7	.622	−.632	−.160	.193	−.054
8	.601	−.636	.293	.146	−.166
9	.661	−.573	.075	.078	.276
10	.677	.272	.366	−.032	−.308
11	.572	−.005	.689	−.066	−.127
12	.775	.185	.409	.035	.170
13	.769	−.226	.015	−.434	.235
14	.695	−.124	−.148	−.446	−.165
15	.792	.046	−.074	−.201	.434

TABLE 17-25
FACTOR LOADING SIMILARITY COEFFICIENTS BETWEEN THE SECOND MZ TWIN FACTORS AND THE SECOND DZ TWIN FACTORS BEFORE ROTATION

	1	2	3	4	5
1	.985	.508	−.157	.274	−.765
2	.087	−.129	−.831	−.216	.104
3	−.012	−.024	.040	.049	−.582
4	.059	−.413	−.448	.546	−.244
5	.210	−.475	.133	−.359	−.185

TABLE 17-26
FACTOR LOADING SIMILARITY COEFFICIENTS BETWEEN THE SECOND MZ TWIN FACTORS AND THE SECOND DZ TWIN FACTORS AFTER ROTATION

	1	2	3	4	5
1	.995	−.023	−.030	.019	.002
2	.056	.859	−.196	−.087	−.151
3	.135	−.137	.660	−.096	−.432
4	.112	.485	.601	.142	.272
5	.156	−.041	.111	−.803	.166

TABLE 17-27
DIZYGOUS TWIN COEFFICIENTS OF CONGRUENCE FOR THE ROTATED DZ FACTORS
($\phi_d = H_1' \delta_d H_2$)

	1	2	3	4	5	$D^2_{\phi d}$	$(\delta^2_m - D^2_{\phi d})$
1	.663	.298	.081	.250	−.310	.439	.310
2	.113	−.403	.058	−.084	−.199	.162	.396
3	.065	−.100	.171	.165	.220	.029	.421
4	−.039	.019	.075	.006	−.042	.000	.376
5	.006	.003	.006	−.014	.005	.000	.171

these tests are not all inherited or (2) our assumed proportion of 0.5 of shared genes in fraternal twins is incorrect. If the second conclusion is drawn, then we can compute a value for the proportion of shared genes in fraternal twins. This (least square) value is 0.694.

Table 17-29 gives the results of a varimax rotation of the genetic difference factors. Only four factors were rotated because no more than four were found in any of the original factor analyses. This table can be compared with Table 17-30, which shows the results of a varimax rota-

TABLE 17-28
FACTOR LOADING MATRIX OF GENETIC DIFFERENCES (F_g)

	I	II	III	IV	V	COMMUNALITY	$h^2 \times 2$	PROP.
1	.360	.247	.107	−.241	−.063	.264	.528	.638
2	.366	.255	.189	−.100	−.091	.253	.506	.696
3	.393	.132	.058	−.267	.040	.248	.496	.645
4	.425	.145	−.168	.153	−.056	.256	.512	.668
5	.445	.119	−.173	.203	−.008	.284	.568	.656
6	.431	.057	−.217	.170	−.024	.266	.532	.694
7	.320	−.358	.038	−.190	−.026	.268	.536	.705
8	.262	−.432	−.041	−.105	−.105	.279	.558	.701
9	.319	−.412	.045	−.135	−.020	.292	.584	.714
10	.339	.013	.278	.276	−.009	.269	.538	.700
11	.215	.027	.369	.321	.076	.292	.584	.706
12	.284	.018	.245	.295	−.025	.229	.458	.752
13	.424	.104	−.032	−.014	.110	.204	.408	.567
14	.316	−.066	−.048	.009	.297	.194	.388	.414
15	.410	.045	−.202	.090	.063	.223	.446	.632
$2 \times (\delta_m^2 - D^2\phi_d)$								
	.620	.792	.842	.752	.342			

TABLE 17-29
VARIMAX ROTATION OF THE GENETIC DIFFERENCES FACTOR LOADING MATRIX

	I. Word Fluency	II. Number	III. Spatial	IV. Verbal
Number				
1 Addition	.077	.504	−.000	.013
2 Multiplication	.222	.440	−.041	.020
3 Three higher	.040	.479	.120	.039
Verbal				
4 Sentences	.168	.229	.023	.414
5 Vocabulary	.204	.200	.043	.448
6 Completion	.147	.174	.099	.451
Spatial				
7 Flags	.032	.148	.495	−.005
8 Figures	.006	.015	.514	.055
9 Cards	.068	.090	.528	.008
Word Fluency				
10 First letters	.498	.097	−.066	.083
11 Four letter words	.534	.011	−.013	−.022
12 Suffixes	.465	.049	.033	.090
Reasoning				
13 Letter series	.151	.325	.098	.233
14 Letter grouping	.105	.158	.192	.182
15 Pedigrees	.097	.204	.119	.392
% Variance	24.834	28.602	23.016	23.548

tion of the first four factors from the factor analysis of first members of monozygous twin pairs (Table 17-7). The conclusion to be drawn is that these four "primary mental abilities" are predominantly inherited.

REFERENCES

Cronbach, L. J. 1958. Proposals leading to analytic treatment of social perception scores. In *Person Perception and Interpersonal Behavior,* ed. R. Tagiuri and L. Petrullo. Stanford: Stanford University Press.

Horst, P. 1963. *Matrix Algebra for Social Scientists.* New York: Holt, Rinehart and Winston.

Sears, R. R. 1951. A theoretical framework for personality and social behavior. *Amer. Psychol.* **9:** 476-83.

Thurstone, L. L. and Thurstone, Thelma G. 1941. *The Primary Mental Abilities Tests.* Chicago: Science Research Associates.

TABLE 17-30
VARIMAX ROTATION OF THE ORIGINAL FIRST TWIN MONOZYGOUS FACTOR LOADING MATRIX
($Q_{m1}\Delta_{m1}$)

	I. Word Fluency	II. Number	III. Verbal	IV. Spatial
Number				
1 Addition	.113	.900	−.005	−.012
2 Multiplication	.354	.765	.024	.005
3 Three higher	.133	.808	.182	.223
Verbal				
4 Sentences	.232	.442	.684	.015
5 Vocabulary	.349	.374	.765	.031
6 Completion	.277	.326	.733	.130
Spatial				
7 Flags	−.052	.280	.182	.801
8 Figures	.045	.094	.294	.823
9 Cards	.192	.178	.192	.844
Word Fluency				
10 First letters	.824	.255	.126	.025
11 Four letter words	.900	.044	.027	−.028
12 Suffixes	.765	−.003	.122	.085
Reasoning				
13 Letter series	.160	.574	.585	.127
14 Letter grouping	.244	.307	.468	.093
15 Pedigrees	.196	.344	.738	.064
% Variance	24.584	29.153	27.933	18.330

Tucker, L. R. 1951. A method for synthesis of factor analytic studies. Washington, D.C.: Dep't of the Army: Adjutant General's Office, Personnel Research Section, Report No. 984.

Vandenberg, S. G. 1965. Multivariate analysis of twin differences. In *Methods and Goals in Human Behavior Genetics,* ed. S. G. Vanenberg. New York: Academic Press.

———. 1966. Contributions of twin research to psychology. *Psychol. Bull.* **66:** 327-52.

WILLIAM MEREDITH
University of California
Berkeley, California

18

FACTOR ANALYSIS AND THE USE OF INBRED STRAINS

INTRODUCTION

I would like to begin by giving a mathematically precise definition of certain terms commonly used in quantitative genetics, terms that I shall have occasion to use frequently. This can be achieved most simply by assuming that we are interested in a unidimensional trait, X, affected by two genetic loci and that in the population of interest there exist only two alleles at the two relevant loci. We must further assume that environmental and genotypic effects enter into the trait in question additively, that mating is random, that there is no linkage and that organisms of a given genotype are exposed to the same distribution of environments as organisms of any other genotype. Using the symbols A and a to denote the alleles at one locus and B and b to denote alleles at the other locus, we may symbolize genotypes by AaBb, AABb, etc. I shall assume that the genotypes AaBb, aAbB, aABb and AabB are all identical and denote all four by the symbol AaBb; that is to say sex-linked effects are specifically excluded, for the time being. This means the population contains nine genotypically different kinds of organisms with respect to the trait in question. Other loci are irrelevant. The relative frequency of any particular genotype may be determined from the relative frequencies of the alleles A, a, B and b if the population is in equilibrium. However, for simplicity I shall merely indicate the relative frequency of a genotype by the notation p (AaBb), etc.

Because of the assumption of additivity we may write

$$X = E + G \tag{1}$$

where E and G denote the effect on X of environment and genotype respectively. This means that

$$\mu_X = \mu_E + \mu_G \tag{2}$$

where μ denotes the expected value or population mean, so that

$$x = (X - \mu_X) = (E - \mu_E) + (G - \mu_G) = e + g\,. \tag{3}$$

Thus, we may work with deviation scores and assume the origins of x, e, and g to be zero without loss of generality.

The genetic effect of a genotype can now be defined as simply the expected value or average of x for a given genotype because of the assumption of independence of environment and genotype. We have, for example,

$$g(AaBB) = \mu_{(x|AaBB)} \tag{4}$$

etc. The effect of a genotype can be further broken down. The effect of the A locus is, for example,

$$g(AA) = \frac{1}{p(AA)} [p(AABB)g(AABB) + p(AABb)g(AABb) + p(AAbb)g(AAbb)] \tag{5}$$

where

$$p(AA) = p(AABB) + p(AABb) + p(AAbb). \tag{6}$$

Similar equations hold for g (Aa), g (aa), g (BB), etc. The genotypic effect can then be expressed as

$$g(AaBB) = g(Aa) + g(BB) + \gamma(AaBB) \tag{7}$$

etc., where γ (AaBB) is simply the deviation of g (AaBB) from the sum of g (Aa) and g (BB) and expresses the across-locus interaction due to the presence of the alleles A, a, B and B. This interaction is usually referred to as epistasis or epistatic interaction.

The various locus effects may be further decomposed. The effect of the A allele is

$$\alpha(A) = \frac{1}{p(A)} [\tfrac{1}{2}p(Aa)g(Aa) + p(AA)g(AA)] \tag{8}$$

where

$$p(A) = \tfrac{1}{2}p(Aa) + p(AA). \tag{9}$$

Similar equations hold for α(a), α(B), and α(b). The within locus interaction or dominance deviation is, for example,

$$\delta(aa) = g(aa) - 2\alpha(A) \tag{10}$$

or

$$\delta(Bb) = g(Bb) - \alpha(B) - \alpha(b). \tag{11}$$

Now we can write the following, utilizing equations (3), (7), (8), (10), and (11):

$$x_{AaBB} = \alpha(A) + \alpha(a) + 2\alpha(B) + \delta(Aa) + \delta(BB) + \gamma(AaBB) + e \tag{12}$$

and similar equations. The sum of the terms $\alpha(A)$, $\alpha(a)$ and $2\alpha(B)$ is called the additive effect of the AaBB genotype; and the sum of the terms $\delta(Aa) + \delta(BB)$ is called the dominance effect of the AaBB genotype so that we could rewrite (12) as, for example,

$$x_{aaBb} = e + \alpha(aaBb) + \delta(aaBb) = \gamma(aaBb)\,. \tag{13}$$

If we compute the variance of x it can readily be shown that

$$\sigma_x^2 = \sigma_e^2 + \sigma_\alpha^2 + \sigma_\delta^2 + \sigma_\gamma^2 \tag{14}$$

because of the assumption of independence of genotype and environment and the definitions of the various effects. The term σ_α^2, which is simply the variance of the additive effects $\alpha(AABB)$, etc., is called the additive genetic variance and is frequently referred to as the principal cause of the resemblance between relatives. The ratio σ_α^2 to σ_x^2 is called the heritability of the trait and is frequently estimated by means of parent-offspring regression or sib analyses. The ratio of σ_g^2 to σ_x^2 is frequently referred to as the coefficient of genetic determination.

Several points should be noted. As can be seen from equation (5), for example, anything that alters gene frequencies (e.g., selection, assortative mating, mutation, migration) will alter the values of the effects defined in equations (7), (8), (10), and (13) and hence the three genetic variances in equation (14). Consequently the coefficients of genetic determination and heritability will also be altered.

An alteration in the distribution of environments will not affect the values of the various genetic effects so long as independence of environment and genotype is maintained, but will alter the coefficients of heritability and genetic determination. It can properly be said that whatever is known about the relative contributions of environment and genotype refers only to a specified population of genotypes existing in a specified population of environments.

The development in equations (1) through (14) can quite readily be generalized to an arbitrary number of loci and an arbitrary number of alleles at each locus. Furthermore the epistatic effects $\gamma(AABB)$, etc., and epistatic variance σ^2_γ can be further broken down (e.g., additive $\times$ additive interaction, etc.). The interested reader can find extensive derivations in Falconer (1960) and Kempthorne (1957).

INBRED STRAINS

From the foregoing it is clear that it would be extremely useful if we could replicate organisms of given, known genotype in a variety of environmental circumstances. It is, of course, impractical to study all possible genotypes if the number of loci and/or alleles affecting a trait is large. Alternatively, we can study additive genetic variance through the correlations between relatives in a random mating population with no association between genotype and environment. However, a random mating population is more difficult to achieve than is generally realized. The typical animal laboratory population proves to be a far cry from a random mating population (McClearn 1968).

One method of attaining genetic control in mammals (i.e., specification of genotypes) that has proven useful is the utilization of inbred strains. An inbred strain is defined as the end result of at least twenty generations of brother-sister mating and such organisms can, for practical purposes, be regarded as homozygous at all genetic loci. In fact, however, there will always remain a small residual of genetic variance within an inbred strain due to mutation and the failure of fixation of a few alleles. A large number of inbred strains of mice are available [1] as well as a few inbred strains of rats.

We can regard all animals within a given inbred strain as genotypically identical except for the fact that males will possess the X and Y chromosomes characteristic of the strain and females two of the X chromosomes characteristic of the strain. Further, since organisms within a given strain can produce only two different kinds of gametes (one with an X and one with a Y chromosome), if we crossbreed animals from two inbred strains only three genotypes can be produced. These can be identified by a knowledge of the type of mating and the sex of the offspring. To make this clear, suppose we reconsider our two-locus, two-allele model. Inbreeding from such a population could produce four kinds of inbred strains with respect to the trait in question. It must not be inferred that only four inbred strains could be produced, for we must consider other genetic loci. However, with respect to the trait in question which is assumed to depend only on the A and B loci we could identify only four different types of organisms. Suppose further that the trait in question was affected by a sex-linked locus with two different X alleles. The Y chromosome is taken to be essentially inert. Then we could obtain the sixteen genotypes AABBX, AABBXX, AABBx, AABBxx, AAbbX,

[1]The Committee on Standardized Genetic Nomenclature for Mice. 1960. Standardized nomenclature for inbred strains of mice. Second listing. *Cancer Research* **20**: 145-169.

AAbbXX, . . ., aabbx, aabbxx by inbreeding. Cross-breeding the ABBXX strain with the AAbbxx strain would yield the genotypes AABbX, AABbx, and AABbXx. The male offspring of an AABBXX dam and an AAbbx sire must have the genotype AABbX, etc. It is perfectly obvious that, given males and females from eight inbred strains chosen so that all sixteen potential homozygous genotypes are represented, we could reconstruct all possible genotypes in the population, with respect to the trait in question. Because of the artificiality of the situation we could not guarantee our ability to reproduce the genotypes in proportion to their rate of occurrence in a "natural" population. This is a small matter, however, insofar as a laboratory population is already an unnatural population.

The foregoing suggests the possibility of defining the population a posteriori by selecting an arbitrary set of N inbred strains and crossbreeding. This would define the basic population of potential genotypes. If the number of loci and alleles affecting a trait in question leads to $N^2 + (N+1)N/2$ or fewer possible genotypes, the inbreds plus the crossbreds would reproduce all the potential genotypes. If the number of potential genotypes is greater than $N^2 + (N+1)N/2$, the inbreds plus the crossbreds would allow replication of a sample (not a random sample) of the potential genotypes.

Before leaving the topic of inbred strains it is only fair to point out that a number of problems arise in connection with the use of inbred strains. A large number of lines are lost in the process of inbreeding due to the fixation of recessive alleles that have a deleterious effect on viability and fertility. Hence the lines that survive cannot be thought of as representing a random sample of the theoretically possible set of inbred strains derivable from a given population. Furthermore, it is not feasible to maintain enough strains adequately to represent all the possibilities if the number of loci and alleles affecting a trait is large.

FACTOR ANALYSIS

Now suppose we consider a multidimensional trait y of n dimensions,

$$y' = (y_1, y_2, \ldots y_i, \ldots, y_n)\,. \tag{15}$$

Without loss of generality we can assume y to be expressed in standard scores, i.e., with means equal to zero and variances equal to unity, in the population of interest. The whole random vector y can be subjected to the kind of analysis described in the introductory section. In particular, if we assume additivity of genotypic and environmental effects, we can write

$$y = e + g \tag{16}$$

where e is now an n-dimensional vector of deviation measures whose ith component consists of the effect of environment on y_i and g is an n-dimensional vector of deviation measures whose ith component consists of the genotypic effect on y_i. We also assume independence of genotype and environment so that

$$R_y = V_e + V_g \tag{17}$$

where R_y denotes the intercorrelation matrix of y, V_e the variance-covariance matrix of e and V_g the variance-covariance matrix of g. Because of the metric imposed on y the diagonal elements of V_g contain the coefficients of genetic determination for each component of y; similarly the diagonal elements of V_e are the proportions of total variance due to environment.

Now suppose that both e and g admit of a factor analytic decomposition. That is, we can write

$$e = F_e z_e + u \tag{18}$$

where F_e denotes an $n \times p$ factor pattern matrix $(n > p)$, z_e a $p \times 1$ vector of common factor scores and u an n x 1 vector of specific and/or error factor scores. As usual z and u are presumed to be independent, and the components of u are assumed to be mutually independent. In addition suppose

$$g = F_g z_g + v \tag{19}$$

where F_g denotes an $n \times q$ factor pattern matrix $(n > q)$, z_g a $q \times 1$ vector of common factor scores and v an $n \times 1$ vector of specific factor scores.

Let us examine further the implications of this model. Equation (18) implies that the environmental portion of the manifest variable can be accounted for by a smaller number of latent variates z_e plus n specific and/or error factors. This merely means that we have chosen the components of y so that features of the environment which affect one component of y affect other components. The components of z_e are to be interpreted as latent variates due to differences in environment. Their nature must be inferred from the factor structure and pattern matrices unless the effects of different kinds of environment can be pin-pointed. In short, such factors are just like ordinary factors except that their locus is known to lie in the environment.

Equation (19) implies pleiotropic gene action, that is to say, the genetic loci that affect one component of y affect still other components. In addition, the presence of the vector v in equation (19) implies that each component of y is affected by a unique genetic locus, or set of loci, not affecting any other component of y. Careful selection of the components of y could reduce v to a null vector for all genotypes. I think it reasonable to assume that in a majority of instances we could simply put

$$g = F_g z_g \quad . \tag{20}$$

Equation (20) will be assumed throughout the rest of this paper. As in the case of environmental factors, the components of z_g are simply latent attributes or factor scores except that the locus is known to be genetic. The components of z_g do not reflect genotypes, but rather genotypic differences and pleiotropy. In short, the genetic factors are to be interpreted as factors usually are, by examination of factor structure and pattern matrices. From equations (17), (18) and (20) it can readily be shown that

$$R_y = F_e \Lambda_e F'_e + F_g \Lambda_g F'_g + \Delta_u \tag{21}$$

where Λ_e and Λ_g denote the variance-covariance matrices of z_e and z_g respectively, and Δ_u denotes the diagonal matrix of variances of u. Another way of writing (21) is

$$R_y = [F_e, F_g] \begin{bmatrix} \Lambda_e : 0 \\ \cdots\cdots \\ 0 : \Lambda_g \end{bmatrix} \begin{bmatrix} F'_e \\ F'_g \end{bmatrix} + \Delta_u \tag{22}$$

which highlights the relationship between this model and the traditional factor analytic model where genetic and environmental effects are not treated separately.

Finally, given the vectors z_g and control of genotype, z_g could be subjected to the kind of analysis defined in equation (13), without the inclusion of an environmental effect. Thus, for each factor we could determine an additive effect, a dominance effect, etc.

THE APPLICATION OF FACTOR ANALYSIS TO INBRED STRAINS

We can now bring together the notions developed in the two previous sections. Suppose we consider a synthetic population derived by crossing inbred strains. The inbreds themselves plus all possible F_1 generations derived by reciprocal crosses provide a total of $N^2 + N(N + 1)/2$ different genotypes if males are assumed heterogametic. We further consider an n-dimensional manifest variable Y. I shall denote a genotype by the notation ijk, $i = 1, \ldots, N$; $j = 1, \ldots, N$; $k = ♀, ♂$. The vector of means or expected values of Y based on males whose sires came from the ith inbred strain and whose dams came from the jth I shall denote by $\mu_{ji}♂$. Note that $\mu_{ij}♀$ is equal to $\mu_{ji}♀$. I shall assume that the effects of genotype and environment are additive with respect to Y, and that within each strain or F_1 generation, organisms are exposed to the various environments in the same relative frequency.

These last assumptions imply that the conditional variance-covariance matrices of Y within genotypes are homogeneous. If we know one of them, the variance-covariance matrix due to environment is available.

However, we cannot determine the population mean vector and variance-covariance without some additional assumptions. The population mean vector, and the various kinds of genetic effects derived in the first section, are in part a function of gene frequency. We can arbitrarily define gene frequency by assuming our synthetic population began with equal numbers of males and females and mated randomly subsequently. If there is no differential fertility or viability gene frequencies would stay fixed. A more serious problem arises in that if the number of loci and/or alleles affecting the manifest variable Y is large in relation to the number of inbred strains, we cannot recover all the possible genotypes, with respect to Y, from the inbreds and F_1 generations.

Since it is somewhat unclear what the concept of population refers to in this context anyway it seems reasonable to ignore the question of population and all it implies for genetic effects and regard the use of inbreds and F_1 generations as a convenient tool for replicating and controlling genotypes. Let us form the supermatrices

$$M♂ = [\mu_{11}♂, \mu_{12}♂, \ldots, \mu_{1N}♂, \mu_{21}♂, \mu_{22}♂, \ldots, \mu_{NN}♂] \tag{23}$$

$$M♀ = [\mu_{11}♀, \mu_{12}♀, \ldots, \mu_{1N}♀, \mu_{21}♀, \mu_{22}♀, \ldots, \mu_{NN}♀] \tag{24}$$

and

$$M = [M♂M♀]. \tag{25}$$

The dimensions of M ♂ are $n \times N^2$, of M ♀ are $n \times N(N+1)/2$ and of M are $n \times [N^2 + N(N+1)/2]$. Let U′ be defined as a $1 \times (N^2 + N(N+1)/2)$ vector given by

$$U' = \frac{1}{2N^2}(11\ldots111\ldots1 \vdots 12\ldots212\ldots1) \tag{26}$$

and

$$\mu = MU. \tag{27}$$

The vector μ is a weighted average with those means corresponding to "unreplicated" genotypes (i.e., μ_{ij} ♀; $i>j$) given double weight. Now let

$$M_0 = M - \mu 1' \tag{28}$$

where 1′ denotes a row vector of ones.

Now let Σ_e denote the within genotype variance-covariance matrix and

$$\Sigma = \Sigma_e + M_0 D_U M_0' \tag{29}$$

where D_U indicates a diagonal matrix formed by arraying the vector U along the main diagonal.

It can readily be shown that if we define y_{ijkh} as

$$y_{ijkh} = D_{\Sigma}^{-1/2} (Y_{ijkh} - \mu) \tag{30}$$

and

$$G = D_{\Sigma}^{-1/2} M_0 \tag{31}$$

where D_{Σ} indicates a diagonal matrix made up of the diagonal elements of Σ, and Y_{ijkh} denotes the value of Y for the hth organism of genotype ijk, we can write

$$y_{ijkh} = e_h + G_{ijk} \tag{32}$$

where G_{ijk} indicates the appropriate column out of G and e_h denotes the environmental effect on Y for the particular organism. The vector e_h is equated to the vector e of equation (16) except for scaling factors; the vector G_{ijk} may be regarded as equivalent to the vector g except for changes in both location and scale of the components due to the arbitrary choice of a mean and variance. If we assume a factor analytic decomposition we can write analogously to (18) and (20)

$$e_h = P_e Z_h + \mu_h \tag{33}$$

and

$$G_{ijk} = P_g \Gamma_{ijk} \tag{34}$$

which leads to

$$G = P_g \Gamma \tag{35}$$

where Γ is a $q \times [N^2 + N(N+1)/2]$ matrix of genetic factor scores formed analogously to the matrix M, and P_g the genetic factor pattern matrix, with the scale defined by (29) and (30).

What the foregoing is predicated on is that if an equation like (20) holds in some larger sense (i.e., with respect to some general population) then a similar equation must also hold for any subset of genotypes. We have chosen genotypes by choosing inbred strains and their F_1's, so that genotype can be controlled. This choice of genotypes introduces arbitrary location and scaling constants but location and scaling constants have no effect on the number of factors or their nature, either environmental or genetic.

It can be seen, however, that the number of genetic factors is limited to $N^2 + \frac{1}{2} N(N+1) - 1$. In practical situations this would not be serious. If no sex-linked loci affect Y then the number of genetic factors is limited by $\frac{1}{2} N(N+1) - 1$ which is still not a serious restriction unless N is small. We are fortunate that the variance-covariance matrix due to environment can be gotten at directly, given the assumptions of additivity and the independence of genotype and environment.

Now we are almost finished. Without loss of generality we can require that the factors be orthogonal. Then we have that

$$D_{\Sigma}^{-\frac{1}{2}} \Sigma_e D_{\Sigma}^{-\frac{1}{2}} = P_e P_e{}' + \Delta_u \tag{36}$$

where Δ_u denotes a diagonal matrix of variances of the components of u, and

$$G D_U G' = P_g P_g{}' . \tag{37}$$

Any arbitrary method of factoring the matrices on the left of (36) and (37) that would yield the desired solutions could be applied. However, we have made an arbitrary choice of scale and should use methods that are invariant with respect to scale. One procedure that has other desirable features would be first to perform an arbitrary factoring.

$$D_{\Sigma}^{-\frac{1}{2}} \Sigma_e D_{\Sigma}^{-\frac{1}{2}} = TT' \quad . \tag{38}$$

The rank of T is n. Next we would form

$$T^{-1} G D_U G' T^{-1\prime} = W \tag{39}$$

and determine the non-zero eigenvalues and associated eigenvectors of W. Let λ denote the $q \times q$ diagonal matrix of eigenvalues arranged in descending order and Q the $n \times q$ matrix of associated eigenvectors. Then

$$P_g = TQ\lambda^{\frac{1}{2}} . \tag{40}$$

This procedure can readily be shown to be invariant with respect to changes in scale and also has desirable features when applied to sample data. This question will be taken up in a subsequent section.

The genotypic factor scores for the $N^2 + \frac{1}{2} N(N+1)$ genotypes can be readily determined by

$$\Gamma = \lambda^{-\frac{1}{2}} Q' T^{-1} G . \tag{41}$$

The components of Γ can be further broken down. What we might consider the autosomal effect of a genotype, keeping the structure of the situation in mind, on the factor scores would be

$$\Gamma_{ij.} = \Gamma_{ji.} = \tfrac{1}{4}\,(\Gamma_{ij♂} + \Gamma_{ji♂} + 2\Gamma_{ij♀}). \tag{42}$$

The "additive" effect of an inbred becomes

$$\alpha_i = \frac{1}{N}\sum_{j=1}^{N}\Gamma_{ij}\ , \tag{43}$$

and the "dominance deviation" for a given cross is

$$\delta_{ij} = \delta_{ji} = \Gamma_{ij} - \alpha_i - \alpha_j\ . \tag{44}$$

The effects due to sex chromosome loci are

$$\Gamma_{..♂} = \frac{1}{N^2}\sum_{i=1}^{N}\sum_{j=1}^{N}\Gamma_{ij♂} \tag{45}$$

and

$$\Gamma_{..♀} = \frac{1}{N^2}\left(\sum_{i=1}^{N}\Gamma_{ii♀} + 2\sum_{i=1}^{N}\ \sum_{j=i+1}^{N}\Gamma_{ij♀}\right). \tag{46}$$

The sex-by-genotype interaction is

$$\gamma_{ijk} = \Gamma_{ijk} - \alpha_i - \alpha_j - \delta_{ij} - \Gamma_{..k} \tag{47}$$

Note that all the effects defined in (42) through (47) are q $\times$ 1 vectors.

Finally we would return to the matrix $D_{\Sigma}^{-1/2}\Sigma_e D_{\Sigma}^{-1/2}$ and refactor, preferably using Rao's (1955) canonical factor analysis to determine the matrices P_e and Δ_u. Rao's method is invariant with respect to scale. The environment factor scores could be determined by regression using

$$e_h = Y_{ijkh} - \mu_{ijk} \tag{48}$$

in the regression equation.

Both P_g and P_e may be rotated without loss of generality and without affecting the development in (42) through (47).

ESTIMATION AND HYPOTHESIS TESTING

The preceding development bears an obvious relationship to multivariate analysis of variance, and the significance tests and estimation procedures appropriate to MANOVA can all be applied here. We can determine unbiased estimates without making any assumption about the form of the joint distribution of e_h. The established significance tests all are based on the assumption that the joint distribution of e_h is multivariate normal.

Suppose we have N inbred strains and their reciprocal crosses and observe the value of Y for m males chosen at random from each genotype. For females we would observe the value of Y for m organisms chosen at random from the ii♀ genotypes and m organisms chosen at random from each of the reciprocal crosses giving the ij♀ genotype. Let the mean vectors be

$$\overline{Y}_{ii\male} = \frac{1}{m} \sum_{h=1}^{m} Y_{ij\male h} \qquad (49)$$

$$\overline{Y}_{ii\female} = \frac{1}{m} \sum_{h=1}^{m} Y_{ii\female h} \qquad (50)$$

and

$$\overline{Y}_{ij\female} = \frac{1}{2m} \left(\sum_{h=1}^{m} Y_{ij\female h} + \sum_{h=1}^{m} Y_{ji\female h} \right) = \overline{Y}_{ji\female}. \qquad (51)$$

The mean vector $\overline{Y}_{ijk}$ is known to be a "best linear unbiased" estimator of μ_{ijk} in the sense that each component of $\overline{Y}_{ijk}$ is a BLU estimator of the corresponding component of μ_{ijk}. Let the matrices M♂, M♀, and M be defined analogously to equations (23), (24) and (25) with the μ_{ijk} replaced by $\overline{Y}_{ijk}$. Then we can estimate M_0 by

$$\hat{M}_0 = \hat{M} - \hat{\mu}1' \qquad (52)$$

where

$$\hat{\mu} = \hat{M}U \qquad (53)$$

The matrix of variances and covariances due to environment may be readily estimated by

$$S_e = \frac{1}{N^2[m-1] + [N(N-1)/2][2m-1]} \left[\sum_{i=1}^{N} \sum_{j=1}^{N} \sum_{h=1}^{m} (Y_{ij\male h} - Y_{ij\male}) \right.$$

$$\left. (Y_{ij\male h} - \overline{Y}_{ij\male})' + \sum_{i=1}^{N} \sum_{j=1}^{N} \sum_{h=1}^{m} (Y_{ij\female h} - \overline{Y}_{ij\female})(Y_{ij\female h} - \overline{Y}_{ij\female})' \right]. \qquad (54)$$

Now let

$$S = S_e + \hat{M}_0 D_U \hat{M}_0' \tag{55}$$

and

$$D_S^{-1/2} S_e D_S^{-1/2} = tt' \tag{56}$$

The likelihood ratio statistic for testing the hypothesis of no genetic effect at all, given that e is mulitvariate normal (Rao, 1952), is

$$L = \prod_r \left(1 + \frac{m_g}{m_e} \hat{\lambda}_r\right)^{-1} \tag{57}$$

where $\hat{\lambda}_r$ is an eigenvalue of the matrix

$$\hat{W} = \frac{2N^2m}{N^2 + N(N+1)/2} \{t^{-1} D_S^{-1/2} \hat{M}_0 D_U \hat{M}_0' D_S^{-1/2} t'^{-1}\} \tag{58}$$

m_g equals $[N^2 + N(N + 1)/2] - 1$, and m_e equals $N^2[m - 1] + [N(N + 1)/2][2m - 1]$.

More importantly, each eigenvalue of W can be tested for significance in a quasi-independent manner using a chi-square approximation suggested by Kullback (1959) and Rao (1952). Now in this case testing λ_r for significance is essentially a test of significance for the number of factors in equation (35). We would retain as many factors as are statistically significant.

The estimated factor pattern matrix is

$$\hat{P}_g = t'\hat{Q}\Lambda \tag{59}$$

where Q is a matrix of normalized eigenvectors of W and Λ is a diagonal matrix of $\{[N^2 + N(N + 1)/2]/[2N^2m]\}^{1/2}$ times the square roots of the significant eigenvalues.

The next steps would be to estimate the genotypic factor scores by

$$\hat{\Gamma} = \Lambda^{-1}\hat{Q}'t^{-1}D_S^{-1/2}\hat{M}_0 \tag{60}$$

and carry out an analysis similar to that described in equations (42) through (47); and finally to factor analyze the matrix $D_S^{-1/2} S_e D_S^{-1/2}$ into $\hat{P}_e$ and $\hat{\Delta}_u$.

An alternative method of analysis would be to break down the matrix $D_S^{-1/2}\hat{M}_0$ into the effects $\hat{\alpha}_i$, $\hat{\delta}_{ij}$, etc., analogously to those described in equations (42) through (47). Here we would be operating directly on the n-dimensional variate y rather than on genetic factor scores. Then mean square dispersion matrices could be calculated for each effect and MANOVA techniques applied to each. I am opposed to this procedure in this case for it seems that what we are looking for are pleiotropic effects; and a particular set of loci may behave additively with respect to one component of y and have a complete dominance-recessive effect on another component of y. In forming the dispersion matrices for additive and dominance effects separately the pleiotropic effect (i.e., correlation) could be completely obscured. Similar problems would occur with the other kinds of effects.

It is also possible to use a test of significance for the homogeneity of variance-covariance matrices to test whether the basic assumptions of additivity of genetic and environmental effects and independence of genotype and environment are met. Rejection of the hypothesis of homogeneity could be due to the failure of either one of these two basic assumptions. Presumably if heterogeneity occurs we would attempt to find transformations of the individual components of Y that would result in homogeneity.

References

Falconer, D. S. 1960. *Introduction to Quantitative Genetics.* New York: Ronald Press.

Kempthorne, O. 1957. *An Introduction to Genetic Statistics.* New York: Wiley.

Kullback, S. 1959. *Information Theory and Statistics.* New York: Wiley.

McClearn, G. E. 1968. Genes, Generality and Behavioral Research. (Unpublished manuscript.)

Rao, C. R. 1952. *Advanced Statistical Methods in Biometric Research.* New York: Wiley.

———. 1955. Estimation and tests of significance in factor analysis. *Psychometrika* **20**: 93-111.

SUBJECT INDEX

AUTHOR INDEX